AF799963

The Hospitality and Tourism Industry in ASEAN and East Asian Destinations

New Growth, Trends, and Developments

Advances in Hospitality and Tourism Series

The Hospitality and Tourism Industry in ASEAN and East Asian Destinations

New Growth, Trends, and Developments

Edited by

Jinlin Zhao, PhD

Lianping Ren, D.HTM

Xiangping Li, PhD

AAP APPLE ACADEMIC PRESS

First edition published 2022

Apple Academic Press Inc.
1265 Goldenrod Circle, NE,
Palm Bay, FL 32905 USA

4164 Lakeshore Road, Burlington,
ON, L7L 1A4 Canada

CRC Press
6000 Broken Sound Parkway NW,
Suite 300, Boca Raton, FL 33487-2742 USA

2 Park Square, Milton Park,
Abingdon, Oxon, OX14 4RN UK

© 2022 Apple Academic Press, Inc.

Apple Academic Press exclusively co-publishes with CRC Press, an imprint of Taylor & Francis Group, LLC

Reasonable efforts have been made to publish reliable data and information, but the authors, editors, and publisher cannot assume responsibility for the validity of all materials or the consequences of their use. The authors, editors, and publishers have attempted to trace the copyright holders of all material reproduced in this publication and apologize to copyright holders if permission to publish in this form has not been obtained. If any copyright material has not been acknowledged, please write and let us know so we may rectify in any future reprint.

Except as permitted under U.S. Copyright Law, no part of this book may be reprinted, reproduced, transmitted, or utilized in any form by any electronic, mechanical, or other means, now known or hereafter invented, including photocopying, microfilming, and recording, or in any information storage or retrieval system, without written permission from the publishers.

For permission to photocopy or use material electronically from this work, access www.copyright.com or contact the Copyright Clearance Center, Inc. (CCC), 222 Rosewood Drive, Danvers, MA 01923, 978-750-8400. For works that are not available on CCC please contact mpkbookspermissions@tandf.co.uk

Trademark notice: Product or corporate names may be trademarks or registered trademarks and are used only for identification and explanation without intent to infringe.

Library and Archives Canada Cataloguing in Publication

Title: The hospitality and tourism industry in ASEAN and East Asian destinations : new growth, trends, and developments/ edited by Jinlin Zhao, PhD, Lianping Ren, D.HTM, Xianping Li, PhD.

Names: Zhao, Jinlin, editor. | Ren, Lianping, editor. | Li, Xiangping, editor.

Description: First edition. | Includes bibliographical references and index.

Identifiers: Canadiana (print) 20200414917 | Canadiana (ebook) 20200415026 | ISBN 9781771889421 (hardcover) | ISBN 9781003082200 (PDF)

Subjects: LCSH: Tourism—Southeast Asia. | LCSH: Tourism—East Asia.

Classification: LCC G155.A743 H67 2021 | DDC 338.4/79159—dc23

Library of Congress Cataloging-in-Publication Data

Names: Zhao, Jinlin, editor. | Ren, Lianping, editor. | Li, Xiangping, editor.

Title: The hospitality and tourism industry in ASEAN and East Asian destinations : new growth, trends, and developments/ edited by Jinlin Zhao, PhD, Lianping Ren, D.HTM, Xianping Li, PhD.

Description: 1st edition. | Palm Bay, FL : Apple Academic Press, 2021. | Series: Advances in hospitality and tourism series | Includes bibliographical references and index. | Summary: "This new volume addresses the growing interest to understand tourism and hospitality in Southeast and South Asia, two regions that have seen tremendous growth in international tourists in recent years. It explores the current development of hospitality and tourism industry in the regions of Hong Kong, Macao and Taiwan as well as other key countries in Southeast and South Asia. The Hospitality and Tourism Industry in ASEAN and East Asian Destinations: New Growth, Trends, and Developments provides updated findings and case studies that highlight opportunities and issues of tourism and hospitality development in ASEAN. Chapters cover such diverse topics as Online marketing strategies Sustainable hospitality development Diversification efforts of the tourism industry Innovations in independent hotels Wildlife tourism in urban destinations The Vietnamese national park system Consumers' positive and negative images of certain destinations Much more While academicians will benefit from the updated research findings summarized by the respected scholars, hospitality professionals will also find the book a valuable source of information as the chapters delve into the most recent topical and industry focused issues"-- Provided by publisher.

Identifiers: LCCN 2020055939 (print) | LCCN 2020055940 (ebook) | ISBN 9781771889421 (hardcover) | ISBN 9781003082200 (ebook)

Subjects: LCSH: Tourism--Southeast Asia. | Tourism--South Asia. | Hospitality industry--Southeast Asia. | Hospitality industry--South Asia.

Classification: LCC G155.E18 H67 2021 (print) | LCC G155.E18 (ebook) | DDC 337.4/79159--dc23

LC record available at https://lccn.loc.gov/2020055939

LC ebook record available at https://lccn.loc.gov/2020055940

ISBN: 978-1-77188-942-1 (hbk)
ISBN: 978-1-77463-794-4 (pbk)
ISBN: 978-1-00308-220-0 (ebk)

ABOUT THE ADVANCES IN HOSPITALITY AND TOURISM BOOK SERIES

Editor-in-Chief:
Mahmood A. Khan, PhD
Professor, Department of Hospitality and Tourism Management,
Pamplin College of Business,
Virginia Polytechnic Institute and State University,
Falls Church, Virginia, USA
Email: mahmood@vt.edu

This series reports on research developments and advances in the rapidly growing area of hospitality and tourism. Each volume in this series presents state-ofthe-art information on a specialized topic of current interest. These one-of-a-kind publications are valuable resources for academia as well as for professionals in the industrial sector.

BOOKS IN THE SERIES:

Food Safety: Researching the Hazard in Hazardous Foods
Editors: Barbara Almanza, PhD, RD, and Richard Ghiselli, PhD

Strategic Winery Tourism and Management: Building Competitive Winery Tourism and Winery Management Strategy
Editor: Kyuho Lee, PhD

Sustainability, Social Responsibility and Innovations in the Hospitality Industry
Editor: H. G. Parsa, PhD
Consulting Editor: Vivaja "Vi" Narapareddy, PhD
Associate Editors: SooCheong (Shawn) Jang, PhD,
Marival Segarra-Oña, PhD, and Rachel J. C. Chen, PhD, CHE

Managing Sustainability in the Hospitality and Tourism Industry: Paradigms and Directions for the Future
Editor: Vinnie Jauhari, PhD

Management Science in Hospitality and Tourism: Theory, Practice, and Applications
Editors: Muzaffer Uysal, PhD, Zvi Schwartz, PhD, and Ercan Sirakaya-Turk, PhD

Tourism in Central Asia: Issues and Challenges
Editors: Kemal Kantarci, PhD, Muzaffer Uysal, PhD, and Vincent Magnini, PhD

Poverty Alleviation through Tourism Development: A Comprehensive and Integrated Approach
Robertico Croes, PhD, and Manuel Rivera, PhD

Chinese Outbound Tourism 2.0
Editor: Xiang (Robert) Li, PhD

Hospitality Marketing and Consumer Behavior: Creating Memorable Experiences
Editor: Vinnie Jauhari, PhD

Women and Travel: Historical and Contemporary Perspectives
Editors: Catheryn Khoo-Lattimore, PhD, and Erica Wilson, PhD

Wilderness of Wildlife Tourism
Editor: Johra Kayeser Fatima, PhD

Medical Tourism and Wellness: Hospitality Bridging Healthcare (H_2H)©
Editor: Frederick J. DeMicco, PhD, RD

Sustainable Viticulture: The Vines and Wines of Burgundy
Claude Chapuis

The Indian Hospitality Industry: Dynamics and Future Trends
Editors: Sandeep Munjal and Sudhanshu Bhushan

Tourism Development and Destination Branding through Content Marketing Strategies and Social Media
Editor: Anukrati Sharma, PhD

Evolving Paradigms in Tourism and Hospitality in Developing Countries: A Case Study of India
Editors: Bindi Varghese, PhD

The Hospitality and Tourism Industry in China: New Growth, Trends, and Developments
Editors: Jinlin Zhao, PhD

Labor in Tourism and Hospitality Industry: Skills, Ethics, Issues, and Rights
Abdallah M. Elshaer, PhD, and Asmaa M. Marzouk, PhD

Sustainable Tourism Development: Futuristic Approaches
Editor: Anukrati Sharma, PhD

Tourism in Turkey: A Comprehensive Overview and Analysis for Sustainable Alternative Tourism
Editor: Ahmet Salih İKİZ

Capacity Building Through Heritage Tourism: An International Perspective
Editor: Surabhi Srivastava, PhD

The Hospitality and Tourism Industry in ASEAN and East Asian Destinations: New Growth, Trends, and Developments
Editors: Jinlin Zhao, PhD, Lianping Ren, D.HTM, and Xiangping Li, PhD

Event Tourism in Asian Countries: Challenges and Prospects
Editors: Shruti Arora, PhD

ABOUT THE SERIES EDITOR

Mahmood A. Khan, PhD, is a Professor in the Department of Hospitality and Tourism Management, Pamplin College of Business at Virginia Tech's National Capital Region campus. He has served in teaching, research, and administrative positions for the past 35 years, working at major U.S. universities. Dr. Khan is the author of several books and has traveled extensively for teaching and consulting on management issues and franchising. He has been invited by national and international corporations to serve as a speaker, keynote speaker, and seminar presenter on different topics related to franchising and services management. Dr. Khan has received the Steven Fletcher Award for his outstanding contribution to hospitality education and research. He is also a recipient of the John Wiley & Sons Award for lifetime contribution to outstanding research and scholarship; the Donald K. Tressler Award for scholarship; and the Cesar Ritz Award for scholarly contribution. He also received the Outstanding Doctoral Faculty Award from Pamplin College of Business. He has served on the Board of Governors of the Educational Foundation of the International Franchise Association, on the Board of Directors of the Virginia Hospitality and Tourism Association, as a Trustee of the International College of Hospitality Management, and as a Trustee on the Foundation of the Hospitality Sales and Marketing Association's International Association. He is also a member of several professional associations.

ABOUT THE EDITORS

Jinlin Zhao, PhD

Dr. Jinlin Zhao is Professor, Director of the Graduate Program, and Director of Asia and Pacific Development, Chaplin School of Hospitality and Tourism Management, Florida International University, Miami, Florida, USA. He came to FIU in fall 2000 and was tenured in 2005. He earned his bachelor's degree in English at Beijing Second Foreign Language Institute in 1979 and remained to teach at the Institute until 1987. After he came to the United States in fall of 1987, he completed his master's degree at Indiana University of Pennsylvania in 1988. He earned his PhD degree in Hospitality and Tourism Management at Virginia Polytechnic Institute and State University in 1994. He started his teaching career in the United States in fall 1992 when he became an assistant professor and coordinator of special studies in hospitality management at Western Carolina University (WCU), North Carolina. He took the initiative in building the hospitality management degree program. Three years later, in 1995, the degree program was approved by the State of North Carolina. He was granted tenure at WCU and promoted to associate professor in 1999.

Dr. Zhao has been an active researcher. His area of research lies in competitive methods, the international environment and impact analysis, and multinational corporate strategy. He edited the book *The Hospitality and Tourism Industry in China: New Growth, Trends, and Developments* (Apple Academic Press, 2019), in which he coauthored two chapters. He also coedited the book *Handbook of Hospitality Strategic Management*, London, Elsevier, in 2008, in which he co-contributed a book chapter to the same book. In 2010, he coauthored the book *Tourism Destination Management: Principles and Practice* (New Delhi: Kanishka Publishers). He was the author of the *instructor's manual to accompany Strategic Management in the Hospitality Industry*, 3rd edition, (Pearson/Prentice Hall Inc. 2008). He has also published many refereed articles in several research journals and five books/chapters, and other publications, and has given more than 70 presentations at international hospitality conferences. Dr. Zhao has guided and mentored graduate students to do research, which resulted in many publications in high-rank academic journals and presentations at international conferences.

Dr. Zhao has much industry experience. He has been involved in pro bono consulting activities in China. He has given hotel management seminars and presentations in China since 1995. He also took the opportunity to work as a foodservice manager during the 1996 Atlanta and 2004 Athens Olympic Games. In 2008, he used his sabbatical leave to work as training manager for the 2008 Beijing Olympic Catering Project, in which he was in charge of recruiting and training 7000 temporary employees. He also worked as a human resource expert for the 2010 Guangzhou Asian Games. He has worked with the Great Miami Convention & Visitors Bureau to conduct research. He has been a visiting professor at several universities, such as Hong Kong Polytechnic University, Macao City University, Jinan University. In 2015, he was named "Municipal International Scholar" by the Shanghai Education Commission at the Shanghai Institute of Tourism. He has been given training courses in many cities in Mainland China as well in Taiwan and Japan. He was named a Fulbright Specialist by the Fulbright Foundation, USA, and was invited to teach at Northern University of Demark in summer 2016.

Lianping Ren, D.HTM

Dr. Lianping Ren (Eve) is Professor and Academic Coordinator for Postgraduate Programs at the Macao Institute for Tourism Studies. She holds a doctorate from the School of Hotel and Tourism Management, Hong Kong Polytechnic University, and a master's degree from the University of Warwick, UK. Dr. Ren is a Certified Hospitality Educator, qualified by the American Hotel & Lodging Educational Institute in 2009; and she obtained certification of Hotel Industry Analyst, jointly issued by the American Hotel & Lodging Educational Institute and the Smith Travel Research in 2013. Her research interest includes strategic management in the hotel and tourism industry, hotel studies, consumer behavior in hospitality and tourism, and hospitality education.

Dr. Ren's academic position also includes acting as an editorial board member of the *Journal of Global Scholars in Marketing Science* and the *Journal of Hospitality and Tourism Cases*. Dr. Ren has served as a committee member for the Quality Tourism Service Accreditation Scheme in Macau since 2016, and now also serves as a committee member for the Guiding Committee of Tourism Programs (undergraduate level) in Guangdong Province, China. Dr. Ren's early career includes two-year full-time work as a tour guide and a tour leader, as well as a cumulative eight months of experience in the hotel sector. This has given her much inspiration for academic research as well as classroom teaching.

Xiangping Li, PhD

Dr. Xiangping Li is Assistant Professor at the Macao Institute for Tourism Studies. She earned her PhD in Hospitality and Tourism Management at Virginia Polytechnic Institute and State University, and a master degree from the University of Hawaii, Manoa. Her research interests include sustainable tourism development, tourist behavior, and destination marketing. She has also won multiple government projects as principal investigator or co-investigator, such as "Compilation of References for the 'Macao Tourism Development Master Plan,'" commissioned by the Macao Tourism Development Committee, and "Malaysian Visitor Profile Study" and "Japanese Visitor Profile Study," commissioned by the Macao Government Tourism Office.

CONTENTS

CONTRIBUTORS

Sandeep Basnyat
Assistant Professor, Macao Institute for Tourism Studies, Macao, China

Nate-tra Dhevabanchachai
Mahidol University International College, Thailand

Leonardo (Don) A.N. Dioko
Professor, Macao Institute for Tourism Studies, Macao, China

Kirti Dutta
Professor of Marketing, Ansal Unviersity, Gurgaon, India

Puvaneswaran Kunasekaran
Senior Lecturer, Faculty of Social Sciences and Leisure Management, Taylor's University, Malaysia

Hung-Sheng (Herman) Lai
Assistant Professor, Dept. of Restaurant, Hotel and Institutional Management,
Fu Jen Catholic University, New Taipei City 24205, Taiwan

Xiangping Li
Assistant Professor, Macao Institute for Tourism Studies, Macao, China

Eliver Cheuk Ki Lin
Associate Professor, Air Transport Management Programme (ATM),
Singapore Institute of Technology (SIT)

Danqing Liu
Professor, School of Management, Tianjin University of Commerce, Tianjin, China, 300134

Zhaoping (George) Liu
Assistant Professor, Hospitality Business Programme (HB), Singapore Institute of Technology (SIT)

Tuan Phong Ly
Visiting Assistant Professor, Tourism College, Macao Institute for Tourism Studies, Macao, China

Kaewta Muangasame
Mahidol University International College, Thailand

Thi Hong Hai Nguyen
Senior Lecturer in Event Management, Department of Marketing, Events and Tourism,
Faculty of Business, University of Greenwich, United Kingdom

Hera Oktadiana
James Cook University, Australia; Trisakti School of Tourism, Indonesia

H. G. Parsa
Professor, Daniels College of Business, University of Denver, USA

Neethiahnanthan Ari Ragavan
Professor and Executive Dean, Faculty of Social Sciences and Leisure Management,
Taylor's University, Maylasia

Lianping Ren
Professor, Macao Institute for Tourism Studies, Macao, China

Wanlanai Saiprasert
Chiang Mai University, Thailand

Eunice Tan
College of Arts, Business, Law and Social Sciences, Murdoch University

Pimtong Tavitiyaman
The Hong Kong Polytechnic University, Hong Kong

Yodmanee Tepanon
Kasetsart University, Thailand

Henry Tsai
The Hong Kong Polytechnic University, Hong Kong

Frank C. Tsai
Assistant Professor; Dept. of Hospitality, Event Management & Tourism, Slippery Rock University of Pennsylvania, USA

Huijun Yang
Assistant Professor, Macao Institute for Tourism Studies, Macao, China

Jinlin Zhao
Professor, Chaplin School of Hospitality and Tourism Management, Florida International University, USA

ABBREVIATIONS

AOR	average occupancy rate
ARR	average room rate
ASTAR	Agency for Science, Technology and Research
B&B	Bed and Breakfast
CAGR	compound average growth rate
CBT	community-based tourism
C&E	Conference and Exhibition
CSR	corporate social responsibility
EVA	E-visitor authentication
FAST	fully automated seamless travel
FDAWU	Food Drinks and Workers Allied Union
GBA	Greater Bay Area
GEG	Galaxy Entertainment Group
GIEI	Global Islamic Economy Indicator
GMTI	Global Muslim Travel Index
GSTC	Global Sustainable Tourism Council
HAN	Hotel Association Nepal
HB	hospitality business
HZMB	Hong Kong-Zhuhai-Macau Bridge
IE	international enterprise
IMDA	Infocomm Media Development Authority
IMTI	Indonesia Muslim Travel Index
INSTO	International Network of Sustainable Tourism Observatories
IPO	initial public offering
IRs	integrated resorts
ISTA	Indonesia Sustainable Tourism Award
ITM	industry transformation map
IVA	international visitor arrivals
MAL	Malayan Airways Limited
MBO	management by objectives
MBS	Marina Bay Sands
MESCOT	Model of Economically Sustainable Community Tourism
MGTO	Macao Government Tourism Office
MICE	meetings, incentives, conventions, exhibitions

MIECF	Macau International Environmental Co-operation Forum and Exhibition
MPD	mobile positioning data
MSA	Malaysia–Singapore Airlines
MTSI	Macau Tourist Satisfaction Index
NGOs	non-governmental organizations
NP	national parks
NPMB	National Park Management Board
NTDP	National Tourism Development Plan
OTA	online travel agents
OTB	online travel booking
PATA	Pacific Asia Travel Association
PNTC	Phong Nha Tourism Centre
PPC	Provincial People's Committee
PPC of QB	Provincial People's Committee of Quang Binh
PWA	progressive web app
Revpar	revenue per available room
RFID	radio frequency identification
SAR	Special Administrative Region
SATS	Singapore Airport Terminal Services
SDGs	sustainable development goals
SEO	search engine optimization
SHA	Singapore Hotel Association
SIA	Singapore Airline
SIT	Singapore Institute of Technology
SiTF	Singapore Infocomm Technology Federation
SNS	social network system
STB	Singapore Tourism Board
STC	sustainable tourism certification
STD	sustainable tourism destination
STO	sustainable tourism observatory
SUF	Special-Use Forest
TAT	Tourism Authority of Thailand
THB	trillion Thai Baht
TR	tourism receipts
UCPN	United Communist Party of Nepal-Maoist
UNEP	United Nations Environmental Program
UNODC	United Nations Office on Drugs and Crime
UNSDSN	United Nations Sustainable Solution Network

UNWTO	United Nations World Tourism Organizations
VFR	visiting friends and relatives
WCTL	World Center of Tourism and Leisure
WEF	World Economic Forum
WTO	World Trade Organization
WTTC	World Travel and Tourism Council

PREFACE

This book follows the success of the book *The Hospitality and Tourism Industry in China,* edited by Prof. Jinlin Zhao (2018), included in the book series on Advances in Hospitality and Tourism (Series Editor-in-Chief: Prof. Mahmood A. Khan). There has been a growing interest to understand the tourism and hospitality development in Greater China, apart from Mainland China, that is, the special administrative regions of Hong Kong and Macao, as well as Taiwan, where there has been huge momentum in the hospitality and tourism industry in recent years, accompanied by changes, challenges, and new developments. The public interest has also been extended to ASEAN countries (and a few South Asian destinations that are not listed in ASEAN, such as Nepal), where there has been tremendous growth in international tourists in recent years as well.

As documented by the UN World Trade Organization (UNWTO), "Asia and the Pacific welcomed 324 million international tourists in 2017, close to a quarter of the world's total. South Asia (+10%) enjoyed the strongest results by subregion, followed by South-East Asia (+8%)." In addition, it is recorded that South Asia and South-East Asia are the fastest growing subregions in the world, with arrivals increasing an average 11% and 8% per year, respectively, in the period 2005 to 2016 (UNWTO, 2017). Tourism growing at such a phenomenal rate has inevitably changed the economic, social, and cultural landscapes in these regions. This book, thus, aims to explore the current development of hospitality and tourism industry in the regions of Hong Kong, Macao, and Taiwan as well as key countries in ASEAN. Nepal, a rising star in South Asia, is also included in this book.

Chapter 1, by Kirti Dutta and H. G. Parsa, explores the online marketing strategies of five countries including Thailand, Malaysia, Singapore, Indonesia, and Vietnam due to their geographic proximity and intense competition among themselves. Using predefined keywords and the Google search engine, they point out that while the search with the government websites does not produce consistent results, consistent search results are presented with online travel agencies. Other online service providers, such as hotels and airlines, are seriously lacking in their presence on the first page. The authors recommend that there is a need for those countries to optimize their search engine marketing strategies.

Chapter 2, by Huijun Yang, offers her insights into the opportunities and challenges to Hong Kong brought up by the development of the Greater Bay Area (GBA). Although it is forecast that GBA could further tourism development in Hong Kong and better facilitate the integration of Hong Kong to the Mainland, challenges have also been acknowledged in materializing the policies provided by development of GBA, particularly in the technological and political aspects.

Chapter 3, by Danqing Liu and Jinlin Zhao, focuses on the sustainable development in Macau's hospitality and tourism industry. The authors use stakeholder theory to illustrate the key players, such as government, enterprises, associations, education, and tourists' engagement in developing Macau as a sustainable tourism destination.

Chapter 4, prepared by Xiangping Li and Lianping Ren, focuses on the diversification efforts of the tourism industry in Macao. With the understanding that it is not sustainable to rely on a single industry and a single source market, diversification has become the priority for the Macao government since 2008 when the National Development and Reform Commission proposed to build Macao into a World Centre of Tourism and Leisure (WCTL). It is acknowledged that with years of diversification, Macao has found some success; however, due to its petite geographic area, regional cooperation should be enhanced for better results.

Chapter 5, by Frank Tsai and Hung-Sheng Lai, reviews the four stages of development of the bed and breakfast (B&B) industry in both rural and metropolitan areas in Taiwan. Nine types of B&B establishments are developed as affordable alternative sharing accommodations for travelers in Taiwan.

Chapter 6, by Yodmanee Tepanon, Wanlanai Saiprasert, and Pimtong Tavitiyaman, explores the positive and negative images of Thailand tourism from tourists' perspectives. They describe the major impacts of tourism on Thai society and present the current position of food/gastronomy and sex tourism in Thailand. They also discuss the managerial implications for Thailand tourism's future development.

Chapter 7, by Nate-Tra Dhevabanchachai, Henry Tsai, and Kaewta Muangasame focuses on the innovations in independent hotels in the Thai hotel industry. It discusses the determinants, approaches, managerial guidelines, and future development of innovation implementation in the Thai hotel industry.

Chapter 8, by Tuan Phong Ly and Nguyen Thi Hong Hai, provides an overview of the Vietnamese national park system, focusing on its past development as well as the current state of tourism operations within the system. The authors list two management models adopted by the Vietnamese national park system, namely, the parastatal model and coexisting model. While most Vietnamese parks are still utilizing the parastatal model, the coexisting model (the new model) seems to be a more promising one as it could bring some benefits by introducing the partnership between the public and private sectors. However, currently, only one national park is taking advantage of the new model; further efforts are needed to penetrate this model to other parks. In addition, the authors discuss three approaches to apply the concept of carrying capacity including (1) a *laissez-faire* approach, (2) a *relaxed* approach, and (3) a *strict* approach. They find that the majority of the national parks implement the first two approaches as overcrowding is not an issue in those parks.

Chapter 9, by Neethiahnanthan Ari Ragavan and Puvaneswaran Kunasekaran, explores tourism and community development in Malaysia toward sustainable tourism practice. It discusses community-based tourism (CBT) as a sociocultural element in sustainable tourism development in Thailand.

Chapter 10, by Zhaoping Liu and Eliver Cheuk Ki Lin, focuses on, hotel and air transport industry in Singapore. It describes innovations in the two industries. It uses two cases to illustrate many innovations encouraged by the government.

Chapter 11, by Eunice Tan, discusses wildlife tourism in the urban destination of Singapore. It reviews key narratives and semantic representations associated with wildlife tourism, natural heritage conservation, and crowd sourced advocacy for nature in Singapore. It provides a conceptual framework for sustainable urban wildlife tourism.

Chapter 12, by Lenardo (Don) A. N. Dioko, reviews the government policies and planning, key issues, and challenges of Philippine tourism development, evaluates the country's tourism performance relative to ASEAN countries, and identifies factors driving the long-term outcome of Philippine tourism. The influence of these factors in accounting for international arrivals is then empirically tested using a time series cross-sectional analysis, the results of which lend some support to the review's arguments.

Chapter 13, by Hera Oktadina, reviews the development of the tourism industry in Indonesia. With the review of the current state of the industry and development initiatives from the government, the author argues that tourism in Indonesia is advancing in various aspects. Despite the advancement, Indonesia has not reached its full potential, particularly in the fields of sustainable tourism and halal tourism development. The author contends that to increase Indonesia's global competitiveness, several pivotal aspects still require improvement, such as caretaking the environment and attending to sustainability, health and hygiene, tourist service infrastructure, safety and security, ICT readiness, and building human resources.

Chapter 14, by Sandeep Basnyat, discusses the trade union issue in the Nepalese hotel industry from hotel manager's perspective. The chapter reviews the roots and nature of employment relationship in the Nepalese hospitality industry, examines political parties' relationships with and influence on trade unions, and consider trade unions' impact on the effective management of hotel operations from hotel of managers' perspectives.

CHAPTER 1

TOURISM AND HOSPITALITY IN SOUTHEAST ASIA: CAPTURING THE CUSTOMERS ONLINE

KIRTI DUTTA* and H. G. PARSA*

G. L. Bajaj Institute of Management & Research, India, and Daniels College of Business, University of Denver, USA

**Corresponding author. E-mail: duttakirti@yahoo.co.in; haragopal.parsa@ du.edu*

ABSTRACT

Asia and the Pacific region account for the second highest share in the global tourist arrivals and tourism receipts with Southeast Asia accounting for 37.3% and 33.6%, respectively, for the same. Five countries namely Thailand, Malaysia, Singapore, Indonesia, and Vietnam, in Southeast Asia, account for 84% of tourism arrivals and 90% of tourism receipts.. The geographic proximity and competitive attractions provided necessitate strategic online marketing as most of the consumers are going online for their purchases including purchases related to travel needs.

The current chapter looks at the online marketing strategy of these five countries and explores how they are trying to capture the online customers. A primary research was carried out on the Google search engine as it has more than 90% of the market share to see how these countries perform for three search results—country name; country name and tourism; country name and tourist destinations. It is interesting to note that the government websites for various countries are not consistently appearing in all the search results. Mediators like MakeMyTrip, Tripadvisor, Hotels.com are consistently appearing in the Southeast Asian countries studied. Hotels, airlines, etc. are conspicuous by their absence on the first page and need to relook their search engine marketing and search engine optimization strategies.

1.1 INTRODUCTION

Economic growth coupled with job creation, environmental protection, development, and cultural preservation are some of the impacts of tourism sector. More specifically travel and tourism accounts for 10% of the World's GDP, 30% of services exports, and 7% of world exports and one in every 10 jobs in the year 2017 (UNWTO, 2018). Upsurge in global economy and robust demand from source markets resulted in an increase in international tourists by 6% during first half of the year 2018, with biggest growth came from Southeast Asia and Southern Europe. China was the top source market for out bound tourism in the world. Growth in Asia was driven by South Korea and India (UNWTO cited in Vietnam Investor Review, 2018).

Southeast Asia includes 11 countries namely Brunei, Malaysia, Cambodia, Myanmar, Indonesia, Philippines, Laos, Singapore, Vietnam, Thailand, and Timor Leste. In 2017, Southeast Asia received 37.3% of international tourist arrivals and 33.6% of international tourism receipts for Asia and Pacific region, which has 24% share of the global pie for tourist arrivals, and 29% for tourism receipts. Southeast Asia also accounted for 24.9% of outbound tourism generation in the year 2017. Globally, Asia and the Pacific region ranks second for both tourism arrivals and outbound tourism. It has the highest growth in Asian subregion due to robust demand, visa exemptions, and improved air connectivity (UNWTO, 2018).

1.2 TOURISM TRENDS IN SOUTHEAST ASIA

Contributions by travel and tourism directly to the GDP in Southeast Asia is expected to rise from US$ 135.8 billion in 2017 to US$ 245.5 billion by the year 2028. The total contribution to GDP is expected to rise from US$ 329.5 billion in 2017 to US$ 598.3 billion by the year 2028. The contribution to the job market will rise from 4.7% to 5.6 % of the total employment in the same time period. The various parameters to capture the long-term growth scenario from the year 2018–2028 (with adjustments made for inflation) (WTTC, 2018) are summarized in Table 1.1. A comparison with the world's growth figures shows that Southeast Asia is growing at a faster rate compared to the world growth rates.

Southeast Asia includes 11 countries namely Brunei, Malaysia, Cambodia, Myanmar, Indonesia, Philippines, Laos, Singapore, Vietnam, Thailand, and Timor Leste, and their percent share of tourism arrivals and receipts for Asia and the Pacific region is detailed in Table 1.2 in descending order.

TABLE 1.1 Southeast Asia's Travel and Tourism Long-Term Growth 2018–2028.

S. No.	Parameter	Southeast Asia % growth per annum	Asia % growth per annum	World % growth per annum
1	Direct contribution to GDP	5.5	5.6	3.8
2	Total contribution to GDP	5.5	5.5	3.8
3	Direct contribution to employment	3	2.2	2.2
4	Total contribution to employment	2.8	2.9	2.5
5	Visitor exports relative to total exports of goods and services	5.4	4.3	4.1
6	Domestic spending relative to whole economy GDP	5.1	6.1	3.9
7	Leisure spending (direct contribution calculated as share of whole economy GDP)	5.5	5.9	4.1
8	Business spending (direct contribution calculated as share of whole economy GDP)	4.7	5.0	3.2
9	Capital invested (relative to whole economy investment)	5.4	5.5	4.3

Source: World Tourism and Travel Council, 2018.

TABLE 1.2 Percent Share of Tourism Arrivals and Receipts by Southeast Asian Countries.

Country	% share of tourism arrivals	% share of tourism receipts
Thailand	11.0	14.8
Malaysia	8.0	4.7
Singapore	4.3	5.1
Indonesia	4.0	3.2
Vietnam	4.0	2.3
Philippines	2.0	1.8
Cambodia	1.7	0.9
Myanmar	1.1	0.6
Laos	1.0	0.2
Brunei	0.1	-
Timor Leste	0	0
Total	37.3	33.6

Source: United Nations World Tourism Organization, 2018.

The top five countries in Southeast Asia that account for 84% of tourism arrivals and 90% of tourism receipts are Thailand, Malaysia, Singapore, Indonesia, and Vietnam. It is interesting to note that not only inbound but also outbound tourism from Southeast Asia is growing with consumers slated to spend US$ 200 billion for online travel booking (OTB) by the year 2025.

1.3 ONLINE MARKETPLACE

1. **Market Size:** Traditionally, online marketplace offers vast number of opportunities for reaching out to consumers anytime and anyplace. The Southeast Asia's internet economy is growing three times faster compared to the United States, and is worth US$ 50 billion (Wan, 2018) in 2017 and the four major contributors are online media, online travel, food and transport, and e-commerce. Among these four, online travel is the largest industry worth US$ 27 billion (2017) and expected to reach US$ 43 billion by 2020. Online travel grew at a compound average growth rate (CAGR) of 18% during 2015–2017 and its future potential can be gauged by the fact that only 35% of bookings are being made online. It is interesting to note that the top five countries mentioned in Table 1.2 are also the constantly evolving countries with huge opportunities with Indonesia being the fastest growing economy with the largest travel market. Its travel market for online booking is expected to reach US$ 12.98 billion by the year 2020.
2. **Digital Traveler:** More than 90% of "digital travelers" in Southeast Asia carry smartphones and spend 3.6 h every day with their cell phones which is more than any other online consumer in the world. About 70% of the consumers are influenced by videos and 50% consumers are influenced to make a purchase by online advertising. Money is not the prime consideration but engaging content and speed, along with personalization are important. Interestingly, a majority of consumers are willing to pay additional money for quality experience. Thus, customer engagement can be done by providing quality brand experience to the customers according to Hermione Joye, Head, Google Travel, Asia Pacific (Wan, 2018).
3. **Women Traveler:** Increasingly, women are venturing out of their comfort zone and undertaking women-only journeys. It is interesting to note that three fourths of cultural, nature, and adventure travelers are women (Muirden, 2018). A study by booking.com shows that this

is because they feel that travelling alone empowers them. Booking accommodation from trusted sites and staying connected with family and friends through social media plays an important role in boosting their confidence (booking.com, 2014)

4. **Technological Advances:** Technological advances have allowed companies to enhance the experience of the customers who are all tech-savvy. For example, the launch of accelerated mobile page of Malaysian Airlines has helped reduce the page load time by 17 s from 20 s previously. This helped increase ROI on generic pages by 93%. Launch of progressive web app (PWA) by MakeMyTrip helped reach consumers with low-end devices who did not have space for an app on their mobile phone, thereby increasing engagement with users by 160% (Wan, 2018).
5. **Selfie Moments**: Social media is here to stay, and it is globally penetrating at a rate of 45% as of January 2019, and is expected to reach a third of the global population by the year 2021. The average time spent daily on social sites is also increasing and is influencing vacation destinations among others (Statista, 2019). Consumers are looking for the dazzling and unconventional selfie moments to post on various platforms and seek "one-of-a-kind" experience (Muirden, 2018). This needs to be kept in mind while landscaping and designing by service providers.
6. **Wi-Fi**: Wi-Fi is free and is more of a hygiene factor now. The internet speed is becoming increasingly more important for consumers, so it's vital to maintain high quality communications.
7. **M-commerce**: Globally, booking and payments through mobile systems is growing and is said to double during the period 2018–2021 (Research and Markets, 2018); hence providing payment options is important for service providers.

1.4 ONLINE TOURISM MARKETING

Online travel market is growing at a robust rate and is expected to reach USD 1,134.55 billion by the year 2023. This is mainly due to growing demand for international travel among consumers and is complemented by the fact that service providers are reaching out to the customers directly (Market Research Future, 2019). A study of the OTB for the five countries shows that (Table 1.3) growth in revenue for all the countries is studied.

TABLE 1.3 Online Travel Booking.

Country	OTB market volume by 2023 (in USD million)	OTB revenue growth rate (CAGR 2019–2023)	OTB revenue growth rate (year-on-year)	Number of users (2019) in million	Users growth rate (year-on-year)
Thailand	3655	11.1%	19.5%	7.2	13.2%
Malaysia	2196	8.5%	12%	4.8	4.2%
Indonesia	5644	7%	10.2%	17.3	7.3%
Vietnam	1930	7.6%	9.9%	9.2	9.1%
Singapore	2009	7%	9.8%	1.7	5.8%

Source: https://www.statista.com/outlook/262/127/online-travel-booking.

To study how the five countries—Thailand, Malaysia, Singapore, Indonesia, and Vietnam compare with each other for online marketing, a search was conducted on Google search engine. To make the study comparable and realistic, analysis was carried out in the same time period for search results coming on the first page. Google was used as it is the number one search engine worldwide as can be seen from Table 1.4 (Statscounter Global Stats, 2019).

TABLE 1.4 Worldwide Market Share for Search Engines.

Search engine	Market share (%)
Google	92.86
Bing	2.41
Yahoo!	1.82
Baidu	0.89
YANDEX RU	0.59
DuckDuckGo	0.41

Source: Statscounter Global Stats, 2019.

1.4.1 ONLINE TOURISM MARKETING—SOUTHEAST ASIA

Search was conducted for studying how Southeast Asian countries compare with each other when a consumer searches randomly and then specifically for the particular country and searched for tourism Southeast Asia. Results on the first page show the link for further Google page (Fig. 1.1) for popular destinations with Bangkok (capital of Thailand) and Singapore are the ones listed from our five countries; sites like Wikitravel, Touropia, Thrillist, TravelTriangle which describe various rankings for best places to visit.

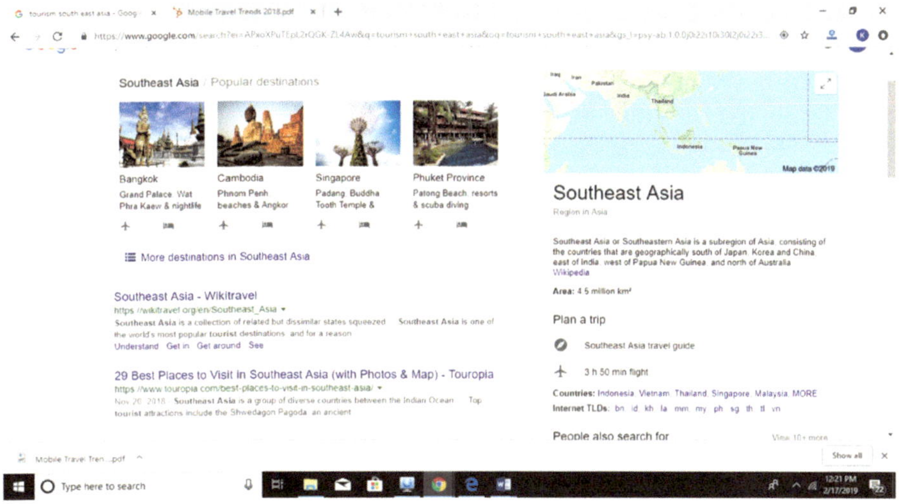

FIGURE 1.1 Google search—tourism Southeast Asia.

Source: Google, 2019a.

When you click on Bangkok and Singapore it takes the visitor to a Google page with further information about the place as shown in Figure 1.2.

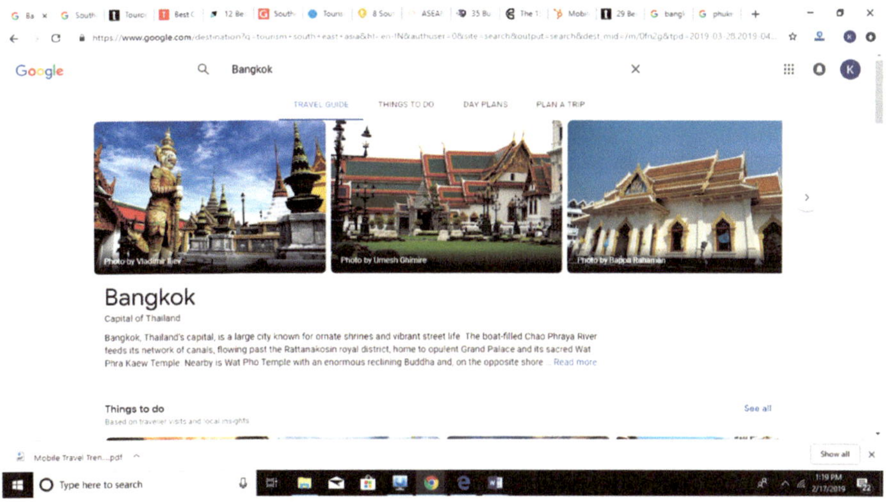

FIGURE 1.2 Google result for Southeast Asia—tourism—Bangkok.

Source: Google, 2019b.

A search for terms like "holiday Southeast Asia," showed top result from traveltriangle.com blog; article from Telegraph for 10 best places to travel in Southeast Asia; MakeMyTrip blog, trip ideas by Audley Travel, etc. Terms like "holiday destinations Southeast Asia"; "travelling Southeast Asia"; "best places to visit in Southeast Asia"; "best country to visit in Southeast Asia"; "top places to visit in Southeast Asia"; "tourist destinations in Southeast Asia" all showed similar results. This shows that none of these countries are using these terms for their search engine marketing. It is interesting to note that a search for "Thailand, Malaysia, Singapore, Indonesia, and Vietnam" showed results for travelindependent.com, Quora, Secom (security company with offices in Thailand, Indonesia, Malaysia, etc.), 7MTours, traveltriangle.com (ad), etc. but no results were seen for any of the government sites.

Web pages on the website are created by using markup language like HTML that uses various tags and the web browsers of search engines read these tags. The meta tags define the page's content and web browsers crawl the website and then list the information in the form of indexing. When someone searches for information on the search engine, the search engine goes through the indexed information and lists the information on the search engine results page. Websites can get top listing in organic searches by influencing the search engines crawling and indexing activities and these techniques are called as search engine optimization (SEO). SEO helps in getting quality customers to the website thereby contributing to leads and revenue (Gupta, 2018). Companies can also pay for various ad words and get top rankings and this is called as search engine marketing. Let us see how these countries and their service providers' fare organically in the country-specific Google search.

1.4.2 ONLINE TOURISM MARKETING—THAILAND

A search for the word "**Thailand**" gives results for further Google pages, Wikipedia link, "Official website of Tourism Authority of Thailand," Wikitravel, TripAdvisor, Lonely Planet, etc. Search result for "**Tourism Thailand**" (Fig. 1.3) shows the Google pages for popular destinations by two results for "Official website of Tourism Authority of Thailand," followed by results from Yatra, blogs, TripAdvisor, etc. Tourism authority of Thailand provides options for destinations, attractions, and numerous options for planning trips also.

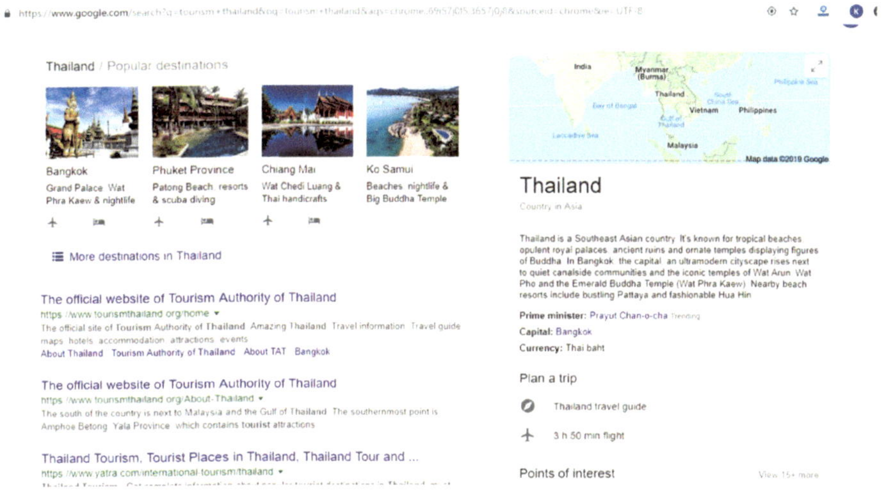

FIGURE 1.3 Google search result—tourism Thailand.

Source: Google, 2019c.

Search for "**Thailand tourist destinations**" shows top result for—Thailand Tour Packages by MakeMyTrip (ad); and Touropia blog; PlanetWare; Holidify, etc. as organic search results but none by official website of "Tourism Authority of Thailand." Touropia blog has details for top 25 destinations to visit and lots of ads like PNB, ICICI, GoDaddy along with an link for "where to stay" next to the destination which takes visitor to a landing site by BookingBuddy—a company for travel search by TripAdvisor that helps in stay booking by comparing sites like Expedia, Trivago, TripAdvisor, and Hotwire. PlanetWare also talks about top 10 destinations to visit with a link for "where to stay" which takes visitor to landing page of TripAdvisor with a lot of hotel deals. Holidify is the first link that shows options for weekends and destinations along with packages and option for customized package as well. TripAdvisor shows multiple options for trips and also has a list of things to do. A search for "holiday Thailand" shows players like Yatra, Thomas Cook, Last Minute, Office holidays, Virgin Holidays, MakeMyTrip, Expedia (in the same order) on the first page and none from "Official website of Tourism Authority of Thailand."

A search for Thailand's capital "**Bangkok**" shows tour packages by MakeMyTrip (ad); and Bangkok.com, Wikipedia, two by TripAdvisor, peninsula.com (five-star luxury hotel) etc. as the top results. Bangkok.com is powered by hotels.com with options for ours, hotels, nightlife etc.—a one stop shop for all travel-related details.

A search of the meta tags shows the keywords that the Tourism Authority of Thailand has linked words like "Thailand, travel Thailand, Thailand tourism, visit Thailand, tourism authority of Thailand, Thailand tourism board, amazing Thailand, Thailand accommodation, Thailand hotels, Thailand holiday, Thailand vacations, Thailand attractions, Thailand restaurants, Thailand map, Thailand guide, Thailand shopping, Thailand events, Thailand festivals, Thailand islands, Thailand culture, Thailand nightlife, travel to Thailand, travel around Thailand."

Managerial implication: Tourism Authority of Thailand website link shows among the top results in searches like Thailand, tourism, etc. They have tagged the meta name of the website title description along with the keywords. They should be adding keywords like "tourism Southeast Asia, Bangkok (and other top cities that draw tourism)" to get a higher ranking when a tourist is searching for tourism in the area.

1.4.3 ONLINE TOURISM MARKETING—MALAYSIA

A search for "Malaysia" shows results for further Google pages, Wikipedia, Wikitravel, Tourism Malaysia Corporate site (government web site to promote tourism in Malaysia); Britannica, Lonely Planet, etc. Search for "Tourism Malaysia" shows Malaysia tourism packages by MakeMyTrip (ad), Google pages, Tourism Malaysia Corporate site, Official website of Tourism Malaysia which is linked to Tourism Malaysia Corporate site (corporate site) but with different URL (Malaysia.Travel) (see figures below); packages by Yatra, MakeMyTrip, Thrillophilia, Wikipedia, etc. Search for "Malaysia tourist destinations" shows results for tour packages by MakeMyTrip (ad), and results from Touropia, Tripadvisor, etc. Touropia shows details for top 23 destinations to visit and lots of ads like ICICI, Adobe, Marriott along with an link for "where to stay" next to the destination which takes visitor to a landing site by BookingBuddy.

A search for "holiday Malaysia" shows public holidays in Malaysia, Thomas Cook, One Stop Malaysia, Yatra, and none from the government.

A search for Malaysia's capital "Kuala Lumpur" shows results for further Google pages, Wikipedia, kualalumpur.ws which is powered by hotels.com, Tripadvisor, and Lonely Planet where you can book hotels, flights, etc.

A study of the website HTML shows that only the meta name has been given "Tourism Malaysia Corporate Site" and no other keywords have been provided.

A study of the official website Tourism Malaysia shows that the following keywords have been added with the meta name "visit Malaysia 2020, travel to Malaysia, holiday in Malaysia, tourist attractions, Malaysia truly Asia, events, destinations, attractions, travel packages, Malaysia travel guide, tourist information, tourism board, flights to Malaysia, travel planner, trip planner, holiday packages, Malaysia travel destinations, Malaysia events, Kuala Lumpur, tours and travels, Asia, ASEAN, official website."

Managerial Implications: The official government website needs to add other keywords related to tourism and search of related information in Southeast Asia if they want to compete with the other countries and service providers in the region.

1.4.4 ONLINE TOURISM MARKETING—INDONESIA

A search for "Indonesia" shows Google pages, Wikipedia, Wikitravel, official website for Indonesia tourism—Indonesia.travel (by Ministry of Tourism, Republic of Indonesia which has destination highlights and travel packages); indonesia-tourism.com (powered by indonesiacommerce.com), Britannica, Lonely Planet, etc. All these provide a number of options for travel-related needs. A search for "tourism Indonesia" shows results for tour packages by MakeMyTrip (ad), and results for indonesia.travel (official website of Indonesia Tourism), Google pages, indonesia-tourism.com, Wikipedia, MakeMyTrip (organic), Tour My India, Yatra, PlanetWare's top 15 destinations, visitindonesia.co.in by Embassy of Republic of Indonesia, tourismindonesia.com (with news, hotels, link to booking.com, and Wikimapia) and Tripadvisor. Search for "Indonesia tourist destinations" shows results for tour packages by MakeMyTrip (ad), and results for Google pages, PlanetWare, Touropia, Tour My India, etc. A search for "holiday Indonesia" shows a number of links for various holidays, last minute, Tripadvisor, MakeMyTrip, and Tripzilla on the first page but none by the government.

Search for Indonesia's capital "Jakarta" shows results for Google pages, Wikipedia, Indonesia.travel, Wikitravel, Tripadvisor, Britannica, Lonely Planet, etc.

A study of the webpage elements shows meta description as "Catch a glimpse of Indonesia's bewitching attractions without having to put on your shoes and discover the ultimate destination that matches your soul. So go ahead and steal a glance at your ideal holiday" and title as "the official website of Indonesia Tourism—wonderful Indonesia."

Managerial Implications: As can be seen that the meta keywords should be included keeping in mind the major keywords that a tourist can be searching for. Only the title containing "Indonesia tourism, wonderful Indonesia" is of help and therefore the weak results in the searches.

1.4.5 ONLINE TOURISM MARKETING—VIETNAM

Search for "Vietnam" shows results for Google pages, Wikipedia, Lonely Planet, Britannica, Wikitravel, Tripadvisor, smarttraveller.gov.au (by Australian Government's Department of Foreign Affairs and Trade). A search for "tourism Vietnam" shows results for tour packages by MakeMyTrip (ad), and results for vietnam.travel (official tourism website of Vietnam) Wikipedia, MakeMyTrip (organic), Tripadvisor, Thrillophilia, vietnamtourism.gov.vn (Ministry of Culture, Sports and Tourism, Vietnam National Administration of Tourism) and Lonely Planet. Search for "Vietnam tourist destinations" (Fig. 1.4) shows results for tour packages by expedia and MakeMyTrip (ads) and two links for Touropia, link for PlanetWare, Tour My India, Vietnam-guide (by hotels.com), culture trip, Tripadvisor, Lonely Planet, etc. A search for "holiday Vietnam" shows links for various holidays followed by results for Tripadvisor, MakeMyTrip, Thomas Cook, etc. but none by the Government.

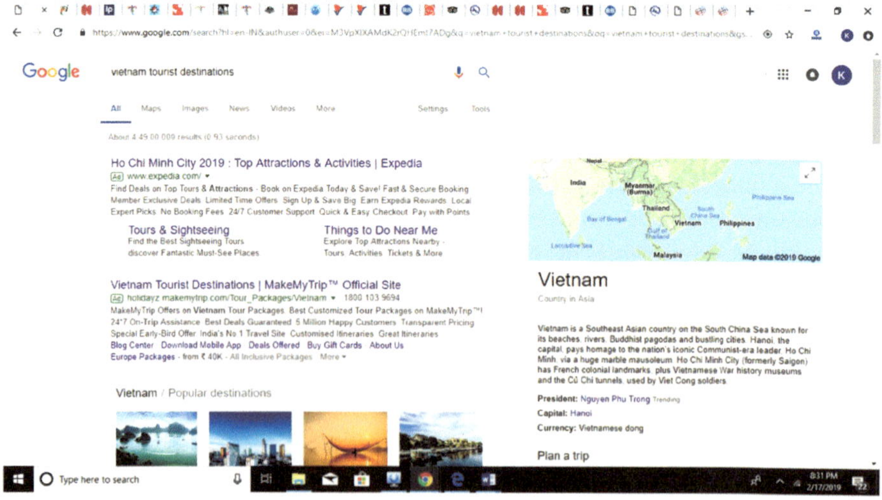

FIGURE 1.4 Google search—Vietnam tourist destinations.

Source: Google (2019d).

A search for Vietnam's capital "Hanoi" shows results for Google pages, Wikipedia, Vietnam-guide, Lonely Planet, two by Tripadvisor, Lonely Planet, etc.

A study of vietnam.travel (official tourism website of Vietnam) shows meta name description "Discover the highlights of Vietnam through the country's official tourism website. Plan your trip with advice on Vietnam's best destinations, what to do in Vietnam, visas for Vietnam, where to go in Vietnam, and much more." Tagging is not done for tourist destinations and "Vietnam" has been repeated five times resulting in "keyword stuffing" which is marked as a negative during SEO ranking hence resulting in absence from the first page of search engine result when searching for Vietnam.

A study of thevietnamtourism.gov.vn (Ministry of Culture, Sports and Tourism, Vietnam National Administration of Tourism) website shows that number of keywords added to the site are "TITC, Trung tam Thong tin du lich, Vietnam, Vietnam tourism, tourist, tourism, travel, Vietnam Travel, travel to Vietnam, festival in Vietnam, hotel, hotel in Vietnam, tourist office, travel agency, tourist company, tourist company in Vietnam, tour, tour in Vietnam, booking, reservation, restaurant, restaurant in Vietnam, destination, destination in Vietnam, cultural tourism, ecotourism, world heritage sites, patrimoine-mondial au Vietnam, entertainment, entertainment in Vietnam, visa, embassy, embassies, information, useful reading, transport, customs, guide, agence, trip, excursion, flight, voyage, Tong cuc Du lich Vietnam, Du lich Viet Nam, Du lich, Du lich Van hoa, Lu hanh, le hoi, diem du lich tai Vietnam, diem thamquan, khu du lich, du lich sinhthai, nghi mat, du lich nghiduong, di san, congty du lich, khach san, nha hang, co so luutru, danhthang, vuichoi, giai tri, Portal, Hanoi, Ha Noi, Saigon, Saigon, Tin tuc, Tintuc, Van ban, VanBan, tin lanhdao, tin hoat dong, Van ban quanlynhanuoc." This shows that the website must be popping up in search results related to all these words.

Managerial Implications: The metaname keywords should be added and meta name description needs to be refined for getting effective visits to the official tourism website of Vietnam. The word Vietnam occurs 5 times in vietnam.travel and 14 times in vietnamtourism.gov.vn. As discussed this results in "keyword stuffing" which scores a negative during SEO ranking thereby not resulting on a high ranking on search engine ranking pages and vietnamtourism.gov.vn scores lower than Vietnam.travel.

1.4.6 ONLINE TOURISM MARKETING—SINGAPORE

A search for "Singapore" shows results for Google pages, Wikipedia, stb.gov.sg (by Singapore Tourism Board which also had a link to Tripadvisor at bottom of page), visitsingapore.com (also a copyright with Singapore Tourism Board); Singapore Air, gov.sg (Ministry of Communications and Information), Wikitravel, Singapore Tourism Board (under Ministry of Trade and Industry), Lonely Planet, and National University of Singapore. It is interesting to note that Singapore Tourism Board website was last updated on 5 Feb, 2018 while gov.sg (last update 8 Feb, 2019) and visitsingapore.com (last update 2019) were more current. A search for "tourism Singapore" showed results for National Gallery Singapore and MakeMyTrip (ads) along with Google pages, visitsingapore.com, Wikipedia, Tripadvisor, Thomas Cook, gov.sg (Singapore Tourism Board), MakeMyTrip (organic), Yatra, NewsAsia-Singapore, etc. A search for "Singapore tourist destination" (Fig. 1.5) shows number of images for various tourist attractions in Singapore followed by packages by national gallery and Aiwa Travels (ads) along with PlanetWare, MakeMyTrip, Singapore Guide (by hotels.com), Touropia, Tour My India, Tripadvisor, Thrillophilia, culture trip, etc. A search for "holiday Singapore" showed a number of links for the dates of public holidays followed by MakeMyTrip, Yatra, Thomas Cook, etc. and none by the government. A separate search for capital was not done as Singapore is a city state.

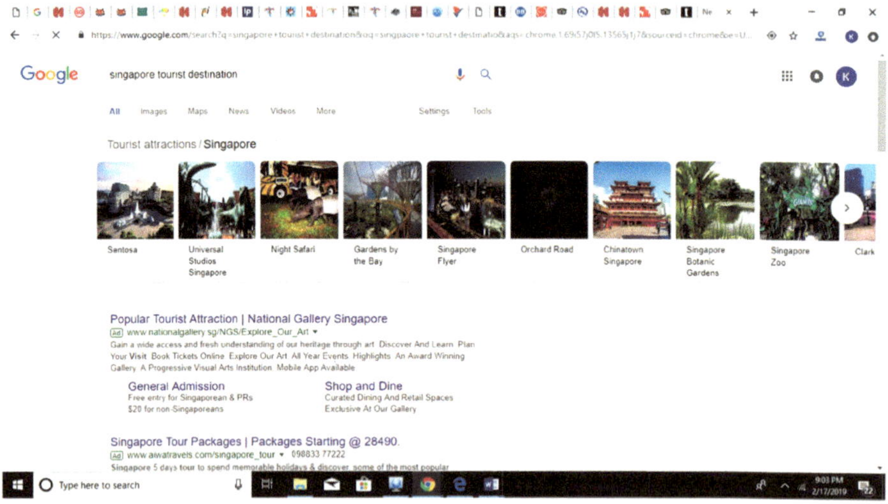

FIGURE 1.5 Google search—Singapore tourist destination.

Source: Google, 2019e.

A study of the various tourism sites by the government likestb.gov.sg (by Singapore Tourism Board) and visitsingapore.com (also a copyright with Singapore Tourism Board) shows the following: stb.gov.sg (by Singapore Tourism Board) shows only title as "Singapore tourism board" and no description or keywords. visitsingapore.com (also a copyright with Singapore Tourism Board) shows keywords like "What to do in Singapore, shopping in Singapore, dining in Singapore, Singapore tourism, Singapore attractions, museum in Singapore, activities in Singapore, Singapore travel guide, nightlife in Singapore, Singapore tourist attractions, Singapore restaurants, Singapore shopping, events in Singapore, travel to Singapore, Singapore tourism board, about Singapore, places in Singapore, Singapore tourist places, tourist spots in Singapore, what to see in Singapore, things to see in Singapore, fun things to do in Singapore, sightseeing in Singapore, places to see in Singapore, travel Singapore" thus resulting in a higher page ranking than the stb.gov.sg website.

Managerial implications: Two websites to promote tourism leads to confusion and division of the traffic and the government might consider diverting traffic from one site to the other and updating one site only (stb.gov.sg site in any case was updated in Feb, 2018).

1.5 CONCLUSION AND RECOMMENDATIONS

According to the Travel and Tourism Competitiveness Report, 2017 by World Economic Forum the global ranking of the countries for tourism are: Singapore—13; Malaysia—26; Thailand—34; Indonesia—42; Vietnam—67. This shows the high level of competition between Southeast Asian countries, along with competition from other Asian destinations and the rest of the world. Tourism Authority of Thailand does not show up in the search for tourism in Southeast Asia and all countries search but shows up in results for keywords like "Thailand," "Tourism Thailand" but not for "Thailand Tourist destinations." Tripadvisor consistently shows up on the first page for all the keywords.

Malaysia Government's website to promote tourism shows up in search for Malaysia but not for tourist destinations. Ministry of Tourism website shows up in the country-specific search and tourism-related search but not for tourist destinations. It shows up for the Jakarta search also. A website by Embassy of Republic of Indonesia also shows up in the tourism-related search. The official tourism website of Vietnam and the website by Ministry of Culture, Sports and Tourism does not come up in the country-specific search. Singapore was the only country showing good results on Google search primarily drawing strength from the fact that it is a city state. There are two websites by Singapore Tourism Board—stb.gov.sg and visitsingapore.com, and a Gov.sg (by Ministry of Communications and Information) showed up for country specific and except for stb.gov.sg also showed up in tourism search but not for Singapore tourist destinations. Different websites by the government and all promoting tourism are confusing and rather there should be one website with all the relevant information. Hotels.com has separate pages for Bangkok, Kuala Lumpur, Vietnam, Hanoi, and Singapore. MakeMyTrip is doing a lot of paid advertising with Google and can be found at the top for all the Southeast Asian countries. Tripadvisor, Touropia, Tour My India are some of the other sites present on the top page for most of the searches related to country-specific tourism and tourist destinations.

It is interesting to note that Peninsula Hotels were the only hotel to show up on the top page (for Bangkok search) amongst all the searches. None of the international hotel chains like Hilton, Four Seasons, Accor, etc., which have properties in Southeast Asia, showed up on any of the searches. Hotels need to be more proactive to capture the online customers as by the year 2030 there will be 1.8 billion tourists (Muirden, 2018) visiting must-see places across the globe. Digital is here to stay and with consumers going on line to search for options, be influenced, and make purchase decisions, marketers cannot ignore this platform or afford to underperform.

KEYWORDS

- **online marketing**
- **tourism**
- **Southeast Asia**
- **search engine ranking**
- **Google**

REFERENCES

Booking.com, Holidays With Me, Myself and I Give Women a Self Esteem Boost. https://news.booking.com/holidays-with-me-myself-and-i-give-women-a-self-esteem-boost/ (accessed June 23, 2019).

Gupta, S. *Digital Marketing;* McGraw Hill Education (India) Private Limited: Chennai, 2018.

Market Research Future. Online Travel Market Driven by Growing Desire for International Travel|Online Travel Industry Analysis by Platform Type, Mode of Booking, Service Type and Region. https://www.globenewswire.com/news-release/2019/05/07/1818412/0/en/Online-Travel-Market-Driven-by-Growing-Desire-for-International-Travel-Online-Travel-Industry-Analysis-by-Platform-Type-Mode-of-Booking-Service-Type-and-Region.html (accessed June 20, 2019).

Muirden, M. Hospitality and Travel Trends Set to Shape 2019. http://www.luxnomade.com/2018/12/20/news-hospitality-travel-trends-set-to-shape-2019/ (accessed Jan 27, 2019).

Research and Markets. Global M-Commerce 2018. https://www.researchandmarkets.com/research/rwqwtm/global_mcommerce?w=4 (accessed Feb 6, 2018).

Statista. Mobile Social Media – Statistics and Facts. https://www.statista.com/topics/2478/mobile-social-networks/ (accessed Feb 6, 2019).

Statscounter Global Stats. Search Engine Market Share Worldwide. http://gs.statcounter.com/search-engine-market-share (accessed Feb 6, 2019).

United Nations World Tourism Organization. UNWTO Tourism Highlights, 2018 Edition, UNWTO, Madrid. DOI: https://doi.org/10.18111/9789284419876.

Vietnam Investor Review. World Tourism Numbers up 6% in First Half of 2018: UN. https://www.vir.com.vn/world-tourism-numbers-up-6-in-first-half-of-2018-un-63013.html (accessed Dec 30, 2018).

Wan, C. WiT Singapore 2018: South-East Asia Travel is Huge with Consumers who Want Speed, Engagement and Personalization. https://www.webintravel.com/wit-singapore-2018-south-east-asia-travel-is-huge-with-consumers-who-want-speed-engagement-and-personalisation/ (accessed Jan 1, 2019).

World Economic Forum. Travel and Tourism Competitiveness Report 2017. http://reports.weforum.org/travel-and-tourism-competitiveness-report-2017/asia-pacific-results/ (accessed Feb 6, 2019).

World Travel and Tourism Council. Economic Impact 2018: Southeast Asia. https://www.wttc.org/-/media/files/reports/economic-impact-research/regions-2018/southeastasia2018.pdf (accessed Jan 21, 2019).

WEB REFERENCES

Google (2019a). https://www.Google.com/search?ei=APxoXPuTEpL2rQGK-ZL4Aw&q=tourism+south+east+asia&oq=tourism+south+east+asia&gs_l=psy-ab.1.0.0j0i22i10i30l2j0i22i30l2j0i22i10i30j0i22i30l4.1817780.1825690..1827555...0.0..3.375.8149.1j0j19j10......0....1..gws-wiz.....0..0i71j0i131j0i67j0i131i67.EI3UlYePSo4 (accessed Feb 11, 2019).

Google (2019b). https://www.Google.com/destination?q=tourism+south+east+asia&hl=en-IN&authuser=0&site=search&output=search&dest_mid=/m/0fn2g&tpd=2019-03-28,2019-04-04&sa=X&sqi=2&ved=2ahUKEwim8uzglcLgAhVj6uAKHQ_QDd0Q69EBKAAwAHoECAQQAw#dest_mid=/m/0fn2g&tcfs=EhwaGAoKMjAxOS0wMy0yOBIKMjAxOS0wNC0wNCAB&tpd=2019-03-28,2019-04-04 (accessed Feb 11, 2019).

Google (2019c). https://www.Google.com/search?ei=lTJpXMHaIofgrQHl5LRo&q=tourism+thailand&oq=tourism+thailand&gs_l=psy-ab.12..0i71l8.0.0..520330...0.0..0.0.0.......0......gws-wiz.DXy9JA7xibI (accessed Feb 11, 2019).

Google (2019d). https://www.Google.com/search?hl=en-IN&authuser=0&ei=M3VpXIXAMdK2rQHEmI7ADg&q=vietnam+tourist+destinations&oq=vietnam+tourist+destinations&gs_l=psy-ab.3..0j0i22i30l9.528276.534204..534509...0.0..1.628.6174.0j11j1j2j1j5......0....1..gws-wiz.......0i71j0i67j0i131i67j0i7i30j0i131j0i22i10i30.Rl2zoTCkJtg (accessed Feb 11, 2019).
Google (2019e). https://www.Google.com/search?q=singapore+tourist+destination&oq=singpaore+tourist+destinatio&aqs=chrome.1.69i57j0l5.13565j1j7&sourceid=chrome&ie=UTF-8 (accessed Feb 11, 2019).

CHAPTER 2

WHAT ARE THE OPPORTUNITIES AND CHALLENGES TO HONG KONG IN THE DEVELOPMENT OF THE GREATER BAY AREA?

Huijun Yang

Institute for Tourism Studies, Macao, China

E-mail: june@ift.edu.mo

ABSTRACT

The Greater Bay Area (GBA) is an initiative of the Chinese government to coordinate co-competition among 11 cities and Special Administrative Regions (SAR) around the Pearl River Delta. Guangzhou, Shenzhen, Hong Kong, and Macao are expected to be drivers of the economy. The GBA concept will take tourism development in the region to a new height, providing rich resources to fuel the further development of multi-destination travel. Hong Kong's tourism has entered a new chapter with two mega projects of completed: the Guangzhou–Shenzhen–Hong Kong Express Rail Link (XRL) and Hong Kong–Zhuhai–Macau Bridge (HZMB), which are pivotal to the long-term integration with the Mainland. How can Hong Kong leverage these new infrastructures to explore new opportunities and enhance its competitiveness? How will the metropolis ensure the improvement of life quality of local residents? This chapter seeks to discuss the opportunities and challenges to Hong Kong's future development of its tourism and hospitality in the GBA.

2.1 INTRODUCTION

Hong Kong is well known in the world for being a destination popular with millions of tourists every year. The combination of eastern and western cultures, world-class shopping centers, and international cuisine are the special features of Hong Kong. Visitors from different countries are attracted to the metropolis. In 2018, Hong Kong received 65.1 million tourists (Hong Kong Tourism Commission, 2019). With the completion of new transportation infrastructures connecting the Mainland and Hong Kong, including the Guangzhou-Shenzhen-Hong Kong High-speed Rail Link (Hong Kong Section) and the Hong Kong–Zhuhai–Macau Bridge (HZMB), the entire Guangdong-Hong Kong-Macau Bay Area (Greater Bay Area) will become a region to provide a mixture of tourism features and experiences.

After over 100 years of being under the rule by Britain, Hong Kong was handed over back to China in 1997. It was designated as a Special Administrative Region (SAR). Even since being occupied and administered by the British, Hong Kong has been a free seaport welcoming traders from all over the world. From 1970s to 1990s, Hong Kong underwent continuous prosperity in its economic development. Since 1990s, the Chinese Mainland has experienced high speed growth in economy, thanks to its "open-door policy." With its labor-intensive manufacturing businesses transferred to the Mainland, Hong Kong has transformed its main economic activities to service-based industries, including finance, logistics, retail, and tourism (Choi, 2007). Hong Kong has been regarded by all as a gateway to China—the world's largest market (Hong Kong Tourism Board, 2019). For eight consecutive years, this metropolis has been on top of the chart of "most visited destinations in the world" (Euromonitor International, 2018). Tourist from over 170 countries may join visa-free visits to Hong Kong for a certain period of time—from 7 to 180 days (GovHK, 2017). As one of the "four pillar industries"—namely finance, logistics, retail, and tourism—tourism/hospitality industry has provided quite a number of job opportunities to the local citizens since Hong Kong's shift to service businesses in 1980s and 1990s. Since 1980, on average, the total number of visitor arrivals has been increasing at around 10% annually and tourism industry is one of the top hard currency earners for Hong Kong (Barrow, 1994). Japan, Taiwan, and ASEAN are the major source markets and in 1996 a total of 11.7 million tourists visited Hong Kong (Mok and Dewald, 1999). In 1997, it showed an overall decrease of 11.1% of tourist arrivals to 10.4 million tourists (Hong Kong Tourist Association, 1997). In 1998, the financial crisis swept over Asia. With the support from the Central Government of China, Hong Kong survived the challenge and the local economy recovered its vitality, thanks to the resilience of tourism development. In comparison with 1997, the number of tourist arrivals in 2002 increased by 47% (Choi, 2007).

In 2003, Hong Kong suffered from a heavy blow inflicted by the breakout of SARS. In 2003, the Hong Kong Government presented the Individual Visit Scheme (IVS) to cope with such a crisis. The IVS enabled a huge increase of tourist arrivals from the Chinese Mainland. In 2004, 35% of the tourists to Hong Kong were from the Mainland; in 2017, this percentage reached 57.1%. By the end of 2017, over 238 million domestic tourists visited Hong Kong since the IVS had been implemented (Hong Kong Tourism Commission, 2018). According to the IVS, citizens from 49 cities in the Mainland may tour Hong Kong as independent travelers. This move drastically drove up the local hospitality.

Despite the numerous economic benefits from the Mainland Chinese tourists, the local communities of Hong Kong have become more concerned about the negative impact of overgrowth of tourism. The problems have become more prominent since the expansion of the IVS in 2009, such as price increase of consumer goods and properties (Chow, 2012), overcrowding and goods supply shortage (Tsang, 2013; Yang et al., 2017), cultural conflicts between tourists and locals, shortage of public medical resources (Ye et al., 2011), and negative impacts of local residents' life quality (Yang et al., 2017; Ye et al., 2014). Some local residents have descended on some areas to protest Mainland visitors and parallel traders since 2013. Parallel traders from the Mainland take advantage of this scheme to transport goods from Hong Kong to the Mainland, causing shortage of household goods in Hong Kong and inconveniencing other passengers by distributing their goods at stations near the border (Ye et al., 2014). Such activities have caused a series of social issues in Hong Kong. In 2014, tourist arrivals to Hong Kong registered a notable growth of 12%, and the year after that, this percentage dropped by 2.5% (including a decline by 3% of the tourists from the Mainland). In 2016, tourism slumped by 4.5% (with a decline of visitation from the Mainland being 6.7%) (Hong Kong Tourism Board, 2015, 2016). After a decline in 2015 and 2016, the total number of visitors increased in 2017, rising 3.2% (GovHK, 2018). Of those, Mainland visitors were 3.9% higher than in 2016.

Hong Kong has about 7.4 million citizens and the limited space received more than 58.4 million tourists from the world (Hong Kong Tourism Board, 2017). Based on a statistical review of Hong Kong Tourism Board (2018a), the short haul markets of Taiwan and Korea experienced a decline of 4.3% and 4.5%, respectively. Visitor arrivals from long haul markets increased by 2.8%, with the United States increasing by 7.3%, United Kingdom by 3.1%, and Australia by 2.2%. As for the Chinese Mainland, visitors grew by 14.8% in 2018. The large number of Chinese tourists come to Hong

Kong via Guangzhou–Shenzhen–Hong Kong Express Rail Link (XRL) and ZHMB and boosted the local retail, service, and other industries related to hospitality. In 2018, Hong Kong's revenue of tourism/hospitality was recorded as 42 billon US dollars, compared with 38 billon in the previous year (CEIC, 2019). The major market area showing the highest growth rate was the Chinese Mainland visitors; therefore, important to Hong Kong's tourism industry.

2.2 THE GREATER BAY AREA (GBA)

In 2017, the Chinese government made the decision to leverage the advantages of Hong Kong and Macao together with nine cities of Dongguan, Huizhou, Foshan, Shenzhen, Guangzhou, Jiangmen, Zhuhai, Zhaoqing, and Zhongshan with the purpose of constructing an economic hub. Up to 2017, the GBA had a population of 70 million people with a GDP translated to 1.51 trillion US dollars. If compared with major economies in the world, in GBA equals Russia, exceeds Australia, and get very close Korea in GDP (2022 Foundation, 2019). Among these 11 cities and metropolises, Hong Kong is the global hub of international air cargo (ACI, 2017), and No. 2 center of initial public offering (IPO) in the world (Yiu, 2017); Macao is the most prosperous gaming city in the world (Casino News, 2018), while Hong Kong, Shenzhen, and Guangzhou are all pivotal container ports (AAPA, 2018). The Chinese government designed the plan to integrate the 11 cities with the purpose to compete with the bay areas in the United States and Japan. Funding and investment will be infused into this area to support the growth of hi-tech, manufacturing industry, and finance. All these efforts are expected to lay a base for "Smart City" in the near future (Shih, 2019).

The GBA's infrastructural construction is marked by the high-speed railway connecting Guangzhou and Hong Kong and a bridge serving Hong Kong, Zhuhai and Macao. In September 2018, Hong Kong joined the network on China's extensive high-speed railway grid. XRL is built to improve life quality, facilitate trade, and provide more flexibility in GBA. A high-speed train covers the distance of 87 miles from Hong Kong to Guangzhou in 47 min, including Shenzhen, Dongguan, and Guangzhou in the 1-h circle for commuters. Through joining the China's massive high-speed railway grid, tourists and businessmen from the Mainland may easily enter Hong Kong for shopping and other economic activities. Meanwhile, Hong Kongers may find it convenient to settle down in the GBA cities and travel even further to explore the vast continent (Lui, 2019).

In 2018, the mega-sized HZMB was completed and put into operation. This longest bridge-tunnel system is designed for use of 120 years and includes Zhuhai and Macao in the 1-h circle for Hong Kongers. It is expected that Hong Kong's industries of finance, logistics, tourism, and professional support will expand to GBA and may even reach other provinces. The HZMB, a milestone in the construction of GBA, will help boost the economic and technological development to compete with the Silicon Valley in California or Keihanshi Region which consists of Tokyo, Osaka, and Kobe in Japan (CGTN, 2018).

2.3 THE CHALLENGES TO HONG KONG TOURISM INDUSTRY

Hong Kong's tourism industry is mainly driven by the global economic recovery, largely steady growth in the Mainland, stable exchange rate of renminbi, and efficient cross-border transportation. Nevertheless, Hong Kong still faces a number of challenges ahead.

Hong Kong is a shopping paradise for tourists, which is well-known for its department stores, magnificent shops, and street markets. As technology and new media are on the rise, retail landscape is being reshaped and consumer behavior is changing. In recent years, online shopping has become a major trend around the world. Alibaba and JD.com have attracted millions of online shoppers who should been customers to Hong Kong or other areas (Hong Kong Business, 2019). Since 2014, cross-border e-commerce has become popular with Chinese netizens with improved online shopping experience and more and more Chinese customers are making purchasing products from overseas merchants (Nielsen Survey, 2015). In China, online retail sales accumulated to 1060 billion US dollars, almost 4.5 folds increase over the past 5 years, and more overseas retailers have established their presence in the Mainland (Choi, 2018). One third of global luxury products are bought by the Chinese consumers who have grown accustomed to online shopping has a deep implication for luxury retailers based in Hong Kong (Hong Kong Business, 2019). If Hong Kong still wishes to maintain high growth rate of tourism/hospitality, it will be hard to put all the hopes on the shoppers who come to Hong Kong. In 2018, the Chinese Mainland remained as the largest source market for Hong Kong, accounting for 78% of the total arrivals, and the figure was 14.8% higher than that in 2017 (Hong Kong Tourism Commission, 2019). Though the number of arrivals has risen from its low, the sale of high-end luxury goods remains much lower than its peak (Choi, 2018). In 2013, Hong Kong sold luxury goods, such as jewelry, wrist watches, diamonds, or valuable gifts worth of HKD118.3, compared with HKD85.33 billion in 2018 (Statista, 2019). The statistic displays the sale of high-valued goods in Hong Kong from 2008 to 2018 (Fig. 2.1). If the number of high-spending tourists continues to decline, Hong Kong's hospitality industry will surely meet challenges when making effort to attract more visitors.

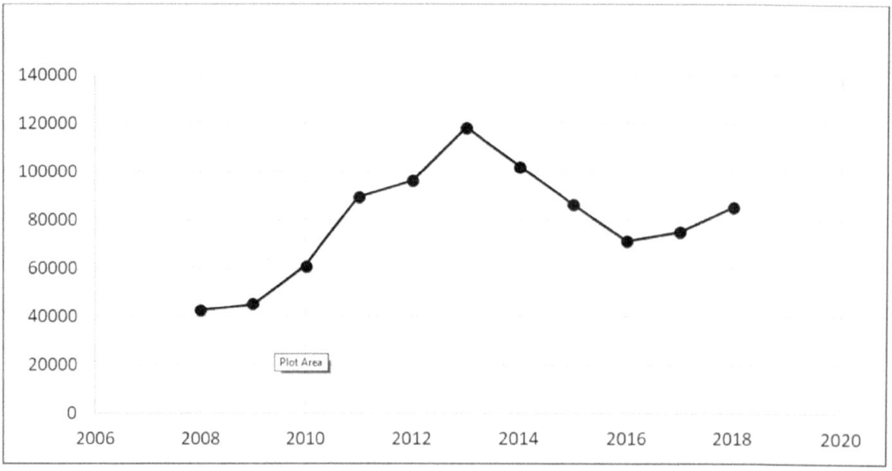

FIGURE 2.1 Value of retail sales of jewelry, watches, clocks, and valuable gifts in Hong Kong 2008–2018 (in millon Hong Kong dollars).

Source: Statista (2019).

Tourism influences local people's quality of life (Andereck and Nyaupane, 2011), and impacts upon the local people with regard to quality of life should be taken into serious consideration. Tourists from the Mainland are crucial for Hong Kong tourism industry. With the increasing number of tourists from the Mainland, however, Hong Kong has to cope with the negative impact from overdevelopment of tourism. Shopping centers are usually for the high-spending tourists instead of local citizens. Local communities are complaining about the shortage of powdered milk for babies, soaring price of property, expensive accommodation, etc. Featured local shops or restaurants are closed because of strong competition from international chains (Brownlee, 2013). While bringing economic benefits, the development of tourism also incur the issues such as waste, pollution, and congestion. When local people perceive that the tourism development happened at the cost of their resources and it outweighs the benefits, they may feel resentment and irritation toward tourists (Ko and Stewart, 2002), such as the incidents of harassing tourists by some local residents, that in turn reduce community satisfaction. Since tourism relies heavily upon the goodwill of the residents, local support is essential to its development, successful operation, and sustainability of the industry in the long term (Vargas-Sánchez et al., 2011). Thus, tourism development should enhance the quality of local people's lives by addressing the economic, social, cultural, recreation, and other benefits of tourism (Peters and Schuckert, 2014). Government decision makers and community developers have to take into account local people's quality of life in tourism planning and development (Uysal et al., 2016).

Quality of life, innovation, economy has been regarded as the main drivers of city success and competitiveness (Burohappold Engineering, 2016). In the GBA, Hong Kong has taken up its role as a leader in international finance, transportation, and logistics (Hong Kong General Chamber of Commerce, 2019b). The metropolis, nevertheless, is relatively slow in terms of technological growth (HKMA, 2019). On the chart of competitiveness rankings in 2018, Hong Kong's place dropped to the 14th, while regional competitors such as Singapore, Japan, the Mainland, and South Korea have all made improvements (Global Innovation Index, 2018). Compared to its northerly neighbor Shenzhen, which competes with the Silicon Valley in California, Hong Kong lags behind in technological development (Barrett, 2019). Technological giants like Huawei and Xiaomi have their headquarters in Shenzhen in which electric buses and taxis run around supported by the technology of Biyadi. And also, mobile payment has dominated all the cities, including Shenzhen. Hong Kong government has been committed to enhancing the competitiveness and vibrancy of the city's economy (CHINADAILY, 2019). Innovation and technology are two drivers of competitiveness (Su, 2018). Although the Hong Kong government has shown dedication to developing the innovation and technology industry, lack of talents or professionals continues to be a challenge that needs to be overcome. The competition for top-notch tech talents between Hong Kong, Macao and other Mainland cities will be fiercer especially in the GBA (HKMA, 2019).

SMART technology is the engine of development and has shown its impact everywhere, especially in the field of tourism. In this sense, hotel as the pillar of hospitality as a whole, calls for innovation and technological improvement to equip itself on the course of being smart. From the perspectives of countries, in the Mainland, both scholars and hotel practitioners are actively developing SMART hotels, such as the Dragon Hotel Hangzhou, "Future Hotel 2.0" by Alibaba Group, JW Marriott Hotel Hangzhou, and Inter Continental Beijing Sanlitun (Guo, 2018). In America, many SMART hotels provide different kinds of SMART rooms services, such as Loews Hotel 1000, Aliante Casino Hotel, Hotel Bel-Air, New York City Hotel, the Wit Hotel, CitizenM Times Square, The James Chicago, and Wynn Hotel & Resort. Japan opened a hotel in which robots are used to carry bags for the guests (Griffiths, 2015). In recent years, the Singapore Tourism Board (STB) has been encouraging the use of technology and local hotels, such as Bayview Hotel, Conrad Centennial Singapore, and Grand Park City Hall (Chow, 2018), are increasingly developing SMART hotel rooms. Compared to the countries where SMART hotel is enjoying rapid development, SMART hotel in Hong Kong is still in its primary stage, relatively falling behind.

Current global trade disputes and economic downturn is putting pressure on Hong Kong, which has been caught in a trade friction between the Chinese Mainland and the United States. Hong Kong's retail and tourism sectors have taken a big hit. In May 2019, retail kept on declining (Leung, 2019). Furthermore, uncertainties remain as protests against the extradition bill grip the city. Hong Kong's political unrest has affected shoppers' motivation for spending, damaged retail sales and added to the city's economic woes (Tsang, 2019). Though Hong Kong has been one of Asia's most popular tourist destinations and a global financial hub, the violent clashes have hurt its international image. Hong Kong protests hit, in particular, the tourism industry. Hotel occupancy rates dropped by 20% in June and by 40% in July. Group bookings from the Taiwan, Korea, and Japan have plunged by 50% (Reed, 2019). Industries such as catering, retail, and transportation have suffered from with a significant loss of revenue. The trade tension, depreciation in renminbi, and incessant demonstrations has combined to lower the tourists' spending and retailers are keenly aware of the global risk recession (The Straitstimes, 2019).

In August 2019, the Chinese government announced that Shenzhen will be designated as "a pilot demonstration area" (Xinhua, 2019), seeking to improve the blueprint of GBA and enhance the "one-country-two-systems" policy. The announcement shows that Shenzhen is to become the key player of the GBA strategy (Chen and Li, 2019). The endless protests, demonstrations, and violence in Hong Kong have formed a sharp contrast to Shenzhen's rising importance in the national strategy. If Hong Kong is still caught in its internal turmoils and riots, the metropolis' prosperity and development would be endangered in the future "while Shenzhen is running at a much faster speed" (Tian, 2019). The protracted tensions could damage the local economy in both short and long runs. Negatively affected will be the SAR's long-term competitiveness, including its established image of Asia's World

City, a popular destination, a global financial hub (Reed, 2019), and as the core city of GBA.

2.4 THE OPPORTUNITIES TO HONG KONG'S TOURISM INDUSTRY

As a core metropolis of GBA, Hong Kong is expected to go beyond its current roles while playing new roles in further collaboration with other cities in the area. The construction and growth of GBA has provided more opportunities for Hong Kong to take advantage of its strengths and competitive edge (KPMG, 2018). Hong Kong may take the lead in developing multi-destination tourism within the GBA framework.

In recent years, Shenzhen and Dongguan—two cities traditionally based on manufacturing—have risen to become powerhouses of innovation. This rise has brought both pressure and opportunities to Hong Kong. Compared with other cities in GBA, Hong Kong has the largest group of world-class universities whose intellectual and technological resources can be drawn for innovation in building smart cities in Hong Kong and the area as well. The construction of a smart city has been on priority for the Hong Kong government, with dozens of initiatives conducted (Yang, 2019) to cover the aspects of mobility, living, environment, people, government, and economy (Smart City Blueprint, 2017). As tourism has been one of Hong Kong's pillar industries, the local government is paying due attention to it in the SMART City Blueprint. Smart tourism is presenting more convenience, excitement to tourists as well as residents by improving tourist satisfaction and daily life of the host communities. Digitalization is not only a trend of tourism but also the base of smart cities. Hong Kong's well-developed infrastructure, once digitalized, will be a leading example in GBA. One of the GBA initiatives is to construct a high-quality "living circle." That means the area is anticipated to establish a complete and streamlined modern system of housing, education, medical care, recreation, and tourism for harmonious working, living, and travelling (Hong Kong General Chamber of Commerce, 2019a). The cities in GBA will have more sports venues, community centers, galleries, MICE center, theme parks, tourism attractions, hotels, and shopping malls (KPMG, 2018). Hong Kong-style private residential areas, such as housing, schools, amenities, and elderly and medical care, are proposed to construction which could be appealing to Hong Kong people to purchase at affordable prices (KPMG, 2018, 2019).

Meetings, incentives, conventions, exhibitions (MICE) tourism is vital to attracting the profitable business visitors and underlining Hong Kong's leading position as an international city in Asia (Hong Kong Tourism Board, 2018a). The major convention and exhibition centers in the GBA, with a total floor space of more than 360,000 square meters, are important to the development process as a whole. Hong Kong's expertise and professionals in MICE management may help other destinations of convention and exhibition in GBA. Co-competition among different destinations is healthy for the overall development of the industry and greater connectivity has opened up the potentials for cross-boundary events (2020 Foundation, 2019). Hong Kong's leading role in MICE will pull up the promotion of the GBA as an international destination.

For years, Hong Kong has been putting efforts into developing its cruise facilities. Kai Tak Cruise Terminal has been under reconstruction and expansion, which shows the local authority's determination to turn Asia's World City into a cruise hub in the region. Cruise travelers may leverage the convenient location of Hong Kong to come into or out of it to try "fly-cruise" or "rail-cruise" through the accessibility of the XRL and HZMB. All this will contribute to the growth of cruise industry in Hong Kong and the region (Hong Kong Constitutional and Mainland Affairs Bureau, 2018).

With the powerhouses emerging in Guangdong and economic integration in GBA, Hong Kong may need a clearer vision of its role so as to make better use its strategic position and new infrastructure construction. XRL and HZMB have shortened the time between Hong Kong and neighboring cities in GBA, making it possible for more co-competition and partnership. Meanwhile, the advancement in GBA's infrastructural construction may enhance Hong Kong's edge to attract more domestic and international travelers. Hong Kong will continue to be a popular destination and buffer area where "East meets West." Domestic tourists are charmed by Hong Kong has a lot of western elements while to the international visitors by joining hands with tourism authorities in other cities, Hong Kong's function in promoting the image of GBA to the world will be better performed (Hong Kong Tourism Board, 2018b).

2.5 CONCLUDING REMARK

The GBA concept will take tourism development in the region to the next level and provide rich resources to fuel the further development of multi-destination travel. If it gets over the unrest and return to the right track of development, Hong Kong will continue to serve as a global tourism gateway for overseas tourists to enter the GBA, offering exotic experience as Asia's World City The completion of XRL and HZMB—two infrastructural projects—may also put the global spotlight on Hong Kong by repolishing its tourism image, which will help encourage overseas visitors to consider starting their journey to the Mainland via Hong Kong. The government's stance on the sociocultural differences that separate the two opposing groups could be difficult to understand, but the metropolis must acknowledge the Chinese connection in order to avoid confrontation with central government and the political system; and the population's ethnicity makes such acknowledgement inevitable. Flows of people and finance should be strategically planned in the GBA so that the professional and high-valued services of Hong Kong will benefit the whole area by diversifying its economy through better connectivity with the Mainland. Both government and communities should seriously consider correct positioning of Hong Kong so as to catch the boat of China's latest development in economy.

KEYWORDS

- **the Greater Bay Area (GBA)**
- **high-speed railway**
- **Hong Kong-Zhuhai-Macao Bridge**
- **opportunities**
- **challenges**

REFERENCES

Airports Council International (ACI) (2017). ACI World Releases Preliminary 2017 World Airport Traffic Rankings. https://aci.aero/news/2018/04/09/aci- world-releases-preliminary-2017-world-airport-traffic-rankings-passenger-traffic-indian- and-chinese- airports-major-contributors-to-growth-air-cargo-volumes-surge-at-major-hubs-as-trade-wars-thre/ (accessed July 29, 2019).

American Association of Port Authorities (AAPA) (2018). *Port Industry Statistics*, 2018. https://www.aapa- ports.org/unifying/content.aspx?ItemNumber=21048 (accessed July 29, 2019).

Andereck, K. L.; Nyaupane, G. P. Exploring the Nature of Tourism and Quality of Life Perceptions Among Residents. *J. Travel Res.* **2011,** *50*(3), 248–260.

Barrett, E. China is Creating a 'Greater Bay Area' to Rival Silicon Valley. What Will it Mean for Hong Kong? https://fortune.com/2019/06/02/china- hong-kong-greater-bay-area/ (accessed July 25, 2019).

Barrow, M. G. The Second Largest Foreign Exchange Earner, Tourism. *Hong Kong Bus. Annu.* **1994,** *13*, 138–143.

Brownlee, I. Regulate Tourist Numbers for Hong Kong's Benefit. South China Morning Post, 2013. https://www.scmp.com/comment/insight-opinion/article/1218510/regulate-tourist-numbers-hong-kongs-benefit (accessed Aug 20, 2019).

Burohappold Engineering (2016). What makes cities competitive? Rankings & Success Factors. ttps://regioncity.se/wp-content/uploads/2017/02/What_Makes_Cities_Competitive-November_2016.pdf (accessed Aug 20, 2019).

Casino News. Las Vegas, Macao Casino Revenue Comparison Report. 21 February 2018. https://www.casinonewsdaily.com/2018/02/21/las-vegas-Macao-casino-revenue-comparison-report-infographic/ (accessed July 29, 2019).

CEIC. Hong Kong SAR tourism revenue. https://www.ceicdata.com/en/indicator/hong-kong/tourism-revenue (accessed July 26, 2019).

CGTN. Hong Kong-Zhuhai-Macao Bridge to be sustainable and promote integration. https://news.cgtn.com/news/3d3d514f3541444d30457a6333566d54/index.html (accessed July 25, 2019).

Chen, Q.; Li. R. China to Build City Into Socialist Demonstration Area. Global Times, August 18, 2019. http://www.globaltimes.cn/content/1161866.shtml (acceseed Aug 20, 2019).

CHINADAILY. HK Government Committed to Enhancing SAR's Competitiveness. May 29, 2019. https://www.chinadailyhk.com/articles/224/38/26/1559125151182.html (accessed July 18, 2019).

Choi, S. W. From Fishing Port to 'City of Life': Hong Kong History According to the Hong Kong Tourism Board. http://commons.ln.edu.hk/mcsln/vol3/iss1/1 (accessed July 18, 2019).

Choi, W. H. The Recent Development of Hong Kong Tourism Industry and its Outlook. Economic Review. https://www.bochk.com/dam/investment/bocecon/SY2018014(en).pdf (accessed July 22, 2019).

Chow, V. Anger at Mainland Visitors Escalates with 'Locust' ad'. South China Morning Post, 1 February 2012.

Chow. P. Smart Hotels on the Rise in Singapore. https://www.ttgasia.com/2018/02/27/smart-hotels-on-the-rise-in-singapore/ (accessed July 25, 2019).

Cutter, S. L. *Rating Places: A Geographer's View on Quality of Life*. Resource Publications in Geography, the Association of American Geographers: Washington, DC, 1985.

Euromonitor International. Top 100 City Destinations 2018. https://go.euromonitor.com/white-paper-travel-2018-100-cities.html?utm_source=blog&utmmedium=blog&utm_campaign=CT_WP_18_12_04_100%20Cities&utm_content=organic (accessed July 29, 2019).

Global Innovation Index. Hong Kong Slips in Global Innovation Ranking as Regional Competitors Edge Ahead. South China Morning Post, 16 Aug 2016.

GovHK. Hong Kong: The Facts–Tourism. https://www.gov.hk/en/about/abouthk/factsheets/docs/tourism.pdf (accessed July 30, 2019).

GovHK. Hong Kong: The Facts Tourism. https://www.gov.hk/en/about/abouthk/factsheets/docs/tourism.pdf (accessed July 20, 2019).

GovHK. Tackling Problems Brought by Increasing Number of Visitors to Hong Kong. https://www.info.gov.hk/gia/general/201405/28/P201405280418.html (accessed Aug 20, 2019).

Griffiths, S. Faulty Towers? Japan Opens Hotel Run by ROBOTS that will Welcome Guests, Carry Bags and Even Clean Your Room. from https://www.dailymail.co.uk/sciencetech/article-2946103/FAULTY-towers-Japan-opens-hotel-multi-lingual-ROBOTS-welcome-guests-carry-bags-clean-room.html (accessed July 23, 2019).

Guo, D. Chinese Hotel uses 'High-Tech'! Face Recognition Check in, Room Facilities 'Free Hands' Voice Control. https://www.limitlessiq.com/news/post/view/id/6641/ (accessed July 25, 2019).

Hong Kong Business. The Changing Face of Hong Kong's Retail. https://hongkongbusiness.hk/retail/in-focus/changing-face-hong-kongs-retail (accessed Aug 15, 2019).

Hong Kong Constitutional and Mainland Affairs Bureau. Greater Bay Area: Tourism. https://www.bayarea.gov.hk/en/opportunities/tourism.html (accessed July 30, 2019).

Hong Kong General Chamber of Commerce (HKGCC) (2019a). Navigating the Greater Bay Area. https://www.chamber.org.hk/en/information/the-bulletin_detail.aspx?id=233&sid=1 (accessed July 30, 2019).

Hong Kong General Chamber of Commerce (2019b). The Voice of Business in Hong Kong. df (Aug 20, 2019).

The Hong Kong Management Association (HKMA). A Study of Trends and Challenges Facing Hong Kong. https://www.hkma.org.hk/pdf/2019_PYA.pdf (accessed July 22, 2019).

Hong Kong Tourism Board. A Statistical Review of Hong Kong Tourism 2015. http://securepartnernet.hktb.com/filemanager/intranet/ir/ResearchStatistics/paper/Stat-Review/StatReview2015/Statistical%20Review%202015.pdf. (accessed July 30, 2019).

Hong Kong Tourism Board. A Statistical Review of Hong Kong Tourism 2016. http://securepartnernet.hktb.com/filemanager/intranet/ir/ResearchStatistics/paper/Stat-Review/StatReview2016/Stat%20Review%202016.pdf (accessed July 30, 2019).

Hong Kong Tourism Board. Monthly Report - Visitor Arrival Statistics: Aug 2017. http://partnernet.hktb.com/filemanager/intranet/pm/VisitorArrivalStatistics/ViS_Stat_E/VisE_2017/Visitor%20Arrivals%20To%20Hong%20Kong%20in%20Aug%202017.pdf (accessed July 31, 2019).

Hong Kong Tourism Board. (2018a). Annual Report. http://www.discoverhongkong.com/eng/about-hktb/annual-report/annual-report20172018/ strategic-initiatives/mice.html#landing (accessed July 30, 2019).

Hong Kong Tourism Board. (2018b). Hong Kong–Zhuhai–Macao Bridge. http://www.discoverhongkong.com/eng/see-do/culture-heritage/modern- architecture/hong-kong-zhuhai-Macao-bridge.jsp (accessed July 25, 2019).

Hong Kong Tourism Board. Discover Hong Kong: History. http://www.discoverhongkong.com/eng/plan-your-trip/traveller-info/about-hong- kong/history.jsp (accessed 22 July, 2019).

Hong Kong Tourism Commission. Tourism Fact Sheets. https://www.tourism.gov.hk/english/papers/papers_fact_sheets_2018.html (accessed July 26, 2019).

Hong Kong Tourism Commission. Tourism Performance in 2018. https://www.tourism.gov.hk/english/statistics/statistics_perform.html (accessed Aug 1, 2019).

Hong Kong Tourist Association. *Visitor Arrivals Statistics: January–May*; 1998.

Ko, D. W.; Stewart, W. A Structural Equation Model of Residents' Attitudes for Tourism Development. *Tour. Manag.* **2002,** *23*, 521–530.

KPMG. Opportunities in the Greater Bay Area. https://assets.kpmg/content/dam/kpmg/cn/pdf/en/2018/11/connecting-opportunities-in-the greater-bay-area.pdf (accessed July 22, 2019).

Leung, K. Hong Kong Retail Sales in May Fall for Fourth Month in Row Despite Surge in Mainland Chinese Visitors for Labour Day Holiday. South China Morning Post, 2 July 2019.

Lui, S. W. Spotlight: Cross-Border Infrastructure Helps Promote Tourism in Hong Kong and Greater Bay Area. http://www.xinhuanet.com/english/2019-02/07/c_137804510.htm (accessed July 22, 2019).

Mok, C.; Dewald, B. Tourism in Hong Kong: After the Handover. *Asia Pacific J. Tour. Res.* **1999,** *3*(2), 32–40. DOI: 10.1080/10941669908722017

Nielsen Survey. China's Cross-Border Online Shippers are Increasingly Mature. https://www.nielsen.com/cn/en/insights/article/2015/Nielsen- Survey-China-s- Cross-Border-Online-Shoppers-are-Increasingly-Mature/(accessed Aug 15, 2019).

Peters, M.; Schuckert, S. Tourism Entrepreneurs' Perception of Quality of Life: An Explorative Study. *Tour. Anal.* **2014,** *19*(6), 731-740.

Reed, J. R. Global Firms are Sounding the Alarm as Escalating Hong Kong Protests Deal a 'Serious Blow' to the City's Outlook. Aug. 14 2019. https://www.cnbc.com/2019/08/14/global-firms-are-sounding-the-alarm-as-hong-kong- protests-escalate.html (accessed Aug 20, 2019).

Shih, G. China Wants to Pull Hong Kong into a 'Greater Bay Area' that Could Compete with Silicon Valley. https://www.washingtonpost.com/world/china-wants-to-pull-hong-kong-into-a-greater-bay- area-that-could-compete-with-silicon-valley/2019/02/19/71f2518c-3412-11e9-946a-115a593 2c45b_story.html?noredirect=on&utm_term=.7a5ec34ce9f7 (accessed July 29, 2019).

SMART City Blueprint. Hong Kong Innovation and Technology Bureau. https://www.smartcity.gov.hk/blueprint/HongKongSmartCityBlueprint_e- flipbook_EN/ (accessed July 27, 2019).

Statista. Value of Retail Sales of Jewelry, Watches, Clocks, and Valuable Gifts in Hong Kong from 2008 to 2018 (in million Hong Kong dollars). https://www.statista.com/statistics/975629/hong-kong-jewelry-watches-clocks-and-valuable-gifts-retail-sales-value/ (accessed Aug 18, 2019).

The Straitstimes. Pain from Hong Kong Protests Spreads as Luxury Brands Get Hit. https://www.straitstimes.com/business/economy/pain-from-hong-kong- protests-spreads-as-luxury-brands-get-hit (accessed July 25, 2019).

Su. X. Hong Kong Still China's Most Competitive City, but Greater Bay Area has a Long Way to go, Study Shows. South China Morning Post. https://www.scmp.com/news/hong-kong/hong-kong-economy/article/2170594/hong-kong- still-chinas-most-competitive-city (accessed Aug 18, 2019).

Tian, F. Told the Global Times. August 18, 2019. http://www.globaltimes.cn/content/1161866.shtml (accessed Aug 21, 2019).

Tsang, E. Parallel Trading of Baby Milk Formula Still Rampant Despite Hong Kong Restrictions. South China Morning Post, 1 April, 2013.

Tsang, D. Hong Kong Protests Hit City Where it Hurts – in the Wallet. South China Morning Post, 8 July 2019.

Uysal, M.; Sirgy, M. J.; Woo, E.; Kim, H. Quality of Life (QOL) and Well-Being Research in Tourism. *Tour. Manag.* **2016,** *53,* 244–261. DOI: 10.1016/j.tourman.2015.07.013

Vargas-Sánchez, A.; Porras-Bueno, N.; de los Ángeles Plaza-Mejía, M. Explaining Residents' Attitudes to Tourism. Is a Universal Model Possible? *Ann. Tour. Res.* **2011,** *38*(2), 460–480. DOI: 10.1016/j.annals.2010.10.004

Xinhua. China to Build Shenzhen into Socialist Demonstration Area. Aug. 19, 2019. http://www.xinhuanet.com/english/2019- 08/18/c_138318854.html (accessed Aug 20, 2019).

Yang, H. J.; Chan S. H. G.; Li, W. J. Tourism Impacts on the Quality of Life in Hong Kong. *Int. J. Mark. Stud.* **2017,** *9*(3), 103–112. DOI: https://doi.org/10.5539/ijms.v9n3p103

Yang, R. Hong Kong: Our Smart City Journey Conference. https://www.ogcio.gov.hk/tc/news/press_releases/2019/01/pr_20190110.html (accessed July 26, 2019).

Ye, H. B.; Qiu, H. Z.; Yuen, P. P. Motivations and Experiences of Mainland Chinese Medical Tourists in Hong Kong. *Tour. Manag.* **2011,** *32*(5), 1125–1127.

Ye, H. B.; Zhang, Q. H.; Shen, H. J.; Goh, C. Does Social Identity Affect Residents' Attitude Toward Tourism Development? *Int. J. Contemp. Hosp. Manag.* **2014,** *26*(6), 907–929.

Yiu, E. Hong Kong Remains in Pole Position for IPOs. South China Morning Post, 24 June 2017. https://www.scmp.com/business/companies/article/2099755/hong-kong-remains-pole-position-ipos (accessed July 29, 2019).

2022 Foundation. Creating the Greater Bay Area of the Future–Opportunities for Hong Kong. http://www.2022foundation.com/images/GBA_MainReport.pdf (accessed July 29, 2019).

CHAPTER 3

STAKEHOLDER ENGAGEMENT IN MAKING MACAU A SUSTAINABLE TOURISM DESTINATION

DANQING LIU[1,2*] and JINLIN ZHAO[1,2]

[1]*School of Management, Tianjin University of Commerce, Tianjin 300134, China*

[2]*Chaplin School of Hospitality and Tourism Management, Florida International University, Miami, FL, USA*

[*]*Corresponding author. E-mail: dliu@tjcu.edu.cn*

ABSTRACT

Tourism industry is the principal industry of Macau, and it is of great significance for Macau's sustainability. This chapter includes a brief introduction to Macau's tourism industry, a statement of the significance of sustainable tourism for Macau, and a description of key stakeholders' engagement in developing Macau as a sustainable tourism destination.

3.1 TOURISM INDUSTRY IN MACAU

3.1.1 INDUSTRIAL STRUCTURE OF MACAU

Macau is the Special Administrative Region (SAR) of China. It was a Portuguese colony and returned to China in 1999. In the past, Macau has developed industries such as textiles, electronics and toys, and gaming industry (MGTO, 2018). In 1999, the major economic sectors of Macau were manufacturing, construction and real estate, financial services, and gaming (Sheng and Gu, 2018). Since its return to China, the economy has grown rapidly, and its industrial structure has changed "with gaming tourism as the engine, complete with other service industries" (Hong Kong Trade Development Council, 2019). According to Macau Statistics and Census Service report (Macau Statistics and Census Service, 2018), the secondary sector accounted for 5.1% in the year 2017, while as in 2016 and 2015, it was 6.6% and 7.8%, respectively; the tertiary sector of Macau accounted for 94.9% in the year 2017, comparing with 93.4% in 2016, and 92.2% in 2015, it has been increasing. Based on Macau Statistics and Census Services report (Macau Statistics and Census Service, 2018), Macau's economic sectors and their proportion in 2017 can be summarized in Table 3.1.

TABLE 3.1 Structure of Economic Activities of Macau in 2017.

Economic activities	**Gaming and junket**	**Real estate**	**Public administration, education, health, and other services**	**Banking, insurance, and pension**	**Wholesale and retail**	**Renting and business**
Percentage	49.1%	10.6%	10%	6.5%	5.6%	4.4%
Economic activities	Hotel and similar	Construction	Transport, storage, and communication	Restaurant and similar	Electricity, gas, and water supply	Manu-facturing
Percentage	4.3%	3.7%	2.7%	1.7%	0.7%	0.6%

The results of statistics demonstrated that Macau's economy heavily relied on gaming industry and gaming tourism was the main driver for its economic growth. According to International Monetary Fund (2019), the growth rate of gaming tourism revenue reached 9.7% in 2017, 4.7% in 2018, and it is projected that the growth rate for 2019 is 4.3%, and over the medium term the growth rate will remain solid at about 4%.

Macau's economy heavily relies on the gaming tourism industry, which raised the concern of the sustainability of its economy and its growth (Greenwood and Dwyer, 2016; Sheng and Gu, 2018; Wan and Li, 2013). To address the sustainability issue, Macau has positioned itself as the World Centre of Tourism and Leisure and has been working toward a quality international tourist destination.

3.1.2 TOURISM INDUSTRY IN MACAU

Tourism is the pillar industry of Macau, it accounts for a significant proportion of Macau's economy. According to World Travel and Tourism Council's

(2018) report, the total contribution of Macau Travel and Tourism to GDP was 61.3% in 2017 and the total contribution of Macau Travel and Tourism to total employment was 53.6% in 2017; the direct contribution of Travel and Tourism to GDP was 29.3% in 2017, and the direct contribution of Travel and Tourism to total employment was 27.6% (WTTC, 2018).

With the opening of Hong Kong–Zhuhai–Macau Bridge in 2018, and more joint efforts for developing the Guangdong–Hong Kong–Macau Greater Bay Area, tourism industry in Macau is expected to have a remarkable growth.

Working toward a world tourism destination, Macau provides a wide choice of hotels, resorts, MICE facilities, restaurants, and casinos. According to the 2018 report of Macau Government Tourism Office, there were 116 hotels including 35 five-star hotels, 17 four-star hotels, 16 three-star hotels, 14 two-star hotels, and 34 guesthouses; and there were 1381 MICEs, 2237 restaurants, 40 casinos, and 210 travel agents in Macau in the year 2018 (MGTO, 2019).

In July 2005, the Historic Centre of Macau was officially recognized as a UNESCO World Heritage List. The major attractions (Table 3.3) of the world heritage are located along a linear route, linking religious, military, and civil monuments of both western and Chinese concepts (Macau Cultural Affairs Bureau, 2018). The major attractions are A-Ma Temple, St. Augustine's Church, Ruins of St. Paul's, Barra Square, Moorish Barracks, "Leal Senado" Building, Na Tcha Temple, Lilau Square, Mandarin's House, Sam Kai Vui Kun (Kuan Tai Temple), Section of the Old City Walls, St. Augustine's Square, St. Lawrence's Church, Holy House of Mercy, Mount Fortress, Senado Square, St. Joseph's Seminary and Church, Cathedral, St. Anthony's Church, Cathedral Square, Dom Pedro V Theatre, Lou Kau Mansion, the Casa Garden, St. Dominic's Square, Sir Robert Ho Tung Library, St. Dominic's Church, the Protestant Cemetery, Company of Jesus Square, and Camões Square.

The unique blend of Chinese and Portuguese cultures has drawn visitors and business people from all over the world (MGTO, 2017). In 2018, Macau visitor arrivals reached 35.8 million, a year-on-year increase of 9.8% (MGTO and DSEC, 2019). The increase of tourist arrivals has been maintained since the Macau's return to China in 1999 (MGTO, 2012, 2018).

3.2 SIGNIFICANCE OF SUSTAINABLE TOURISM DEVELOPMENT FOR MACAU

Tourism can have both positive and negative impacts on the environment, culture, and host populations' well-being (UNWTO and UNEP, 2005). Tourism development in Macau raised sustainability issues.

Macau records the highest population density per square kilometer (Macau News, 2016; MGTO, 2015). Based on Worldometers, elaboration of the latest United Nations data, as of October 3, 2019, the population of Macau is 642,719, and its population density is 21,348 per km^2 (Worldometer, 2019).

Only permanently settled population of Macau was involved in the density calculation, the millions of tourist arrivals each year were excluded. In fact, the tourist arrivals stood at 35.8 million in 2018, and the inbound tourism of Macau is expected to maintain the growth rate in the future, and more people will squeeze to this most densely populated place. Consequently, Macau has sustainability challenges including the increasing of pollution, motor vehicle density, garbage, and deaths from respiratory diseases (Chua, 2015).

Sustainable tourism practices for the development of a sustainable tourism destination are essential for Macau, since Macau has positioned to be a World Centre of Tourism and Leisure (MGTO, 2017). In Macau, gambling tourism played a vital role for its economy, the revenue it generated accounted for approximately 50% of total revenue, therefore, the industry practices exert huge impacts on the environment (Chua, 2015). Increasing electricity consumption and carbon emissions from casinos and hotels caused more environmental problems. Therefore, sustainable actions including design and construction, procurement management, waste management, reuse, and recycle initiatives are required (MGTO, 2010).

Tourism has contributed to the social–cultural development in Macau, such as Macau's cultural heritage preservation, and convenient transportation for both tourists and residents. However, the development of tourism, especially gaming tourism, has caused sustainable issues, such as land availability and land cost due to increased use for tourism resorts and real estate; inadequate local human resource, especially for top tourism positions which seem to normally require tertiary education due to increasing dropout rate in secondary schools, as the gaming industry employs people with a low educational level for better paid jobs (Atamhenwan, 2013).

Wan and Li (2013) conducted a study on sustainable tourism development in Macau and summarized the impacts of tourism development in Macau (Table 3.2).

To take the benefits of tourism development in Macau, and to minimize its negative effects, sustainability is therefore the primary concern for Macau.

TABLE 3.2 The Impacts of Tourism Development in Macau (Wan and Li, 2013).

	Economic impacts	**Social impacts**	**Environmental impacts**	**Visitor satisfaction**	**Community participation**
Positive findings	Increased in governmental revenue, overall employment, and household income	Improved public transportation system and social welfare	Increased in public expenditure on environmental protection	Tourist satisfaction monitored	
Identified issues	Human resources issue (e.g., shortage of labor) and economic diversification issue (overly rely on gaming industry)	Increased school dropout rate, problem gamblers, crimes, and gambling-related crimes	Issue of public land reduction, traffic congestion, and increased electricity consumption	Only an average satisfaction was achieved between 2004 and 2008	The need for a master plan in tourism development and public involvement in tourism decision-making

3.3 STAKEHOLDERS' ENGAGEMENT IN DEVELOPING A SUSTAINABLE DESTINATION

Sustainable development is "a process to meet the needs of the present without compromising the ability of future generations to meet their own needs" (United Nations, 1987). And sustainable tourism, according to United Nations World Tourism Organizations (UNWTO), is "tourism that takes full account of its current and future economic, social and environmental impacts, addressing the needs of visitors, the industry, the environment and host communities" (UNWTO and UNEP, 2005). UNWTO and United Nations Environmental Program (UNEP) identified 12 aims for sustainable tourism which include "economic viability, local prosperity, employment quality, social equity, visitor fulfillment, local control, community well-being, cultural richness, physical integrity, biological diversity, resource efficiency and environmental purity" (UNWTO and UNEP, 2005). Therefore, to achieve sustainable tourism goals, all interest groups' participation is required. With the support from stakeholders, tourism destination thus can be developed in a sustainable manner (Andereck and Vogt, 2000; Andriotis, 2005; Kong et al., 2015). In 1992, the first United Nations Conference on Environment and Development emphasized the importance of "the participation of all concerned citizens" for sustainable development (UNESCO, 1992). Based on Freeman's (1984) stakeholder theory, Goeldner and Ritchie (2008) identified "tourists, residents, entrepreneurs and local governments" as key stakeholders for a tourist destination. Stakeholders in tourism development all assume the responsibility of sustainability, and governments should take actions in policy-making for sustainable tourism development and policy implementation, which can "enable and encourage all stakeholders to respond to sustainability issues" (UNWTO and UNEP, 2005).

With strong support from the central government, Macau SAR government works with all stakeholders including regional partners, environmental associations, higher education institutions, tourism industry, residents, and tourists to develop Macau a sustainable tourism destination (Fig. 3.1).

FIGURE 3.1 Key stakeholders for the sustainable tourism development of Macau.

3.3.1 GOVERNMENT AGENCIES: LEADING MACAU FOR SUSTAINABLE DEVELOPMENT

The Macau government has been working on creating an image of green tourism destination for Macau, and making its pillar industry, the tourism industry far more sustainable (Chua, 2015).

3.3.1.1 STRATEGIC PLANS IN PLACE

3.3.1.1.1 2016–2020 Plan

The Five-Year Development Plan of the SAR (2016–2020) states that Macau is developing into a world tourism and leisure center (referred to as "One Centre"). Therefore, a better environment is important for Macau and great efforts have been made to manage air pollution, water pollution, food safety, and public order. According to Macau's Five-Year Plan, the sustainable development of Macau is in line with the 12th and the 13th National Five-Year Plans. Aiming to be a tourism and entertainment friendly city, Macau is working on establishing an overall business form for the tourism and leisure industry (Macau Government, 2016).

3.3.1.1.2 Environmental Protection Planning 2010–2020

In an effort to protect the valuable ecological environment and natural resources in Macau, and also enhance the quality of living environment, the Macau Environmental Protection Bureau (DSPA), upon its inception, has initiated the first people-oriented environmental protection planning, thus develop Macau into a "Global center of tourism and recreation" (DSPA, 2012). Chapter 3 of the plan focuses on "optimizing the environment suitable for living and tourism," key aspects are improvement of air quality, enhancement of water environment quality, treatment and disposal of solid waste, control of noise pollution, conservation of ecological environment, prevention and control of light pollution, and protection of environmental radiation. In Chapter 4, energy conservation, recycling of waste and water resources, emission reduction, and low carbon production were highlighted. This planning primarily aims for the sustainable development of Macau.

3.3.1.1.3 Macau Tourism Industry Development Master Plan (September 2017)

In 2015, the Macau Government Tourism Office (MGTO) initiated the "Macau Tourism Industry Development Master Plan" (the Master Plan) project. The Master Plan exists under the overall framework of the "Five-Year

Development Plan" (2016–2020) of the Macau SAR Government. It provides a road map for the future development of Macau's tourism industry and helps to steer tourism-related developments to be in line with the Macau SAR's vision to become the "World Centre of Tourism and Leisure." Major subject areas of the plan include tourism resources and products system, tourism quality and services system, tourism branding and marketing strategies, tourism industry and city development, tourism carrying capacity, tourism and the application of innovative technology, Tourism Industry Cooperation System, International and Regional Tourism Cooperation and Development (MGTO, 2017).

3.3.1.2 QUALITY TOURISM SERVICES ACCREDITATION SCHEME

Starting from 2014, the Macau Government Tourism Office launched the "Quality Tourism Services Accreditation Scheme" to recognize quality service, and to encourage and support the industry to further improve service quality and to enhance service culture, with the ultimate goal of developing Macau to a "World Centre of Tourism and Leisure."

TABLE 3.3 Quality Tourism Services Accreditation Scheme.

	Participants	Criteria for assessment	Awards
Food and beverage industry	All licensed restaurants and F&B establishments in Macau	The assessment is composed of two parts: "Mystery Shopping Assessment" and "Service Management System Audit," whereas the frontline service standard and management system will be assessed, respectively	Two types of awards: "Star Merchant Award" and "Service Star Award"
Travel agency industry	All licensed Travel Agency Headquarters/ Branches in Macau	The assessment consists of two parts, namely "Mystery Shopping Assessment" and "Service Management System Audit," whereas the frontline service standard and management system will be assessed, respectively. "Traveler Satisfaction Survey" is added to cope with the business nature of travel agencies that provide inbound services in Macau	

3.3.1.3 ENVIRONMENTAL PROGRAMS

Major governmental programs for environmental protection are listed in Table 3.4.

TABLE 3.4 Environmental Programs (DSPA, 2018).

Environmental Protection Fun Program	Promote the continuous implementation of various environmental protection practices and let the public turn environmental protection behavior into daily living habits, enjoy green life together
Macau Green Hotel Award	The award was launched to raise the awareness of the importance of environmental management in Macau's hotel industry and commend those hotels, which have adopted environmental measures. Thirty-eight hotels have received the "Macau Green Hotel Award," accounting for about 35% of the total number of hotels in Macau
Green Enterprises Partnership Program	Large shopping malls, supermarkets, convenience stores, bakeries, and bookstores, these enterprises actively involve in environmental management, waste reduction, and energy conservation and can help enhance their corporate image
Green School Partnership Program (Green Campus Award; Green School Award; Green Teaching Award)	Schools integrate environmental management and environmental education into their comprehensive development plans, policies, and education practices and continuously use all internal and external resources and opportunities for sustainable education, thus enhance the quality of environmental education of the school

3.3.2 REGIONAL PARTNERS: COOPERATING WITH REGIONAL GOVERNMENTS FOR GREENING

Being a major party of the Guangdong–Hong Kong–Macau Greater Bay Area, Macau has been working closely with regional partners of the Greater Bay Area to achieve sustainable development goals and also has been capitalizing on the opportunities brought about by the Belt and Road Initiative, in an effort to develop Macau into a well-known center for tourism and leisure.

3.3.2.1 OUTLINE DEVELOPMENT PLAN FOR THE GUANGDONG–HONG KONG–MACAU GREATER BAY AREA

The Pearl River Delta Bay Area (PRD Bay Area) has become a major focus of future development and growth. On February 18, 2019, China Central Committee of the Communist Party and the State Council released Guangdong–Hong Kong–Macau Greater Bay Area Development Blueprint, to guide the current and future cooperation and development of the Greater

Bay Area, covers the period from the current year to 2022 in the immediate term and up to 2035 in the long term. One of the objectives of the plan is about conservation and ecology; the plan aims to dramatically improve the efficiency of resource conservation and effectiveness of environmental protection and to fully develop Macau to be an international tourism center (China Daily, 2019).

3.3.2.2 GUANGDONG–HONG KONG–MACAU SUSTAINABLE DEVELOPMENT CONFERENCE

Guangdong Hong Kong Macau Sustainable Development Conference was founded in 2002 by Guangdong Provincial Association for Science and Technology and the Civil, Environmental and Building Divisions of the Hong Kong Institution of Engineers. This conference has brought together scholars, policymakers, industry practitioners, and other stakeholders in the Great Bay Area to discuss sustainable issues including sustainable economy, green transportation, smart city, clean energy, eco-friendly development, climate change, and sustainable social and cultural development, and so on. As a collaborative platform and a communication channel of the Great Bay Area, the conference has provided stakeholders opportunities of sharing, discussing, and cooperating toward international standardization and management in sustainable development (Hong Kong Institution of Engineers, 2019).

3.3.2.3 MACAU INTERNATIONAL ENVIRONMENTAL COOPERATION FORUM AND EXHIBITION

The Macau International Environmental Cooperation Forum and Exhibition (MIECF), first held in 2008, is an annual event and has become a platform fostering the exchange and collaboration on environmental development in the region of Pan-Pearl River Delta. 9+2 provincial and regional governments of the Pan-Pearl River Delta are participated members of the MIECF, which include Fujian, Jiangxi, Hunan, Guangdong, Guangxi, Hainan, Sichuan, Guizhou, Yunnan, Macau, and Hong Kong. Joint efforts of these governments have been made to push sustainable development in the region (Macau MIECF, 2019). Main themes of the nine conferences are listed in Table 3.5.

TABLE 3.5 Macau International Environmental Cooperation Forum and Exhibition (MIECF).

Year	Dates	Main theme
2019	March 28–30	Promoting ecological civilization and green development
2018	April 12–14	Shaping of Eco: cities for inclusive green economy
2017	March 30–April 1	Innovative green development for sustainable future
2016	March 31–April 2	Green economy: opportunities for waste management
2015	March 26–28	Green economy: solutions for clean air
2014	March 27–29	Energizing green business growth
2013	March 21–23	Sustainable cities: the way toward a green future
2012	March 29–31	Green economy: new engine for growth
2011	March 31–April 2	Green opportunities: low carbon urban development
2010	April 8–10	The green gateway: green knowledge, green technology, green finance
2009	April 2–4	A global green gathering, environment–business–connection
2008	April 23–25	Water, energy, and sustainable development

3.3.3 ENVIRONMENTAL ASSOCIATIONS: JOINING EFFORTS FOR SUSTAINABLE WELL-BEING

Macau is called an "association society," as there are more than 8000 nongovernmental social groups or associations, which meet people's different social needs or political interests. The culture of association is a political characteristic of Macau as associations have been playing an important role for making Macau a better society (Macau Monthly, 2018). According to Macau statistics, there are 113 registered environmental associations in Macau. Although these associations have different areas of focus in environmental protection, their ultimate goal is the same: making Macau greener or more sustainable. See Table 3.6 below, which summarizes the 113 registered environmental associations.

3.3.4 HIGHER EDUCATION: PRODUCING TALENTS AND MAKING ACADEMIC CONTRIBUTION

Higher education institutions take coresponsibility for sustainable development (Walter and Brandli, 2016). In Macau, there are 10 tertiary educational institutions offering doctorate, master, and bachelor's degree programs,

TABLE 3.6 Registered Environmental Associations in Macau (Macau Government Printing Bureau, 2019).

1	Macau Federation of Green Building and Building Materials Enterprises	58	Macau Source Pulp Biotechnology Research Association
2	Macau Energy Conservation and Environmental Protection Technology Association	59	Macau Institute of Urban Development and Environmental Management
3	Macau Research Society of Green Resources	60	Macau Green Sense Youth Association
4	Green Energized Association	61	Macau Ecological Society
5	The Art of Plant Association	62	Green Future
6	Ut. Sin Association	63	Green City Construction Association
7	Macau Green Energy Association	64	Macau Forest Protection Association
8	Green friends Association	65	Macau Ecological Harmony Society
9	Macau Green Industry Chamber of Commerce	66	Community Environmental Promotion Association
10	Cherish Macau Association	67	Macau Food Waste Management Association
11	Macau New Energy Development Promotion Association	68	Macau Cross border Environment Concern Association
12	Macau Pest Control Association	69	Macau Free Cycle Association
13	Macau Environmental Energy Remote Control Sports Association	70	Macau Environmental Protection Technology Association
14	Macau Environmental Protective Materials and Technology Association	71	Macau Primary Ecology Science and Technology Institute
15	Association for the Promotion of Landscaping and Greening in Macao	72	Macau Institute of Water, Power and Environmental Engineering
16	Macau Environmental Protection Institute	73	Macau Society of Oceanography
17	Macau Health and Water Culture Association	74	Environmental Parenting Conservation Association
18	Macau Association of Desalination Technology Development Research	75	Macau Environmental Exhibition and Convention Industry Association
19	Macau Environmental Protection Development Association	76	Macau Water Resources Society and Macau Hydraulic Engineering Society

TABLE 3.6 *(Continued)*

20	Macau Wasted Papers Collectors Industries Association	77	Macau Ecology and Landscape Planning and Design Association
21	Macau International Environmental Car Association	78	Macau Environmental Culture Communication Association
22	Macau Nature Education and Promotion Association	79	Macau Tire Retreading, Repairing and Recycling Association
23	Macau Association of Ecological and Environmental Conservation	80	Macau Environment Protection Volunteer Association
24	Macau Wasted Electrical and Electronic Collectors Industries Association	81	Sino-Macau Energy Preservation and Environmental Protection Technology Research Center
25	Macau Marine Development Association	82	Macau Indoor Air Quality Association
26	Macau (New Energy) Green Environmental Protection Association	83	Macau Environmental Protection Equipment Chamber of Commerce
27	Macau Love Rabbit Association	84	Macau International Green Industry Alliance
28	Macau Association for Chinese Ecological Development and Promotion	85	Macau Green Environmental Protection Motor Vehicle Association
29	Karuna Association	86	Macau Environment Monitor Center
30	Macau Sanitation Management Service Industry Development Promotion Association	87	On Earth – Association of Education for Sustainable Development
31	Environmental Protection Development Concern Committee of Macau	88	Macau Environmental Materials Recycling Association
32	Environment Protection Federation Macau	89	Macau Environmental Education Association
33	The International Environmental Protection Equipment Professional Association of Macau	90	Green Environment Protection Association of Macau
34	Macau Recycling of Vehicles Association	91	Green City Construction Consultation Association
35	Macau Happy Swords Sports Association	92	Macau Energy Saving Association
36	Macau Promotion of Renewable Energy and Environmental Protection Association	93	Macau Youth Environmental Ambassadors Association

TABLE 3.6 *(Continued)*

37	Guangdong and Macau Association of Environmental Protection	94	Macau Natural Resources Conservation Promotion Association
38	Macau Green Life Concept Association	95	Macau Environmental Engineering Association
39	Green Monday Association	96	Macau Water Quality Association
40	Macau Green and Peaceful Association	97	Green Care Association
41	Macau Environment Protection Charity Association	98	Macau Gaseous Fuel Technology and Safety Association
42	Macau Environmental Protection Concern Association	99	Chinese (Macau) Wildlife Conservation Association
43	Macau International General Federation of Environmental Protection Groups	100	Macau Environmental Protection Substances and Technology Invention Society
44	Macau Green and Organic Plant Association	101	Macau Forestry Society
45	Macau Environmental Protection and Management Association	102	Macau Environmental Impact Assessment Association
46	Macau Federation of Ecology and Environment Protection Unions and Macau Union for Conservation of Nature	103	Macau, China—Research and Development Center of Renewable Energy Industry between Asia-Pacific and Latin America
47	Macau Urban Landscape Research Association	104	Macau Society of Environmental Sciences
48	Macau Endangered Species Conservation Society	105	Macau Wildlife Conservation Association
49	Macau Hydroponics Association	106	Macau Wetland Society
50	China Green Building and Energy Saving (Macau) Association	107	Macau Electromagnetic Radiation Protection Association
51	Macau Environmental Protection Association	108	Macau Greening Promotion Association
52	Macau Association for Ecological Study	109	Green Environment Institute
53	Macau Radiation Protection Association	110	Macau Research Society of Biological Diversity
54	Weng Chok Ecological Civilization Association	111	The Macau Society for Panda Conservation
55	Macau Food and Environmental Protection Association	112	Macau Youth Environmental Promotion Association
56	Blue Planet Association	113	Macau Green Power
57	Macau Youth Ecological Tourism Protection Society and Macau Youth Ecological Protection Society		

higher diplomas, postgraduate certificates, and diploma programs (Macau Higher Education Bureau, 2019). Five of the higher institutions have programs in tourism and hospitality management, and research units specialized in research for sustainable development. They produce professionals for the tourism and tourism-related industries; conduct research projects for sustainable tourism practices such as policy development, sustainable technology, sustainable planning, marketing, and management; provide academic support for the government, communities, the tourism industry, and other stakeholders of sustainable development for Macau. Tables 3.7–3.11 list the five institutions of tourism studies and their focuses.

3.3.4.1 MACAU INSTITUTE FOR TOURISM STUDIES

TABLE 3.7 Institutes and Their Areas of Focus.

Institutes	Areas of focus
Macau Institute for Tourism Studies	**Macau Institute for Tourism Studies** offers degree and professional programs in tourism-related disciplines such as hospitality, tourism business, heritage, events, retail and marketing, leisure and entertainment, sports and recreation, creative and cultural studies, and culinary arts. The institute also delivers various programs at the executive level in partnership with leading tourism schools in the world. The Institute conduct policy level research related to tourism planning and development for local authorities
IFT Tourism Research Centre	**IFT Tourism Research Centre** is funded by the Macau Institute of Tourism Studies and the Macau SAR government. It serves as a management and public policy think-tank and a knowledge provider for tourism and hospitality related research, and conducts policy and policy-review studies, monitoring and evaluation studies, polling and opinion surveys, support for academic research; communication via organizing seminars, workshops and conferences, publication, communication and dissemination of research findings, as well as research design and administration
Global Centre for Tourism Education and Training	**Global Centre for Tourism Education and Training** was established by the Institute for Tourism Studies, Macau, with the support of the Macau SAR Government and in collaboration, through an MOU, with the United Nations World Tourism Organization to enhance the quality of human capital and increase the competitiveness of tourism destinations (particularly in the Asia Pacific Region) in order to achieve sustainable tourism development in the long run

Source: Macau Institute for Tourism Studies.

3.3.4.2 UNIVERSITY OF MACAU

TABLE 3.8 Institutes and Their Areas of Focus.

Institutes	Areas of focus
Environmental Engineering Laboratory	**Environmental Engineering Laboratory** is under the program of Civil Engineering in Faculty of Science and Technology, University of Macau. The laboratory can serve students for experimental study in four-year Bachelor's Degree Program and research activities in Final Year Projects. The laboratory can contribute its pilot study facilities and chemical analytical equipment as well as experimental technology to the environmental studies and continuing learning in the environmental professions in Macau
International Integrated Resort Management Group	**The International Integrated Resort Management (IIRM) Group,** formerly **Hospitality and Gaming Management (HGM)**, is a unique academic group within the Faculty of Business Administration, University of Macau. It offers a Bachelor of Science in International Integrated Resort Management, a Master of Science in International Integrated Resort Management, and a Ph.D. in Business Administration with specialization in Hospitality and Gaming Management
Institute for the Study of Commercial Gaming	**The Institute for the Study of Commercial Gaming (ISCG)** is committed to the advancement of gaming knowledge that will benefit regulators, businesses and, most importantly, society as a whole. ISCG provides continuing public education and high-quality research in gaming and responsible gaming

Source: University of Macau.

3.3.4.3 MACAU UNIVERSITY OF SCIENCE AND TECHNOLOGY

TABLE 3.9 Institutes and Their Areas of Focus.

Institutes	Areas of focus
Faculty of Hospitality and Tourism Management	**The Faculty of Hospitality and Tourism Management (FHTM)** at Macau University of Science and Technology offers high-quality undergraduate and graduate programs in the fields of hospitality and tourism management. Undergraduate programs include International Tourism Management, Hotel Management, and Food and Beverage Management; The Master's Degree Program in International Tourism Management includes two areas of specialization: Meetings, Incentives, Conventions and Exhibitions (MICE) and Event Management, and Hotel Management; The Doctor of Philosophy in Tourism Management. Gained "TedQual" (Tourism Education Quality) certification from the United Nations World Tourism Organization (UNWTO)

TABLE 3.9 *(Continued)*

Institutes	Areas of focus
Institute for Sustainable Development	**The Institute for Sustainable Development (ISD)** of MUST aims at "objective, scientific and constructive research on Macao society," and has carried out a series of research projects, executive education programs, lectures, and forums since its establishment. ISD has set up two independent centers to meet the research demand of Macao society, "Market Research Center" and "Contemporary Macau Study Center."
Institute for Social and Cultural Research	**The Institute for Social and Cultural Research** is founded in December 2010. It rises as a nexus of communication and cooperation with academic institutions in Macau and in various regions of the world. With the community of Macau in mind, the institute organizes Macau Society Seminar Series and Ambassadorial Lecture Series to share research findings, and to broaden vision
Macau Environmental Research Institute	**Macau Environmental Research Institute (MERI)** is established in September 2015. MERI is committed to a research strategy in accordance with the social, economic, and cultural significance of Macau. The institute organizes full-scale academic communication and cooperation with research institutions in Macau, from periphery and around the world. With these capacities combined, MIRI hopes to further carry out the systematic researches on the new environmental topics in Macau, and endeavors to build an advantaged research platform for interested scholars and students and to offer original and meaningful research findings for local and international academics
Macau Institute of Smart City	**Macau Institute of Smart City** carries out smart city research, applications, and education and provides smart city products and services to the stakeholders. Smart city initiatives afford Macau a new opportunity to start its next round of highly efficient and sustainable growth toward economic prosperity and social well-being, and to turn Macau into the top tourist destination in the world

Source: Macau University of Science and Technology.

3.3.4.4 CITY UNIVERSITY OF MACAU

TABLE 3.10 Institutes and Their Areas of Focus.

Institutes	Areas of focus
Faculty of International Tourism and Management	**Faculty of International Tourism and Management** was established in September 2010 and has been approved by the Macau government to offer Bachelor, Master and Doctor of International Hospitality and Tourism Management programs. Gained "TedQual" (Tourism Education Quality) certification from the United Nations World Tourism Organization (UNWTO)

TABLE 3.10 *(Continued)*

Institutes	Areas of focus
Institute of Tourism and Gaming	**The Institute of Tourism and Gaming (ITG)** aims to serve the society and promote research in social, economic, and cultural fields that are deeply rooted and focused on Macau. The projects of the ITG are committed to the sustainable development of Macau and the enhancement of the distinction of the university as a world-class leader in education

Source: City University of Macau.

3.3.4.5 MACAU POLYTECHNIC INSTITUTE

TABLE 3.11 Institutes and Their Areas of Focus.

Institutes	Areas of focus
School of Business	**The School of Business** is to promote academic excellence that empowers students to advance their careers and life-long learning. Two of seven bachelor's degree programs are Bachelor of Social Science in Gaming and Recreation Management and Bachelor of Business Administration in Gaming and Recreation Management
Social, Economic and Public Policy Research Centre	**Social, Economic and Public Policy Research Centre** is united by The Centre of Social Economic Research and The Centre of Research and Planning of Human Resources. Its main research deals with the gambling industry of Macau, the development of Macau's economy and society, the government and public policy of Macau. More specifically, to conduct studies on Macau's social and economic problems; to strengthen academic development and to enhance teaching quality; to provide consultancy service for MSAR Government; to promote and assist in Macau's economic and social development
Cultural and Creative Industry Teaching and Research Centre	**The Cultural and Creative Industry Teaching and Research Centre** is devoted in responding to the real need in mid- and long-term teaching and training for the development of cultural and creative industries in Macau and enhancing academic research and the research and development of the cultural and creative profession. Work for diversified long-term sustainable development of regional economy and bring Macau up to the level of a high-quality society

Source: Macau Polytechnic Institute.

3.3.5 TOURISM ENTERPRISES: TAKING SOCIALLY RESPONSIBLE ACTIONS

In the tourism industry, sustainable development has gained momentum. Tourism enterprises have to play a role in the exercise of good governance practices (Martínez et al., 2013). In Macau, business entrepreneurs are encouraged to implement environmental measures. Other than participating in government and association initiatives, more and more Macau enterprises have included social and environmental commitments as key components of their corporate social responsibility (CSR). Practices of Macau international airport and two leading gambling operators (Sociedade de Jogos de Macau, S.A. "SJM" and Galaxy Casino) are briefly described. Tables 3.12–3.14 summarize all the sustainable actions taken by tourism enterprises in Macau.

3.3.5.1 MACAU INTERNATIONAL AIRPORT

Macau International Airport takes sustainability as its priority to make Macau International Airport an environmental-friendly green airport. Green practices including better energy efficiency, environmental quality, and cost-saving are integrated components of its planning, development, and daily operations.

TABLE 3.12 Macau International Airport in Action.

Environmental Management 5-Year Plan (2014–2018)	Vision: Develop and implement environmentally responsible activities with a view to approach the objective of "Green Airport"
	Slogan: Green the Life; Blue the Sky
	Objectives: better environment; social responsibility; financial benefit; sustainable development
	Targets: Reduce carbon emission and power consumption, increase waste recycle
Green projects	*Green wall*
	Hybrid/electric vehicles: replacement of cars to reduce carbon dioxide emissions
	Food waste recycling scheme: Introduced the food waste disposer, a microbial decomposition technology, degrade the majority of food waste into water, carbon dioxide, and organic fertilizer
	Green the Airport with various plants: 670 potted plants are put in different areas of the terminal building
	Replacement of airfield ground light: increase safety for aerodrome operation, higher capacity for airport utilization, reduce electricity consumption and CO_2 emission
	Macau International Environmental Cooperation Forum and Exhibition 2015: presented the plan and results of the airport environmental protection, to achieve the goal of "Green Airport."
Carbon emission	With the implementation of various environmental initiatives, the Greenhouse Gas Emissions per aircraft movement in 2017 was reduced by 28.7% as compared to that of 2012, exceeding the target of 20%

Source: www.macau-airport.com.

3.3.5.2 SJM HOLDINGS LIMITED

SJM Holdings Limited is a Macau casino gaming concessionaire. It was the only authorized gaming company in Macau before 2002, and now, one of the six companies to operate game business. The company's core value is "from society, to society;" therefore, they have been given back to the society and community in the way of supporting education, arts and culture, sports, and other charitable events.

TABLE 3.13 SJM in Action.

Education: nurturing talents	Annual scholarships to the University of Macau and Macao Polytechnic Institute
	Annual scholarship to employees' children for their study for up to five years
	Sponsors of local educational institutions
	Sponsors of school staffs to study at Macau Millennium College
	Supports employees to complete secondary school education at Millennium Secondary School
	Organize staffs participating in patriotic education in Jinggangshan, China, since 2014
Arts and culture	Regularly sponsors cultural events in Macau.
	Supports cultural activities at the Kam Pek Community Centre in Macau's Inner Harbor neighborhood and music performances at Macau Military Club
Sports	SJM is the title sponsor of the Formula 3 racing team "SJM Theodore Racing" and the "SJM Macau GT Cup—FIA GT World Cup"
	SJM is the title sponsor of the SJM Macau Derby at Macau Jockey Club, and a long-term sponsor of the Macau Jockey Club Charity Day
	SJM supports the sports events organized by Macau Sports Bureau
	SJM sponsors or subsidizes Hong Kong/Macau ferry tickets for sports organizations in Macau
Responsible gaming	Organizes a series of Responsible Gaming promotions every year
	Provides Responsible Gaming training for its employees
Charity and community activities	SJM has been the host of the Charity Auction and Dinner at Grand Lisboa since 2007
	SJM donates to the "Walk for a Million in Macau" charity event annually
	SJM is a regular sponsor of the events organized by the Macau Government Tourism Office
	SJM established the SJM Volunteer Team to help at Macau Social Services Centre, Macau Special Olympics, Macau Association of the Hearing Impaired, Macau Youth Volunteers Association, Macau Holy House of Mercy, Caritas de Macau and Bosco Youth Service Network

TABLE 3.13 *(Continued)*

Support for Macau's SMEs	SJM is a long-term supporter of local SMEs by prioritizing purchases from local SMEs, and jointly launched the "SJM and Macau SME Procurement Partnership Program" with the Macau Chamber of Commerce in 2016
	SJM launched the "SJM provides benefits to SME restaurants in San Ma Lou and Inner Harbor Area, promoting Macau as the City of Gastronomy" campaign
	SJM supported the "2018 Macau Year of Gastronomy" initiative
Commitments beyond Macau's borders	SJM sponsors art and culture groups from China and overseas for exhibitions and performances in Macau
	SJM contributes transportation and Macau hotel accommodation to charity groups and associations involved in sports, arts, and culture from Hong Kong, China, and overseas
	SJM donated extensively to various organizations on the Mainland in time of need

Source: www.sjmholdings.com.

3.3.5.3 GALAXY ENTERTAINMENT GROUP

Galaxy Entertainment Group (GEG) is a one of the leading companies of hospitality and gaming business in the world, and one of the major authorized concessionaires to operate gaming business in Macau. It strives to be a responsible corporate by investing heavily in social development, talent management, environmental conservation and resource, and supply chain management.

TABLE 3.14 GEG in Action

Stakeholders engagement	To achieve sustainable business growth and sustained value for its customers, employees, and the society, GEG engaged stakeholders through a variety of channels (e.g., an independent third party to facilitate the engagement exercise; all stakeholders were invited to rate the importance of sustainability issues online)
Social responsibility	***Contribution to community*** by volunteering in community activities and services and by providing immense relief efforts for the recovery of Macau from natural disasters
	Galaxy Entertainment Group Foundation to nurture young people, enabling them to become responsible citizens in their communities; *the GEG Sign Bilingualism Co-enrolment Pilot Scheme* to help hearing-impaired children to study in inclusive classrooms; *the GEG Chinese Literacy Development Pilot Scheme* to help students with language and literacy problems

TABLE 3.14 *(Continued)*

	Education and culture support Macau SAR Government's programs and policies on youth and talent development by working with various parties to empower through learning and education
	Charitable activities: Support to various charity groups (e.g., Charity Fund from the Readers of Macau Daily News, Macau Holy House of Mercy, Macau Special Olympics, etc.)
	Sports Development: Working with Macau SAR Government initiatives on enhancing cultural diversity, tourism, sports development as well as social harmony within the city. Sponsored various sports such as the "FIVB Volleyball World Grand Prix™ Macau"
Compliance with laws and regulations	***Anti-Money Laundering and Corruption***, An Anti-Bribery and Anti-Corruption Policy is in place and provides team members with guidance; ***Protecting Customer Privacy***, in full compliance with the Personal Data (Privacy) Ordinance of the government
Employees engagement	***Diversity:*** GEG embraces inclusivity and equal opportunity; ***working environment:*** GEG promotes a healthy work-life balance and is a recognized "Family-Friendly" Employer; ***labor standards:*** GEG strict comply with Government rules and regulations, and provides employees benefits and support; ***team member health and safety:*** Safety procedures and controls have been established and are in full compliance with relevant laws and regulations; ***talent management:*** provides training and development opportunities for employees for their career development.
Guests engagement	Satisfy customer needs by delivering exceptional experiences to guests; follows corporate policies and government rules and regulations to protect the health and safety of customers
Supply management	To work collectively with the suppliers in driving sustainable growth: control and mitigate adverse environmental impacts; prevent pollution and reduce waste production; and follow all relevant legal obligations and codes of practice; GEG also prioritizes local SMEs in its procurement process
Environmental protection and sustainability	Develop a culture to be an environmentally responsible company; conduct energy and water conservation; comply with all applicable environmental requirements and legislations; reduce, recycle, reuse and replace; educate, train and encourage employees to be environmentally responsible
Responsive gambling	GEG works closely with the Social Welfare Bureau, the Gaming Inspection and Coordination Bureau, the Institute for the Study of Commercial Gaming of the University of Macau (ISCG), local gaming operators, and nongovernmental organizations to develop a wide range of initiatives to enhance Macau's Responsible Gaming culture for employees and patrons

Source: www.galaxyentertainment.com.

3.3.6 *LOCAL COMMUNITY: VOICING MATTERS FOR COCREATING*

Macau residents' voice matters in tourism development in recent years, and residents' participation is required for tourism planning and development. Therefore, residents play a cocreating role for the development.

3.3.6.1 MACAU TOURISM INDUSTRY DEVELOPMENT MASTER PLAN

Part 3 of the plan described the study process, which includes site visits, international forums, surveys, interviews with key stakeholders, and two-month public consultation. In particular, a survey was conducted by 508 residents. A public consultation version of the plan was publicly available in 2016. After all stakeholders' concerns being considered, the master plan was then released in 2017.

3.3.6.2 MACAU TOURISM CARRYING CAPACITY STUDY

IFT Tourism Research Centre conducts research in tourism with a specific focus on the carrying capacity of Macau tourism. Studies on tourists' and residents' opinions and satisfactions are also conducted, and service providers are also involved to assess the tourism demand and tourism capacity. The research center generates an annual report and an executive summary to help the better management of tourism industry in Macau (Macau IFT Tourism Research Centre, 2019).

3.3.7 *TOURISTS: BEING A CORE CONSIDERATION*

High importance has been attached to visitors' experience in Macau. How well tourism services and organizations in Macau perform in satisfying visitors over time has been studied quarterly by IFT Tourism Research Centre. The result is the Macau Tourist Satisfaction Index (MTSI), which measures visitors' satisfaction across 10 tourism-related sectors. This includes (a) hotels, (b) restaurants, (c) transportation, (d) heritage attractions, (e) non-heritage attractions, (f) retail shops, (g) public services (e.g. customs and immigration), (h) events, (i) tour operators and guides, and (j) gaming. An overall satisfaction index for the tourism industry in Macau is then estimated. This MTSI thus provides industry with a regular and updated assessment of the "health" and effectiveness of Macau's tourism and hospitality services in serving the need of its guests and visitors (Macau IFT Tourism Research Centre, 2019).

3.4 CONCLUSION

With the support of China central government, Macau SAR government has been proactively taking the leading role in developing Macau to a World Centre of Tourism and Leisure, various strategies for the sustainable development of Macau have been proposed and put into practice. The SAR government also closely collaborates with regional partners for sustainable development. Environmental associations, higher education institutions, tourism enterprises, local community, and tourists, all these key stakeholders have their contributions, respectively, toward the sustainability of Macau tourism industry. As one can see from the above discussion, stakeholders of Macau active pursue their common goal for the sustainable development of Macau. The results of consistent growth in economic activities and tourist arrival reflect the payoff of sustainable practices and efforts by the stakeholders in Macau.

Sustainable development is a global trend. Macau sustainable development and its practices provide a great example. To develop sustainability, it needs all stakeholders to participate in any region, country as well as the whole world. A government plays a key role for establishing policies and regulations, governing sustainable economic development toward a sustainable tourism destination. The industry leaders should shoulder CSR, lead the companies for economic development, being corporate citizens, taking care of their employees and local community and implementing green practices. Education sector in a region or country should play an important role to educate students' sustainable development theories and practices to protect the Earth for generations to come. Universities and institutions should also assistant the local government for sustainable development research. Local community and associations should be in sustainability evocates. Tourists are more sensitive to a destination sustainable development. They are willing to participate and engage in the destination sustainability development. Every human being should take an active role to protect our mother Earth, so that generation by generation will enjoy a happy and healthy life on the Earth.

KEYWORDS

- **sustainable tourism destination**
- **stakeholder engagement**
- **tourism industry in Macau**
- **sustainable tourism development**
- **sustainable tourism practice**

REFERENCES

Andereck, K. L.; Vogt, C. A. The Relationship between Residents' Attitudes Toward Tourism and Tourism Development Options. *J. Travel Res.* **2000,** *32* (4), 1056–1076.

Andriotis, K. Community Groups' Perceptions of and Preferences for Tourism Development: Evidence from Crete. *J. Hospitality Tour. Res.* **2005,** *29* (1), 67–90.

China Daily. HK Welcomes Greater Bay Area Development Blueprint. 2019. Retrieved from https://www.chinadailyhk.com/articles/122/159/176/1550489831137.html?newsId=62185.

Chua, J. Betting on a Greener Macau. 2015. Retrieved from http://www.macaomiecf.com/miecf2015/.

Countrymeters. Macau Population Density Is 20,935.2 People Per Square Kilometer (54,221.9/mi^2) as of February 2019. n.d. Retrieved from https://countrymeters.info/en/Macau#population_density.

Goeldner, C. R.; Ritchie, J. R. B. *Tourism: Principles, Practices, Philosophies*, 11th ed., Wiley: New Jersey, 2008.

Greenwood, V. A.; Dwyer, L. Reinventing Macau Tourism: Gambling on Creativity? *Curr. Issues Tour.* **2016,** *20* (6), 580–602.

Hong Kong Institute of Engineers. 2019th Guangdong Hong Kong Macau Sustainable Development Conference. 2018. Retrieved from http://ev.hkie.org.hk.

Hong Kong Trade Development Council. Macao Special Administrative Region: Market Profile. 2019. Retrieved from http://hktdc.com/research.

International Monetary Fund (IMF). People's Republic of China-Macau Special Administrative Region. 2019. Retrieved from http://www.imf.org.

Kong, W. H.; du Cros, H.; Ong, C. E. Tourism Destination Image Development: A Lesson from Macau. *Int. J. Tourism Cities* **2015,** *1* (4), 1–17.

Macau Cultural Affairs Bureau (MCAB). Attractions of the World Heritage: The Historic Centre of Macao. 2018. Retrieved from http://www.icm.gov.mo.

Macau Economy Profile. 2018. Retrieved from https://www.indexmundi.com/macau/economy_profile.html.

Macau Environmental Protection Bureau (DSPA). n.d.a. https://www.dspa.gov.mo/.

Macau Environmental Protection Bureau (DSPA). Environmental Programs, n.d.b. Retrieved from https://www.dspa.gov.mo/topic.aspx.

Macau Environmental Protection Bureau (DSPA). Environmental Protection Planning 2010–2020. 2010. Retrieved from https://www.dspa.gov.mo/Publications/.
Macau Government Printing Bureau. Environmental Associations in Macau. 2019. Retrieved from: https://www.io.gov.mo/cn/entities/priv/cat/environment/?p=1.
Macau Government Tourism Office (MGTO). Arrivals of Total Visitors & International Visitors 1995–2011. 2012. Retrieved from: http://macaotourism.gov.mo.
Macau Government Tourism Office (MGTO). Macau's Population Density on Top of the World. 2015. Retrieved from http://mtt.macaotourism.gov.mo/201506/en/contents/3/172.html.
Macau Government Tourism Office (MGTO). Macau Tourism Industry Development Master Plan. 2017. Retrieved from http://zh.macaotourism.gov.mo/.
Macau Government Tourism Office (MGTO). Quality Tourism Services Accreditation Scheme. 2018a. Retrieved from http://industry.macaotourism.gov.mo/en/page/content.php?page_id=219.
Macau Government Tourism Office (MGTO). Hotels, MICEs, Restaurants, Casinos & Travel Agents in Macau. 2018b. Retrieved from http://macaotourism.gov.mo/.
Macau Government Tourism Office (MGTO). Hotels in Macau. 2018c. Retrieved from http://macaotourism.gov.mo/.
Macau Government Tourism Office. Tourism Statistics. 2019. Retrieved from https://dataplus.macaotourism.gov.mo/Publication/document/ENG/Other/TUR/E_TUR_PUB_2018_Y.pdf.
Macau Higher Education Bureau. 2019. https://www.daes.gov.mo/.
Macau IFT Tourism Research Centre (ITRC). 2019. http://itrc.ift.edu.mo/.
Macau MIECF. Main Theme of 2019 MIECF: Promoting Ecological Civilization and Green Development. 2019. Retrieved from: http://www.macaomiecf.com/miecf2019/.
Macau Monthly. Focusing on the Phenomenon of Macau Association Culture. 2018. Retrieved from: http://www.macaumonthly.net/yuekan/biandu/2018-10-29/128701.html.
Macau News. Macau Records the Highest Population Density Per Square Kilometer. 2016. Retrieved from https://macaunews.mo/macau-records-highest-population-density-per-sq-km/.
Macau News. Record 35.8 Million Visits Macau Last Year. 2019a. Retrieved from https://macaunews.mo/category/tourism/.
Macau News. Hotels Report Record 14.1 Million Guests in 2018. 2019b. Retrieved from https://macaunews.mo/category/tourism/.
Macau Statistics and Census Services. Macao Industrial Structure 2016. 2016. Retrieved from https://www.dsec.gov.mo.
Macau Statistics and Census Service. Industrial Structure of Macao 2017. 2018. Retrieved from https://www.dsec.gov.mo/Statistic.
Sheng, M.; Gu, C. Economic Growth and Development in Macau (1999–2016): The Role of the Booming Gaming Industry. *Cities* **2018,** *75*, 72–80.
UNESCO. The Rio Declaration on Environment and Development 1992. 1992. Retrieved from http://www.unesco.org/education/pdf/RIO_E.PDF.
UNESCO. Sustainable Development. n.d. Retrieved from https://en.unesco.org/themes/education-sustainable-development/what-is-esd/sd.
United Nations. Our Common Future: Report of the World Commission on Environment and Development. 1987. Retrieved from https://sustainabledevelopment.un.org/content/documents/5987our-common-future.pdf.
UNWTO. Sustainable Tourism for Development. 2013. Retrieved from http://www2.unwto.org/publication/sustainable-tourism-development.

UNWTO and UNEP. Making Tourism More Sustainable—A Guide for Policy Kakers. 2005. Retrieved from https://www.e-unwto.org/doi/book/10.18111/9789284408214.

Walter, L. F.; Brandli, L. Eds. *Engaging Stakeholders in Education for Sustainable Development at University Level.* World Sustainability Series. Springer International Publishing: New York, 2016.

Wan, Y. K. P.; Li, X. Sustainability of Tourism Development in Macao, China. *Int. J. Tourism Res.* **2013,** *15*, 52–65.

Worldometer. Macau Population. 2019. Retrieved from: https://www.worldometers.info/world-population/china-macao-sar-population/.

World Travel and Tourism Council (WTTC). Economic Contribution of Macau Travel and Tourism 2017. 2018. Retrieved from https://www.wttc.org/.

CHAPTER 4

DIVERSIFICATION OF MACAO'S TOURISM DEVELOPMENT: A REVIEW

XIANGPING LI* and LIANPING REN

Macao Institute for Tourism Studies, China

Corresponding author. E-mail: xiangping@iftm.edu.mo

ABSTRACT

This chapter summarizes the diversification, initiatives, and evaluates their effectiveness in Macao. Diversification in source market as well as of tourism products are introduced and assessed. Statistics between 2008 and 2018, when applicable, from Bureau of Statistics and Census Service (DSEC) are utilized for the analysis and evaluation.

4.1 INTRODUCTION

Chinese economic surge, coupled with Macao's gaming liberalization in 2002 and the Individual Visit Scheme in 2003, has pushed the economy in Macao to a level that has never been envisioned. Macao's GDP increased from MOP51.9 billion to MOP440.3 billion, an 8.5 times of growth during the period between 1999 and 2018 (Table 4.1). Over the same period, GDP per capita also saw an upsurge from MOP121,363 to MOP666,893, a 549.5% of increase, which makes Macao one of the richest cities in the world (Meneses, 2018). It has been well recognized that the municipal economy is heavily reliant on tourism, which has been closely linked to its gaming. Since 2006, the city's gaming revenue accounted for almost 50% to its GDP with an ever-increasing proportion over the past decades (Table 4.2). Macao surpassed Las Vegas for the first time in 2006 with a gaming revenue of MOP57.5 billion (USD7.2 billion); and in 2013, the city recorded a gaming revenue of over MOP360 billion (USD45 billion), almost seven times that of Las Vegas, further cementing Macao's status as "the undisputed heavy-weight champion of the gambling industry" (Riley, 2014). Another feature of tourism development in Macao is its overreliance on a single-source market, namely, Mainland China (Greenwood and Dwyer, 2017).

TABLE 4.1 Macao's GDP and GDP Per Capita (1999–2018).

	Total GDP			GDP per capita		
	MOP (billion)	USD (billion)	Change (%)	MOP	USD	Change (%)
1999	51.9	6.5	–	121,363	15,186	–
2000	53.9	6.7	4.0	125,271	15,608	3.2
2001	54.7	6.8	1.4	126,107	15,698	0.7
2002	58.8	7.3	7.5	134,181	16,703	6.4
2003	65.7	8.2	11.7	148,182	18,473	10.4
2004	84.9	10.6	29.2	186,776	23,281	26.0
2005	96.9	12.1	14.1	204,607	25,541	9.5
2006	118.3	14.8	22.2	238,057	29,755	16.3
2007	147.4	18.3	24.5	282,962	35,212	18.9
2008	167.8	20.9	13.8	312,149	38,918	10.3
2009	171.5	21.5	2.2	318,611	39,905	2.1
2010	225.1	28.1	31.3	419,153	52,380	31.6
2011	294.3	36.7	30.8	536,178	66,867	27.9
2012	343.8	43.0	16.8	603,525	75,536	12.6
2013	411.9	51.6	19.8	692,501	86,680	14.7
2014	442.1	55.3	7.3	710,895	89,005	2.7
2015	362.2	45.4	–18.1	564,635	70,712	–20.6
2016	362.4	45.3	0.0	561,053	70,177	–0.6
2017	405.8	50.6	12.0	625,254	77,902	11.4
2018	440.3	54.5	8.5	666,893	82,609	6.7

Source: DSEC (1999–2018).

Therefore, even with the exponential growth of the economy, the dominance of gaming in Macao's economic structure is problematic, as this makes the city's economy rather volatile and easily subject to external influences (Choi et al., 2019). One such influence was witnessed in 2013 and two years forward when Chinese government was determined to fight against corruption, which was hurting the high spending market of the casinos (Tao and Li, 2018). June 2013 saw the first monthly decline in gaming revenue by 3.7% since 2009 (DSEC, 2014). To combat against overdependence on gaming industry and a single source market as well as to ensure more sustainable economic progress, both central government and Macao government have made economic diversification a priority for its development (Choi et al., 2019; Sheng, 2017). As far as tourism industry is concerned, the endeavors have been directed to transform Macao into a World Center of Tourism and Leisure (WCTL) by restructuring and diversifying tourism products beyond gaming and tourism source market beyond Mainland China.

TABLE 4.2 Game Revenue and Its Contribution to GDP (2002–2018).

	Gross gaming revenue		Tax revenue from Gaming		GDP	Contribution of gaming to GDP
	MOP (billion)	% change	MOP (billion)	% change	MOP (billion)	
2002	23.5	–	7.8	–	58.8	39.9
2003	30.3	29.02	10.6	36.23	65.7	46.1
2004	43.5	43.53	15.2	44.03	84.9	51.2
2005	47.1	8.33	17.3	13.66	96.9	48.7
2006	57.5	22.04	20.7	19.80	118.3	48.6
2007	83.8	45.77	31.9	53.85	147.4	56.9
2008	109.8	30.98	43.2	35.36	167.8	65.5
2009	120.4	9.61	45.7	5.76	171.5	70.2
2010	189.6	57.49	68.8	50.50	225.1	84.2
2011	269.1	41.92	99.7	44.90	294.3	91.4
2012	305.2	13.45	113.4	13.77	343.8	88.8
2013	361.9	18.55	134.4	18.53	411.9	87.9
2014	352.7	−2.53	136.7	1.73	442.1	79.8
2015	231.8	−34.28	89.6	−34.48	362.2	64.0
2016	224.1	−3.31	84.4	−5.80	362.4	61.9
2017	266.6	18.95	99.8	18.33	405.8	65.7
2018	303.9	13.98	113.5	13.69	440.3	69.0

Source: DSEC (2002–2013) (no statistics of gross gaming revenue provided before 2002).

4.2 BUILDING MACAO INTO A WORLD CENTER OF TOURISM AND LEISURE

The notion of building Macao into a WCTL surfaced in 2008 when the National Development and Reform Commission announced it for the first time in the "Outline of the Plan for the Reform and Development of the Pearl

River Delta (2008–2020)." In 2011, the positioning of Macao as a WCTL was integrated into the national development strategy and supported by the 12th National Five-Year Plan (Liu, 2011). In early 2015, Macao government submitted a specific proposal for the 13th National Five-Year Plan, and in October same year, the Chief Executive established the "Committee for Development of the World Centre of Tourism and Leisure" (MGTO, 2017). Although the national policies have been in place for years, it was not until 2015 when Macao Government Tourism Office (MGTO) took the initiative and compiled a "Macao Tourism Industry Development Master Plan" (the Master Plan) to fully leverage Macao's WCTL position and formulate relevant objectives, strategies, and action plans (MGTO, 2017). Based on the strategies stated in the Master Plan, this chapter focuses on initiatives to spur the diversity in terms of the source market as well as the tourism products. Therefore, diversification efforts to be discussed in this chapter include:

1. Source markets
2. Cultural and creative tourism
3. MICE tourism
4. Marine tourism
5. Accommodations
6. Retailing

Though the diversification initiatives were just put into the Master Plan in 2017, Macao government has acknowledged the importance of diversification and endeavors have been exerted since 2008. Accordingly, this chapter will evaluate the effectiveness of the diversification efforts between 2008 and 2018, when applicable.

4.2.1 DIVERSIFICATION OF SOURCE MARKETS

It has been realized that overreliance on one single source market, namely Mainland China, is not viable for Macao's sustainable tourism development, as visitor arrivals are susceptible to the policy changes in Mainland (Sheng and Gu, 2018). Therefore, MGTO has been seeking to develop potential markets and attracting visitors from other source markets. Two major actions have been taken to achieve this goal. First, MGTO representatives have been sent to the key source markets, including eight countries in Asia and Pacific regions, USA, and two countries in Europe (Table 4.3). They are responsible for tailor-devising promotion schemes and activities for each market. For example, to showcase Macanese cuisine to the Americans,

MGTO representatives organized a series of mega promotional campaigns of "Experience Macao Food Truck" in New York City in 2017, Los Angeles in 2018, and San Francisco in 2019. Second, MGTO has been working hard to strengthen its connections with overseas travel trade as well as regional and international tourism organizations to increase awareness of Macao in the international and regional arenas. For instance, MGTO attended PATA Travel Mart in Kazakhstan to enhance Macao's destination profile in September 2019 (MGTO, 2019b). In addition, MGTO received the Grand Award in the Marketing Category for the campaign "Experience Macao Food Truck" staged in Los Angeles in 2018.

TABLE 4.3 MGTO Overseas Representatives.

	Country
Asia/Pacific	Australia
	India—Mumbai
	India—New Delhi
	Indonesia
	Japan
	Korea
	Malaysia
	New Zealand
	Thailand
America	USA—California
	USA—New York
Europe	Russia
	United Kingdom and Republic of Ireland—London
	United Kingdom and Republic of Ireland—Oxford

However, an examination of the visitor arrivals and market shares of the key source markets suggests that Mainland China still enjoys the lion's share, which has increased from 50.6% in 2008 to 70.6% in 2018 (Table 4.4). Nonetheless, it does not mean that MGTO's efforts have not been paid off at all. One notable pattern is that Korea has become the number one international source market, surpassing Japan, since 2011. The number of tourists from Korea has more than tripled from 279,794 in 2008 to 874,253 in 2017. In addition, 2018 also witnessed a rise in tourist counts from several long-haul markets. Specifically, the number of visitor arrivals increased by 8.3%, 2.2%, and 2.1% from the USA, Australia, and UK respectively. Russia, the new

TABLE 4.4 Visitor Arrivals (,000) and Market Share (%).

	Mainland China	Hong Kong SAR	Taiwan	Korea	Japan	Philippines	Malaysia	USA	Thailand	Indonesia	India	Singapore	Australia	Canada	UK	France	Russia
2018	25,260.6	6327.9	1061.0	812.8	325.8	312.1	227.9	201.8	181.4	173.8	147.9	134.8	90.9	74.3	58.3	38.9	30.8
	70.6%	17.7%	3.0%	2.3%	0.9%	0.9%	0.6%	0.6%	0.5%	0.5%	0.4%	0.4%	0.3%	0.2%	0.2%	0.1%	0.1%
2017	22,196.2	6165.1	1060.1	874.3	329.0	307.1	218.3	186.4	198.2	197.1	148.1	143.1	89.0	74.3	57.1	40.4	27.0
	68.1%	18.9%	3.3%	2.7%	1.0%	0.9%	0.7%	0.6%	0.6%	0.6%	0.5%	0.4%	0.3%	0.2%	0.2%	0.1%	0.1%
2016	20,454.1	6419.8	1074.5	662.3	300.6	287.0	222.8	190.9	236.2	182.5	165.3	155.8	93.3	75.2	61.3	42.7	25.1
	66.1%	20.7%	3.5%	2.1%	1.0%	0.9%	0.7%	0.6%	0.8%	0.6%	0.5%	0.5%	0.3%	0.2%	0.2%	0.1%	0.1%
2015	20,410.6	6534.5	988.1	554.2	282.2	276.8	229.1	182.5	180.8	163.4	167.6	158.8	92.4	71.0	60.0	41.0	22.8
	66.5%	21.3%	3.2%	1.8%	0.9%	0.9%	0.7%	0.6%	0.6%	0.5%	0.5%	0.5%	0.3%	0.2%	0.2%	0.1%	0.1%
2014	21,252.4	6426.6	953.8	554.5	299.8	262.9	250.0	181.5	175.9	189.2	167.2	196.5	105.9	70.6	60.8	40.0	31.9
	67.4%	20.4%	3.0%	1.8%	1.0%	0.8%	0.8%	0.6%	0.6%	0.6%	0.5%	0.6%	0.3%	0.2%	0.2%	0.1%	0.1%
2013	18,632.2	6766.0	1001.2	474.3	290.6	274.1	291.1	179.5	238.6	208.5	160.0	189.8	109.6	74.2	61.3	43.4	30.5
	63.5%	23.1%	3.4%	1.6%	1.0%	0.9%	1.0%	0.6%	0.8%	0.7%	0.5%	0.6%	0.4%	0.3%	0.2%	0.1%	0.1%
2012	16,902.5	7081.2	1072.1	444.8	396.0	283.9	301.8	188.7	231.3	209.1	150.8	205.7	113.3	83.5	59.5	42.5	26.8
	60.2%	25.2%	3.8%	1.6%	1.4%	1.0%	1.1%	0.7%	0.8%	0.7%	0.5%	0.7%	0.4%	0.3%	0.2%	0.2%	0.1%
2011	16,162.7	7582.9	1215.2	398.8	396.0	268.7	324.5	196.1	196.4	220.4	169.7	280.6	111.8	79.8	61.6	42.7	16.5
	57.7%	27.1%	4.3%	1.4%	1.4%	1.0%	1.2%	0.7%	0.7%	0.8%	0.6%	1.0%	0.4%	0.3%	0.2%	0.2%	0.1%
2010	13,229.1	7466.1	1292.7	331.8	413.5	247.8	338.1	188.3	212.4	208.4	169.1	257.2	111.8	79.1	60.4	41.4	11.7
	53.0%	29.9%	5.2%	1.3%	1.7%	1.0%	1.4%	0.8%	0.9%	0.8%	0.7%	1.0%	0.4%	0.3%	0.2%	0.2%	0.0%
2009	10,989.5	6727.8	1292.6	204.8	379.2	247.5	332.5	179.5	242.5	191.4	107.5	256.5	120.4	74.7	63.4	39.2	7.8
	50.5%	30.9%	5.9%	0.9%	1.7%	1.1%	1.5%	0.8%	1.1%	0.9%	0.5%	1.2%	0.6%	0.3%	0.3%	0.2%	0.0%
2008	11,613.2	7016.5	1315.9	279.8	366.9	283.0	428.6	203.2	252.1	180.3	82.4	263.6	139.6	82.1	76.9	40.5	7.1
	50.6%	30.6%	5.7%	1.2%	1.6%	1.2%	1.9%	0.9%	1.1%	0.8%	0.4%	1.1%	0.6%	0.4%	0.3%	0.2%	0.0%

Source: DSEC (2008–2018).

long-distance market, also contributed 30,784 tourists in 2018. Although the absolute number is not impressive, that is a 13.9% growth from previous year. All these are indicative of the effectiveness of MGTO's diversification endeavors to some extent.

4.2.2 DIVERSIFICATION WITH CULTURAL AND CREATIVE TOURISM

Macao is blessed with a rich mixture of Chinese and Western cultural history and traditions. While its history can be traced back in the Qin Dynasty of China back in 221 BC; it was also the first European settlement in Asia since 1553. As a former colony of Portugal and with its root in China, Macao has managed to blend these two cultural forces. The harmonious coexistence and fusion of the two cultures over the past hundreds of years have made Macao a very unique city of both oriental and western styles. The highlight of the rich Chinese and Western cultures is embodied by the Historic Center of Macao, which was inscribed on the UNESCO World Heritage List in 2005. The Historic Center of Macao was described by UNESCO as: "with its historic street, residential, religious and public Portuguese and Chinese buildings, the historic center of Macao provides a unique testimony to the meeting of aesthetic, cultural, architectural and technological influences from East and West. The site also contains a fortress and a lighthouse, the oldest in China. It bears witness to one of the earliest and longest lasting encounters between China and the West, based on the vibrancy of international trade." More equally important legacies from the cultural fusion are noteworthy as well. In addition to these tangible heritage architectures, Macao is similarly known for its churches, temples, museums, intangible cultural heritages, cultural festivals and events, and gastronomy. Three aspects are discussed in this section: cultural events and festivals, culinary tourism, and cultural and creative industries.

4.2.2.1 CULTURAL EVENTS AND FESTIVALS

MGTO has been aware of Macao's abundant cultural and heritage resources. It has been MGTO's priority to promote what Macao has to offer to the tourists by celebrating and staging various cultural events and festivals year round (Table 4.5). In addition to celebrate traditional Chinese festivals, holy days of Catholic, Buddhism, Taoism, and other Chinese folk religions are commemorated as well in Macao. To develop Macao into an internationally "eventful" city, MGTO has also added a series of international events and festivals to its event/festival-scape. For example, "Art Macao" is a newly staged mega-event in 2019, which will last five months starting June. This mega international arts and cultural event will present and display visual artworks at various exhibition venues in Macao along with a feast of diverse musical and theatrical performances (Macao, 2019). While "Art Macao" was only organized in 2019, other art and music festivals have already been present in Macao for decades.

TABLE 4.5 List of Cultural Events and Festivals in 2019.

Month	Festival	Category
January	18th Macao City Fringe Festival (January 11–27)	
February	Chinese New Year (Parade for the celebration)	Traditional Chinese festival
March	Feast of the God Toutei	Taoism festival
	Procession of the Passion of our Lord, The Good Jesus	Catholic festival
April	Ching Ming Festival	Traditional Chinese festival
	Feast of Pak Tai	Chinese folk religious festival
	Easter Holidays	Catholic festival
	A-Ma Festival	Chinese folk religious festival
	HUSH! Full Music (April 28–May 1)	
May	30th Macao Arts Festival (May 3–June 2)	
	Feast of Buddha	Buddhism festival
	Feast of the Drunken Dragon	Chinese folk religious festival
	Tam Kong Festival	Chinese folk religious festival
	Procession of Our Lady of Fatima	Catholic festival
June	Dragon Boat Festival (Macao International Dragon Boat Races)	Traditional Chinese festival
	Art Macao (June–October)	
	Feast of Na Cha	Taoism festival
	Arts and Cultural Festival between China and Portuguese-speaking countries	
July	International Youth Music Festival (July 17–21)	
	International Youth Dance Festival (July 20–25)	
	Fest of Kuan Tai	Chinese folk religious festival

TABLE 4.5 *(Continued)*

Month	Festival	Category
August	2019 Wushu Masters Challenge (August 1–4)	
	Feast of Maidens	Traditional Chinese festival
	Feast of Hungry Ghosts	Buddhism/Taoism festival
September	30th Macao International Fireworks Display Contests	
	Mid-Autumn Festival	Traditional Chinese festival
October	33rd Macao International Music Festival (October 4–30)	
	Festival of Ancestors (Chung Yeung Festival)	traditional Chinese festival
	17th Macau A-Ma Cultural and Tourism Festival (October 8–11)	
	22nd Lusofonia Festival (October 18–20)	Macanese festival
November	All Souls Day	Catholic festival
	19th Macau Food Festival (November 8–24)	
December	Macao Light Festival (December 1–31)	
	4th International Film Festival & Awards (December 5–10)	
	Feast of Immaculate Conception	Catholic festival
	Macao International Parade	
	Winter Solstice	Traditional Chinese festival
	Christmas	Catholic festival

4.2.2.2 CULINARY TOURISM

Macao has been recognized by UNESCO for its 450-plus-year-old culinary legacy, Chinese, Portuguese, and Macanese, and has gained an emerging status as one of Asia's leading culinary destinations since Macao was officially designated as a Creative City of Gastronomy by UNESCO in 2017 (Macau Daily Times, 2017; Springer, 2018). Long before Macao became the Creative City of Gastronomy, Macao has been well known for its varieties of delicacies. For example, Macao has been hosting the International Food Festival since 2000. The Festival has been expanding ever since due to the continuous support from Macao government. The Festival managed to attract an audience of over 800,000 in 2016. The diversification efforts in food area have also been evidenced by a wide array of restaurants and food establishments in Macao (Tables 4.6 and 4.7). Table 4.6 shows that there is an overall increasing trend in total number of restaurants and food establishments. The growth was particularly observable in the years of 2013, 2014, and 2015. The upward trend has also been noted in number of employees in the restaurant industry, as well as the revenues from the industry. In addition to the overall growth, increases have been presented in different types of restaurants, such as Chinese restaurants, Western restaurants (Table 4.7).

TABLE 4.6 Indicators of Restaurants and Establishments.

	Restaurants and establishments		Employees		Revenues	
	Number	% change	Number	% change	(billion MOP)	% change
2008	1568	–	18,974	–	4.495	–
2009	1615	3.0	19,738	4.0	4.790	6.5
2010	1648	2.0	22,172	12.3	5.505	14.9
2011	1714	4.0	21,213	−4.3	6.088	10.6
2012	1678	−2.1	21,987	3.6	6.899	13.3
2013	1918	14.3	26,079	18.6	8.403	21.8
2014	2112	10.1	30,360	16.4	9.602	14.3
2015	2284	8.1	32,313	6.4	10.039	4.6
2016	2265	−0.8	32,551	0.7	10.633	5.9
2017	2306	1.8	32,678	0.4	11.053	3.9
2018	2345	1.7	33,737	3.2	11.829	7.0

Source: DSEC (2008–2018).

Since Macao was designated as a Creative City of Gastronomy in 2017, further efforts have been exerted to promote and conserve Macao's culinary heritage. A four-year action plan with initiatives in six categories was framed in January 2018 to transform Macao into a City of Gastronomy (MGTO, 2019a). A review of tasks performed in 2018 suggested that MGTO was working hard to achieve the goal. For instance, MGTO activated the "2018 Macao Year of Gastronomy" to raise the awareness and strengthen the promotion of Macao's gastronomic culture, and attempted to construct a Macanese cuisine database (MGTO, 2019a). MGTO was also actively engaged with gastronomy-related international events and exchanges, such as attending the 12th UNESCO Creative Cities Network Annual Conference in Poland.

TABLE 4.7 Categories of Restaurants and Establishments.

	Total	Chinese restaurants	Congee and noodle shops	Western restaurants	Japanese and Korean restaurants	Other Asian restaurants	Fast-food restaurants	Cafe	Bars and lounges	Cooked-food stalls	Others
2018	2345	622	858	210	140	106	45	126	34	78	126
2017	2306	615	839	207	132	116	42	123	41	72	119
2016	2265	598	824	192	152	115	38	125	35	110	76
2015	2284	563	921	166	146	113	36	129	30	105	75
2014	2112	504	815	177	132	114	35	117	39	69	110
2013	1918	443	794	153	109	95	34	72	39	75	104

Source: DSEC (2013–2018).

4.2.2.3 CULTURAL AND CREATIVE INDUSTRIES

Although Macao is endowed with abundant tangible and intangible cultural and heritage resources, to deepen the development of cultural tourism products, including cultural festivals and events, and gastronomic tourism, it is understood that new cultural and creative elements are needed. Therefore, development and expansion of cultural and creative industries are essential for providing updated cultural and creative experiences. As a consequence, the Committee of Cultural Industries was founded in 2000 to formulate policies and strategies for the development of the cultural industries. With policies in place, Macao has witnessed the development and growth of the cultural and creative industries in four categories: creative design, cultural exhibitions and performances, art collection, and digital media (Table 4.8). It is demonstrated that a total number of 2088 organizations were in operation in 2017, a 9.1% increase from previous year. The growth mainly came from the category of "creative design," followed by "digital media." In addition, revenues from cultural industries also rose by 4.8%, which was contributed mainly by "creative design." Although the industry has been experiencing a rapid growth, it is still on its infancy, and there is ample room for further expansion (Yong et al., 2018).

TABLE 4.8 Indicators of Cultural Industries.

		Total	Creative design	Cultural exhibitions and performances	Art collection	Digital media
Number of organizations	2015	1708	919	168	109	512
	2016	1913	1021	218	110	564
	2017	2088	1149	220	107	612
Number of employees	2015	10,192	3336	1829	435	4592
	2016	10,996	3368	2429	390	4809
	2017	11,702	3454	2454	427	5367
Revenues (million MOP)	2015	6244	1821	1346	88	2989
	2016	6752	1805	1456	128	3363
	2017	7076	2090	1438	105	3443

Source: DSEC (2015–2017).

4.2.3 DIVERSIFICATION WITH MICE TOURISM

The MICE Sector, the Conference and Exhibition (C&E) in particular, is relatively newer in Macao, but it has been one of the major government efforts in diversifying the economic structure in Macao. Macao has a few advantages in developing the MICE Sector, including fine hotel facilities and conference venues, attractive heritage sites, and good public security. Also thanks to the various incentive schemes the government has been providing, the MICE Sector in Macao has seen quick development (IPIM, 2017). While the MICE sector draws heavily on the capacity of other sectors such as hotels and restaurants, transportation, retail, and advertisement, just to name a few, the capacity growth of this sector in Macao can be seen from the increase in the number of conference venues (establishments) and number employees working for this sector shown in Table 4.9.

TABLE 4.9 Growth Trend of the MICE Sector.

Year	Number of events	Growth (%)	Number of participants	Growth (%)	Number of establish-ments	Growth (%)	People engaged	Growth (%)
2008	–	–	–	–	16	60.00	64	93.94
2009	1195	–	571,287	–	20	25.00	85	32.81
2010	1393	16.57	805,777	41.05	24	20.00	118	38.82
2011	1035	−25.70	1,277,309	58.52	26	8.33	134	13.56
2012	1015	−1.93	1,612,596	26.25	33	26.92	130	−2.99
2013	1024	0.89	2,033,619	26.11	38	15.15	178	36.92
2014	1050	2.54	2,614,473	28.56	61	60.53	274	53.93
2015	1263	20.29	2,516,092	−3.76	67	9.84	373	36.13
2016	1276	1.03	1,721,527	−31.58	80	19.40	344	−7.77
2017	1381	8.23	1,901,674	10.46	98	22.50	426	23.84
2018	1427	3.33	2,117,883	11.37	112	14.29	432	1.41

Source: DSEC (2019).

Under the policy of "Prioritizing Conferences," a total of 1427 conferences and exhibitions were held in Macao in 2018, which was a 3.3% increase compared to that in the previous year, including 1342 conferences and 60 exhibitions. Over 2.1 million people participated in MICE activities in Macao, representing a year-on-year increase of 11.6%. Table 4.9 shows the growth trend of the C&E Sector in Macao by number of C&Es and participants. In terms of number of

C&Es, there has been a steady growth from 2013 to 2018. The land area used for the C&Es has been increasing as well generally, reflecting the government's effort in developing and expanding this sector. The number of participants has grown over the years, except for 2016 and 2017, when economy in all sectors was not optimistic, but 2018 has seen a return.

In terms of the themes of the conferences and exhibitions, a wide coverage is noted, including business trade and management, art, IT and other technology, education, tourism, training, and medical and health (MGTO, 2017). To further push up the C&E sector, the government has plans to further expand and improve the C&E facilities. For example, MGTO (2017) mentioned developing more multifunctional C&E centers, and a committee has been set up solely for the development of the C&E sector, apart from the many promotional activities and incentive schemes carried out by the IPIM (Macao Trade and Investment Promotion Institute), under which the C&E sector is now.

4.2.4 DIVERSIFICATION WITH MARINE TOURISM

Marine tourism is a new tourism element in Macao. This concept has not been visualized until the year 2015 when the central government demarcated an area of 85 square kilometers of waters under Macao's jurisdiction (Marine and Water Bureau, n.d.). The defined water area, along with the future reclamation areas, has provided a strong foundation for Macao to develop water-related leisure facilities and activities. The Master Plan, thus, has outlined a list of action plans within a short-term, medium-term, and long-term timeframe to incorporate marine tourism products into the portfolio (Table 4.10) (MGTO, 2017). To utilize marine resources, the short-term (0–5 years) strategy is to lay the foundations of marine tourism development by cultivating and furthering water sightseeing programs and boat-related activities. Three products are discussed in this section, including Macao cruise sightseeing, yacht tourism, and "Reel Fun" fish boat tours.

4.2.4.1 MACAO CRUISE SIGHTSEEING

In 2018, several Macao sightseeing routes were developed and put into operation as a result of the negotiation and cooperation between MGTO and industry partners around Macao Peninsula, Taipa, and Coloane (Marine and

Water Bureau, 2019) (Table 4.11). As marine tourism is a new form and the routes were only introduced in late 2018, related statistics are not available yet. However, a recent interview with Liang Ye, the managing director of Yuet Tung Shipping, disclosed that the number of passengers embarking on the cruise ships was below the expectation. It was estimated that the occupancy rates of the ships averaged at 50% during holidays, and 20% during weekdays (*Macao Daily News*, 2019). Nevertheless, Mr. Liang was optimistic of the prospects. When the routes are given more promotion, the increased awareness will induce more participation.

TABLE 4.10 Introduce Marine Tourism Products.

	Proposed action plans
Short-term (0–5 years) —laying the foundations	Support the development of water sightseeing programs
	Support boat-related entertainment projects
	Continually develop more water sports
Medium term (6–10 years) —Facilitating productivity and efficiency	Upgrade and optimize the facilities for yacht tourism
	Continually encourage and assist to promote international water sports events and festivals
Long-Term (above 10 years) —achieving sustainable tourism growth	Develop Macao to be the core city of short-haul multidestination trips in the Pearl River Delta through the comprehensive development of marine tourism products

TABLE 4.11 Macao Cruise Sightseeing.

Time	**Route**	**Company/Organization**
14:00	Inner Harbor, Macao → Route 1	Macau Cruise Development
14:00	Inner Harbor, Macao → Route 1	Yuet Tung Shipping
15:00	Taipa Ferry → Coloane	Shun Tak China Travel
16:00	Coloane → Taipa Ferry	Shun Tak China Travel
19:00	Inner Harbor, Macao → Route 1	Macau Cruise Development

4.2.4.2 YACHT TOURISM

Yacht tourism was initiated when the Individual Travel Scheme for Yachts was signed between Guangdong and Macao in July 2014 to deepen cooperation in tourism between the two parties. The first sailings of individual yacht traveling started in November 2016 between Macao and Zhongshan. Although yacht tourism was proposed to diversify Macao tourism product options, the

Scheme seems to be unable to attract yacht owners from either Macao or Mainland as indicated in Table 4.12. Only seven yachts from Mainland in 2016 and one in 2019 from Mainland took advantage of this Scheme. No Macao yacht has undertaken any trip to Mainland. It was pointed out that the Scheme was not working well due to the high deposits, customs, and docking fees that need to be paid by Macao yachts to enter Mainland (Macao, 2019). Therefore, the Scheme is severely underutilized and measures to reduce the cost and facilitate entry are demanded. This is of high significance when the cooperation of yacht travel between Macao and Mainland has been extended to other cities beyond Zhongshan.

TABLE 4.12 Application for Individual Travel Scheme for Yachts.

Year	Quarter	Yachts from Mainland	Yachts from Macao
2016	4.0	7.0	0.0
2017	1.0	0.0	0.0
	2.0	0.0	0.0
	3.0	0.0	0.0
	4.0	0.0	0.0
2018	1.0	0.0	0.0
	2.0	0.0	0.0
	3.0	0.0	0.0
	4.0	1.0	0.0
2019	1.0	0.0	0.0

Source: Marine and Water Bureau (2016–2019).

4.2.4.3 "REEL FUN" FISH BOAT TOURS

Another watered-related entertainment project worth mentioning is "Reel Fun" fish boat tour, which was initiated in 2011 by the Macao Fisherman's Mutual Help Association and supported by Marine and Water Bureau and MGTO. This program was introduced to take advantage of the fishing moratorium period for local fishermen to provide leisure boat tours for residents, and later for tourists as well (MGTO, 2018, 2019a). This program has been well received by residents (Marine and Water Bureau, 2012, 2017), and most seats were occupied (Table 4.13). However, statistics on tourist's participation are not available. It is suggested to collect data from both resident and tourist for further evaluation.

TABLE 4.13 "Reel Fun" Fish Boat Tour.

Year	Period (Saturday and Sunday)	Boat frequency	Carrying capacity/boat	Total carrying capacity
2011	June 11–July 24	70.0	25	1750
2012	May 26–July 29	80	25	2000
2013	June 22–July 28	48	26	1248
2014	May 24–July 13	64	26	1664
2015	June 13–July 13	48	26	1248
2016	June 11–July 11	48	26	1248
2017	June 03–July 10	50	26	1300
2018	May 26–July 08	64	26	1664
2019	May 25–July 07	66	26	1716

Source: Marine and Water Bureau (2011–2019).

4.2.5 DIVERSIFICATION WITH ACCOMMODATIONS

The hotel sector is one of the most important supporting sectors to the tourism industry in any destination. Therefore, the growth of the hotel sector is more or less in parallel with the growth of the tourism industry in Macao. According to the tourism statistics, by the end of 2018, there were 116 hotels altogether in Macao (an addition of 5 hotels from that in 2017), with 38,809 rooms, up by 5.8%. This has included the 34 nonstandard lodging facilities. The average occupancy rate was 91.1%, which was a 3.9% increase compared to that in 2017. Among all, the Five-Star hotels accounted for the biggest percentage of rooms—63%, and the Four-Star hotels accounted for 19.9%, which means that the hotel industry in Macao is more toward the upscale and luxury segments currently. Figure 4.1 shows the details on the number of hotel rooms by category. In 2018, the hotel revenue (the 82 hotels alone, excluding the non-standard hotel operations) has reached 37.29 million MOP in 2018, up by 14.5% compared to that in 2017. This shows a strong upward development signal. In addition, there is a 3.7% increase of manpower dedicated to the hotel sector. In the end of 2018, there were a total of 52,976 employees working for different hotels in Macao, majority working for the four and five-star hotels (87.6%).

With the growing trend of tourist arrivals, there is still ample room for hotel increase in the near future. According to MGTO (2017)'s prediction, the number of hotel rooms will reach 48,600–51,900 by 2025 depending on low and high expected growth rates, which is from 29% to 38% increase compared to the number of hotel rooms in 2016. To cater to the diversified needs of the tourists and to extend their stay, newly developed hotel properties in the Cotai Strip area typically bear more leisure venues, show areas, retail shops, and a greater variety of restaurant choices, compared to the hotels established in the Macao peninsular earlier. There are a greater range of tourist activities in this area. For example, some of the shows such as the House of Dancing Water and the Monkey King have become main attractions. The regular celebrity concert (usually once in every other week) has further contributed to the increase of tourist arrivals. The above effort has been particularly effective with the family tourists from Mainland China, who have an increasingly higher demand for family-oriented hotel/resort product (HVS, 2018). However, as Figure 4.1 shows, majority of the hotel rooms in the current stock belongs to the upscale and luxury categories, accommodating mostly the luxury segment and the gambling tourists, and is therefore not attractive to the budget-conscious tourists. Table 4.14 shows that the increase of the hotels is mostly in the five-star category, while the rest four categories of hotels do not have as much increase. To attract different types of tourists, the government is now planning to increase the choices of hotels by creating more categories (MGTO, 2017).

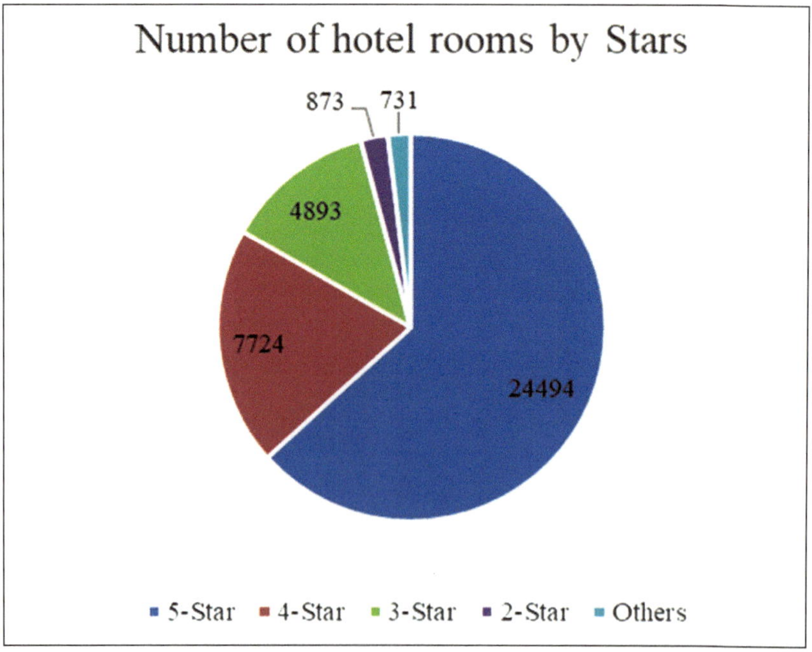

FIGURE 4.1 Number of hotel rooms by category.

TABLE 4.14 Growth of the Accommodation Sector by Category.

Year	Total	Total change	5 Star	5 Star change	4 Star	4 Star change	3 Star	3 Star change	2 Star	2 Star change	Guest house	Guest house change
2008	84	2.44	19	26.67	12	0.00	12	−7.69	11	−8.33	30	0.00
2009	90	7.14	22	15.79	13	8.33	12	0.00	12	9.09	31	3.33
2010	91	1.11	23	4.55	13	0.00	12	0.00	12	0.00	31	0.00
2011	95	4.40	26	13.04	13	0.00	12	0.00	12	0.00	32	3.23
2012	100	5.26	28	7.69	14	7.69	12	0.00	13	8.33	33	3.13
2013	98	−2.00	27	−3.57	14	0.00	12	0.00	12	−7.69	33	0.00
2014	98	0.00	27	0.00	14	0.00	13	8.33	12	0.00	32	−3.03
2015	106	8.16	32	18.52	17	21.43	13	0.00	12	0.00	32	0.00
2016	107	0.94	32	0.00	17	0.00	15	15.38	11	−8.33	32	0.00
2017	111	3.74	33	3.13	17	0.00	16	6.67	12	9.09	33	3.13
2018	116	4.50	35	6.06	17	0.00	16	0.00	14	16.67	34	3.03

Source: DSEC (2019).

4.2.6 *DIVERSIFICATION WITH RETAILING*

Shopping, as a major tourist activity, has always been an important component shaping tourists' experience during their visit, and it also contributes to the extension of the length of their stay. Having a comprehensive portfolio of retail outlets is thus desirable. Therefore, Macao government has included the growth of the retail sector as one of the major strategic efforts, which can be seen in some key government strategic reports, such as the Analysis Report of Statistical Indicator System for Moderate Economic Diversification of Macao released in 2017 and the Comprehensive Report of the Overall Planning of Tourism Industry in Macao (MGTO, 2017). In fact, turning Macao into a "shopping paradise" has been one of the six most important tourism strategies in Macao (MGTO, 2017). This strategic effort has so far seen some good results, which is reflected in the growth in both retail venues and employees. In 2017, there were a total of 7, 017 retail outlets in Macao, employing 42,313 employees, increased by 3% and 3.4% respectively compared with the statistics in the previous year. Generally, both the numbers of retail venues and employees had been increasing at a steady growth rate from 2013 to 2017. Table 4.15 shows the details.

TABLE 4.15 The Growth in Retail Venues and Employees (DSEC, 2018).

	2013	2014	2015	2016	2017
Retail venues	6051	6497	6489	6808	7017
Retail venue growth	1	7.4	−0.1	4.9	3
Employees	30,910	35,516	38,215	40,939	42,313
Employee growth	8	15	7.6	7.1	3.4

In addition, the retail sales volume has seen growth over the years as well. Figure 4.2 shows the growth of the retail sales over the years. Generally speaking, there has been an upward growing trend, except the years of 2015 and 2016, when most economic sectors in Macao suffered from an economic downturn due to various reasons including the implementation of the anticorruption policy in Mainland China.

The government is confident in developing and further expanding the retail sector in Macao. There are a few favorable conditions to do so. First of all, Macao is a free-trade port, which is an important and favorable condition. For example, quite a few retail categories, such as the Portuguese wine, souvenirs, and apparels, have established good reputation among the tourists. Second, the range of high-end international brands is another attraction to tourists. Third, for the Mainland tourists, currency exchange rate between MOP and RBM creates a favorable discount. At a more micro level, there have been efforts in improving tourists' shopping experience in Macao. For example, MGTO (2017) has noted a significant improve in the overall shopping experience rated by 500 tourists in 2016. However, Macao is constrained by the size of the land area and the high rent of the retail shops, and therefore is currently only offering limited categories of retail places. For example, majority of tourists spending in retail is within a limited number of categories including cosmetics and perfumes (30.6%), souvenir (27.1%), apparel (13.4%), watch and jewelry (11.3%), and bags and shoes (9.1%). Therefore, more strategies and initiatives to increase the choices has been included in the government strategic planning (MGTO, 2017). For example, the government is now considering and planning a new comprehensive shopping center in the new living zone.

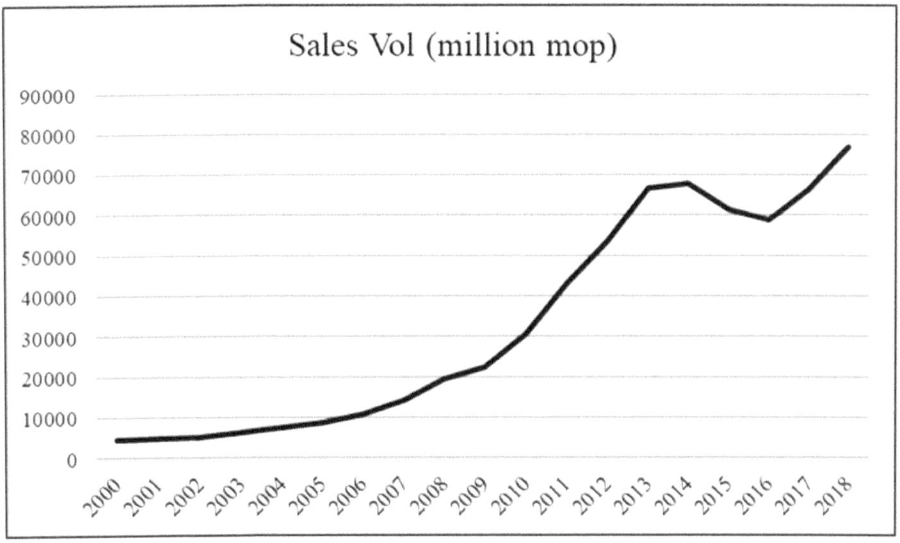

FIGURE 4.2 Growth trend of the sales volume in the retail sector in Macao (DSEC, 2019a).

4.3 CONCLUSION

With the understanding that it is not sustainable to rely on a single industry and a single source market, diversification has become the priority for Macao government since 2008 when the National Development and Reform

Commission proposed to build Macao into a WCTL. Diversification strategy has become ever more pressing when the positioning of Macao was further integrated into the national development strategy by the 12th National Five-Year Plan and written into the Master Plan in 2015. With years of diversification efforts, Macao has found some success in the MICE, hotel, and retailing sectors. In addition, in the field of cultural tourism, inventory of relevant products such as cultural events and festivals has been expanded. With the addition of gastronomy, the future of cultural tourism seems rather promising. However, when it comes to the source markets, although visitor arrivals from some countries such as Korea and the USA have been increasing, the contribution of Mainland market still ranges from 60% to 70%. As marine tourism is a relatively new concept in Macao, it requires more promotion to raise the awareness among visitors. While Macao's diversification initiatives are constrained by its petite geographic area, regional cooperation could be enhanced to address this limitation (Kan, 2017; Zhao, 2017).

KEYWORDS

- **World Centre of Tourism and Leisure**
- **diversification in source market**
- **diversification of tourism products**
- **gaming**
- **Macao**

REFERENCES

Choi, F.; Hon, C. T.; Mao, Y. H.; Lai, I. K. W. Sustainable Development for Small Economy and Diversification from a Dominant Industry: Evidence from Macao. *Sustainability* **2019,** *11* (6), 1626.

Greenwood, V. A.; Dwyer, L. Reinventing Macau Tourism: Gambling on Creativity? *Curr. Issues Tour.* **2017,** *20* (6), 580–602. doi:10.1080/13683500.2016.1187585

HVS. *An HVS Quarterly Macau Update*. 2018. Retrieved from https://www.hvs.com/article/8250-the-hvs-quarterly-macau-update.

IPIM. *A Study on Conference and Exhibition Sector Development Blueprint in Macao*. 2017. Retrieved from https://www.ipim.gov.mo/wp-content/uploads/2017/12/MICE-Extract-_CN.pdf.

Kan, X. The Current Situation and Development Trend of Tourism Cooperation Between Macau and Hengqin. In *Green Book of Macau: Annual Report on Development of Travel and Leisure in Macau (2016–2017)*; Hao, Y., Lin, G., Eds.; Social Sciences Academic Press: Beijing, 2017; pp 73–83.

Liu, Z. *Empirical Investigation of Moderate Economic Diversification in Macao: Development of Tourism Clusters in Macao*; Macao Association of Economic Sciences: Macao, 2011.

Macao, A. (Producer). Art Macao, 2019. Retrieved from http://www.artmacao.mo/2019/en/.

Macao Daily News. Uneven Distribution of Tourists on Cruise [Haishang youkeliang bujun chengxiao weiji yuqi]. (2019, October 12). *Macao Daily News*. Retrieved from http://www.macaodaily.com/html/2019-10/12/content_1387844.htm.

Macau Daily Times. Macau Designated 'Creative City of Gastronomy'. UNESCO. 2017. Retrieved from: https://macaudailytimes.com.mo/macau-designated-creative-city-gastronomy-unesco.html.

Marine and Water Bureau. Residents Responses Positive towards the First-Day Ticket Sale of the Fishing Boat Tours [Xiuyu Yujiale Shouri Shoupiao, Shimin Fanying Rongyue], 2012, May 24. Retrieved from 休漁漁家樂首日售票 市民反應踴躍: https://www.marine.gov.mo/news_detail.aspx?a_id=1438067443.

Marine and Water Bureau. Fishing Boat Tours Ended Successfully with Active Participation from Residents [Xiuyu Yujiale Yuanman Jieshu, Shimin Rongyue Canjia], 2017, October 7. Retrieved from 休漁漁家樂圓滿結束 市民踴躍參與: https://www.marine.gov.mo/news_detail.aspx?a_id=1499676008.

Marine and Water Bureau. Macau Cruise, 2019. Retrieved from https://m.marine.gov.mo/seawayScheduled/Sightseeing/DataTour.aspx?lang=ENG.

Marine and Water Bureau. (n.d.). Waters under Jurisdiction of Macao. Retrieved from https://www.marine.gov.mo/subpage.aspx?a_id=1520388246#.

Meneses, J. P. 2020—Macau Slated to Be Richest Place in the World. *Macau Business,* 2018.

MGTO. *Macao Tourism Industry Development Master Plan Comprehensive Report*. Macao Government Tourism Office (MGTO), 2017.

MGTO. Press Conference of Macao Government Tourism Office, 2018. Retrieved from industry.macaotourism.gov.mo/files/download.php?file...pdf.

MGTO. 2018 Annual Report of MGTO, 2019a. Retrieved from https://industry.macaotourism.gov.mo/cn/ebook/2018_Annual_Report/#p=31.

MGTO. MGTO Attends PATA Travel Mart in Kazakhstan, 2019b. Retrieved from https://industry.macaotourism.gov.mo/en/pressroom/index.php?page_id=7&sp=0&id=3158.

Riley, C. Macau's Gambling Industry Dwarfs Vegas, 2014. Retrieved from http://money.cnn.com/2014/01/06/news/macau-casino-gambling/.

Sheng, L. Explaining Urban Economic Governance: The City of Macao. *Cities* **2017,** *61*, 96–108. https://doi.org/10.1016/j.cities.2016.08.011.

Sheng, M.; Gu, C. Economic Growth and Development in Macau (1999–2016): The Role of the Booming Gaming Industry. *Cities* **2018,** *75*, 72–80. https://doi.org/10.1016/j.cities.2018.01.003.

Springer, K. How Macau Became China's Hub of Haute Cuisine: The 'Las Vegas of Asia' Has Evolved into a Diverse Foodie Destination. *Independent,* 2018, July 2. Retrieved from https://www.independent.co.uk/travel/asia/macau-china-best-restaurants-food-cuisine-holidays-foodie-destinations-a8427171.html.

Tao, L.; Li, S. Corruption and Travel: Effects of China's Anti-Graft Campaign on Macao. *DIE ERDE—J. Geographical Soc. Berlin* **2018,** *149* (1), 44–51.

Yong, X.; Xiuju, X.; Zhao, K. Executive Summary [Zongbaogao]. In *Analysis of the Development of Cutlural Industries in Macao [Aomen Wenhua Changye Fazhan Fenxi I]*; Yong, X., Xiuju, X., Eds.; Committee of Cultural Industries, Macao Special Administrative Region: Macao, 2018.

Zhao, W. Tourism Leisure Economy in Macau: Status Quo, Characteristics and Trends. In *Green Book of Macau: Annual Report on Development of Travel and Leisure in Macau (2016–2017)*; Hao, Y., Lin, G., Eds.; Social Sciences Academic Press: Beijing, 2017; pp 33–44.

CHAPTER 5

THE BED AND BREAKFAST INDUSTRY IN TAIWAN

FRANK C. TSAI[1*] and HUNG-SHENG (HERMAN) LAI[2]

[1]*Slippery Rock University of Pennsylvania, USA*

[2]*Fu Jen Catholic University, Taiwan*

[*]*Corresponding author. E-mail: frank.tsai@sru.edu*

ABSTRACT

Bed and Breakfast (B&B) service was introduced to the Taiwan lodging market as "Minsu" in the Kenting National Park area in the early 1980s. Initially, it was defined as a part-time business operated in a private residence, and the nature of operations was much different from traditional hotels. Over the past four decades, four major development stages in the B&B industry have emerged, and nine types of B&Bs have been identified based the transitions and unique features and services that B&Bs provide. As B&Bs developed, the Taiwanese government implemented B&B regulations that reinforced the industry, but it still does not meet the industry's needs that have arisen from its rapid growth. Overall, the Taiwan B&B industry is a successful accommodation-sharing concept that has successfully been promoted to overseas tourists through multiple channels.

5.1 INTRODUCTION

Bed and Breakfast (B&B) services were introduced to the lodging market in Taiwan about 40 years ago and have continually evolved as "Minsu." By offering a unique accommodation experience to tourists in both rural and metropolitan areas, these B&B establishments have become an affordable alternative for travelers. Due to the nature of B&B operations, guests have a better chance of knowing the owners of a B&B, thus having easier, more direct access to local culture, history, and cuisines. These characteristics are main reasons B&Bs have become one of the more attractive business options in Taiwan.

FIGURE 5.1 Natural resources and culture in Taiwan.

Source: Taiwan Tourism Bureau (https://eng.taiwan.net.tw/).

According to Taiwan's Tourism Bureau (2019a), the number of arriving visitors to Taiwan increased from 1,393,254 people in 1980 to 11,066,707 people in 2018, indicating a 794% increase in tourist arrivals. While the demand for B&Bs is increasing, many potential B&B owners are trying to seize a piece of the market, resulting in a huge B&B boom in the rural areas. Some owners are even trying to position their operations to go beyond B&Bs. B&B owners are relying on their own housing for business, creating a 3-in-1 service (Wang et al., 2012) that incorporates local life style, natural resources, and local produce. Lodging accommodations, dining, and tourist amenities are included as part of the overall service in order to compete both in operation and marketing. Some tourist destination studies (Hsiao and Chuang, 2015; Chen and Zhuang, 2009) and reports from the Taiwan Tourism Bureau (2019b) indicate that Nantou, Hualien, and Yilan Counties currently account for the largest part of B&B operations in Taiwan. In response to the market shift over the years and the success of B&B operations in rural areas, more and more similar operations have been offering short-term stays, called "mini accommodations" in the metropolitan areas since the early 2000s.

Most B&B owners see their operations as a side job during the early stage of ownership, and many B&Bs go through a difficult learning process. Additionally, the B&B industry in Taiwan has caused managerial difficulties for the government. Deviating from the government's intent to decide the management approach of the lodging industry, it also creates an impact and rebound in the lodging industry. After two decades of B&B expansion, Taiwan's first B&B Act, the "Management of Bed & Breakfasts Guidelines," was established by the legislature in 2001 and revised in 2017 in response to many serious issues arising from B&B operations such as sanitation, fire safety, water pollution, zoning code, and building violations, so these issues are in need of attention and discussion.

According to August 2019 statistics from the Taiwan Tourism Bureau (see Chart 5.1), legally registered B&B operations increased dramatically from 36 to 8464 properties between 2001 and 2018, with a total of almost 35,067 rooms in 2018. The B&B sector is a promising component of rural tourism due to its positive contribution to the innovation of tourism services and its small scale (Chen et al., 2013; Tourism Bureau, 2019b). Although the hotel and resort industry debates B&B legalization in order to defend their own businesses, the number of B&Bs is still increasing and creating fiercer competition in Taiwan's lodging industry.

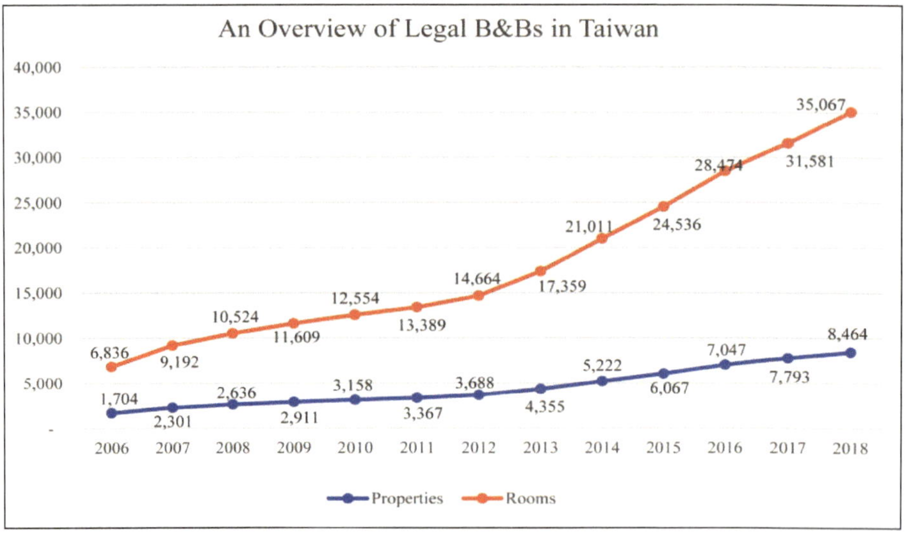

CHART 5.1 An Overview of Legal B&Bs in Taiwan.
Source: Taiwan Tourism Bureau, 2019 (https://eng.taiwan.net.tw/).

The B&B industry in Taiwan presents diverse, unique accommodation services in the lodging industry. B&Bs come in all shapes and sizes, but their services are generated from the specific characteristics of the owners and the integration of local resources. This chapter looks at Taiwan's B&B industry in three ways: (1) identifying the initial services B&Bs provide and how those have evolved along with market progress, (2) accessing the B&B Act and associated legal issues, and (3) illustrating challenges in Taiwan's B&B industry. To provide in-depth information about the uniqueness of the B&B industry in Taiwan, we include anecdotal examples for further discussion and study.

5.2 B&B DEVELOPMENT IN TAIWAN

5.2.1 INTRODUCTION STAGE (1980s–1990s)

In 1980s Taiwan, B&Bs were called "Minsu."At the time, there were a few local residents who rented out the spare bedrooms of their own houses to tourists who stayed overnight in Nanwan Kenting. The services provided were basic and not a real competitive threat to hotels. This was the initial development of B&Bs in Taiwan, and it slowly grew as a new lodging service product. B&B service in Taiwan fits well into the definitions of B&Bs from Hsiao and Chuang (2015), Hotch and Glassman (1992), and Taylor (2004). Although B&Bs are a part-time business for many owners in Nanwan Kenting, they provide tourists more than just a place to sleep and eat; they also offer a warm, family-like feeling during guests' stays.

This unique service offers a promising opportunity for regenerating and extending unique lifestyles and cultural images within local communities (Sharpley, 2002). It also serves as a mechanism affecting the success of the early Taiwan B&B industry during the 1980s and1990s, and it has created a higher market demand and supply phenomenon in the B&B industry for the next stage of development. Following Nanwan Kenting's B&B development, Alishan, Jiufen, Ruifang, Sitou, Penghu, and Ilan also entered the B&B market and quickly transformed many residential homes to businesses. Due to the uniqueness and freedom of B&Bs' operations, the B&B industry has attracted younger generations from various professional fields to invest and develop new B&B operations. As a result, modern B&B operations have entered a diversified and professional generation which leaves the older B&Bs facing the challenge of transformation.

5.2.2 EARLY GROWTH STAGE (2001–2010)

Following the Labor Law on weekly working hours in the late 1990s, Taiwan's government enforced the 2-day weekend policy in 2001, promoting leisure agricultural tourism and opening up the B&B industry to a new stage. Along with the massive reconstruction and recovery after a 7.6-magnitude earthquake rattled the country on September 21, 1999, the government initiated many alternatives and grant opportunities for the most seriously damaged areas in Taiwan. In the legislature, the Post-Earthquake Reconstruction Act raised NT$212.3 billion in reconstruction funds (Taipei Times, 2004) for those communities and individuals suffering from earthquake damages. During this early growth stage, the available funds and contributions of the Labor Law benefited many small villages in rural areas and helped develop many B&Bs and leisure agriculture farms which attracted tourists to those areas and helped generate revenue for economic development.

Hsiao et al. (2018) in a study of "Factors for development of B&B management in Taiwan" and Zheng (2010) indicated that the government began to counsel and promote leisure agriculture programs and form counseling teams for the development of B&Bs in the mountainous and aboriginal regions in early 2002, especially after the Post-Earthquake Reconstruction Act was announced. As a result, B&Bs began to appear in many aboriginal tribal and rural areas. During this period, Taomi Village became an iconic project of Taiwan's B&B development history. The 921 Earthquake reformation showed that by adding unique services, the B&B industry promoted awareness of eco-friendly concepts in the local tourism industry and enriched the appearance of the community to travelers. By providing home-sharing experiences to guests, travelers could enjoy authentic local experience and contribute positively to the innovation of tourist products on smaller scales. The Taiwanese government also strategically encouraged the transformation of traditional agriculture into a new type of agricultural tourism economy by reducing economic impact on the agriculture industry after joining the World Trade Organization (WTO) in 2002. Furthermore, with the economic progress, the Labor Law, a series of tourism campaigns (Tourist Multiplier Program, 2002–2007; Tour Taiwan Years, 2008–2009; Vanguard for Excellence in Tourism, 2009–2014), and B&B Act endorsements in Taiwan, B&B operations have developed rapidly in many rural areas. From 2001 to 2010, legal B&B properties increased dramatically, from 36 to 3158 (Taiwan Tourism Bureau, 2011). During this period of development in the B&B industry, B&B operations have been constantly transforming from the original home-sharing concept into alternative accommodation service styles as well.

5.2.3 GROWTH STAGE (2011–2015)

The government's implementation of tourism rejuvenating campaigns during the early growth stage served as a driving force in attracting more foreign tourists to Taiwan. Based on statistics from the Taiwan Tourism Bureau (2014), the proportion of tourism-oriented foreign visitors to Taiwan increased from 59.69% in 2011 to 68.35% in 2013 (Taiwan Tourism Bureau, 2019b). In 2015, there were 10,439,785 annual inbound visitors from different countries, the highest record in history (Taiwan Tourism Bureau, 2019b). As the demand for accommodations at tourism destinations have increased at the same time, B&Bs have become important resources for the tourism industry in Taiwan.

With the popularity of B&Bs and operators' creativity during this late growth stage, the B&B industry increased its legally registered properties to 6067 B&Bs by the end of 2015 (Taiwan Tourism Bureau, 2019b). During this late growth stage, the industry also successfully integrated social media, online reservation platforms, alternative leisure, and recreational experiences into its services. These value-added options have become increasingly favorable to tourists wanting to stay overnight in a tourist destination. At the same time, many B&Bs have detached from a home-style accommodation service, and their architectural design and operational formats have become increasingly diverse, combining exotic, cultural, and natural elements as well as owner characteristics into the services provided.

FIGURE 5.2 Sky Yard B&B, Nantou County (https://eng.taiwan.net.tw/m1.aspx?sNo=0002114).

Source: Pei-jun Ho.

5.2.4 MATURITY SAGE (2016–CURRENT)

The enhancement of Taiwan's living standard and the practice of a 2-day weekend have made travel and tourism essential in people's lives. B&Bs are a niche segment of the travel and tourism industry and have played a successful role in attracting tourists to many rural areas. After almost four decades of evolution, the two most important factors of B&Bs remain their uniqueness and sense of social environment. Chou et al. (2018) indicate that Taiwan B&Bs have been classified into two major service types—upscale and budget. Though B&B property development has slowed down, legal operations reached 8464 by the end of 2018. The number of annual inbound tourists has also slowed down, growing only from 10,439,785 tourists in 2015 to 11,066,707 tourists in 2018. Therefore, at this stage of B&B development, instead of increasing and expanding B&B operations, owners are focusing more on managerial and operational efficiency of their B&Bs.

Huang and Perng (2018) and (Hsueh, 2015) stated that the rising environmental awareness in past years has people beginning to think about coexisting with the Earth. Ideas about green consumption and ecotourism have become prevalent and gradually integrated into B&B's services in Taiwan. More efficient operations and low environmental impact practices have been introduced and current B&Bs are encouraged to adopt energy saving and carbon-reducing concepts into their operations. Both business and consumers' environmental responsibility presents notable effects on the perceived value of tourism. Huang and Perng (2018) further suggested that environmental protection, environmental responsibility, perceived value, and local culture and natural ecology are the best strategies for sustainable green tourism. Green consumption refers to the management strategies and plan designs of B&Bs in Taiwan. Furthermore, unique landscapes, building patterns in ecotourism areas, and green operations have also become distributive touring spots. In comparison with other touring spots, ecotourism areas may better attract the international market and foreign tourists.

After more than 30 years of development in Taiwan's B&B industry, there are many B&Bs with unique buildings and locations that offer all kinds of experiences to their guests that differ greatly from the usual anonymous, uniform hotel room. There are plenty of unique and interesting B&Bs to choose from in Taiwan according to offers and sharing on social media. Facebook, Instagram, and Twitter pages highlight a variety of picks for Taiwan's best alternative and unique B&Bs, and these lists are constantly being changed and updated on social media through shares and comments made by guests. In recent years, the trend toward information transmission and service functions via the Internet has played an increasingly important role in the B&B industry in Taiwan. B&B owners are shifting their attention to social media and online travel agents (OTA) as their primary marketing tools, and they have shifted their advertising to provide a more visible web presence that attracts oversea tourists. Getting the most out of social media and OTAs for a B&B business is a complex topic for many owners, but it is worth exploring how social networks can help these operations.

5.3 B&B ACT AND LEGAL ISSUES

5.3.1 THE FIRST B&B ACT IN TAIWAN

Taiwan's first B&B Act, the Management of Bed & Breakfasts Guidelines, was endorsed in 2001, after almost 20 years of B&Bs being on the market. The B&B Act helped the B&B industry flourish across the country; however, the guidelines mainly referred to architectural regulations, such room number and building dimensions. In addition to the B&B Act, Taiwan's government also promoted agricultural tourism via the Leisure Agricultural Development Regulations in 2002 in response to the economic impact (Chiang et al., 2004) from joining the WTO, ultimately helping many famers and non-famers invest in B&B operations in the rural agriculture landscape. As B&Bs cooperated with leisure agriculture and utilized rural landscapes, they also created experience-oriented agricultural operations as a lifestyle destination for tourists.

Based on the B&B Act, this type of lodging facility incorporates local culture, natural landscape, ecological environment, environmental resources, and agricultural, forestry, fishery, or livestock farming activities. B&Bs are categorized into nine categories based on their main features:

1. Farm B&B;
2. Beach B&B;
3. Spa B&B;
4. Sports B&B;
5. Traditional Architecture B&B;
6. Gastronomy B&B;
7. Western Farm B&B;
8. Aboriginal Cultural B&B; and
9. Exotic B&B.

5.3.2 REVISION OF B&B ACT

The B&B Act of 2001 was revised in November 2017. Over the past decade or so, there have been major changes to the urban and rural landscapes across Taiwan. The development of economy, transportation, and science and technology has changed our daily life in dramatical ways. The B&B Act endorsed in 2001 was not adjusted for more than a decade, and B&B businesses were unable to function as they should have under certain older restrictions until the revised B&B Act was passed and announced in November 2017.

Since 2010, the number of legal B&B's in Taiwan increased by more than 4,600 properties to a total of 7,793 properties in 2017 (Tourism Bureau,

2019b). The B&B Act of 2001 was an imperfect regulation because it did not consider the environmental impacts associated with the integration of B&B into the local environment (Wu et al., 2008). More seriously, the existing B&B Act was not thorough enough to handle the overdevelopment of B&Bs in rural areas, mountain villages, and even in metropolitan areas.

Both legal and illegal B&B operations have contributed to the rapid development of the B&B industry which has created positive economic value; however, it has also created negative impacts. The overdevelopment of B&Bs in the mountain villages and environmentally sensitive zones has become especially controversial because of environmental damages and pollution concerns to the areas (Chen and Zhuang, 2009). Many illegal B&Bs are violating zoning codes and cannot successfully obtain a legal license for their operations. A more serious problem, though, is some B&Bs offer more rooms than their capacity allows, raising concerns about safety issues.

Regulatory restrictions changed in November 2017 when the Taiwan Tourism Bureau revised the B&B Act. The amendment approved new zoning restrictions and development parameters for B&B operations that include:

1. All new developments must be compatible with the standard of existing developments under the existing B&B Act, and must be complimentary to existing development and responsive to the character of the allowed development in nonurban and national park areas; and
2. Subdivisions of the scenic designated areas, tourist destinations, aboriginal areas, remote areas, islands, leisure agriculture areas, and cultural or historical preservation areas must be permitted along the original subdivision pattern;

At the same time, many tourist destination developments regulated by the Cultural Assets Preservation Act are also approved for planning by local authorities. The revised B&B Act of 2017 has led to a historical high in B&B development with more than 9,000 legal B&Bs in 2018 having been started in Taiwan.

5.3.3 LEGAL ISSUES

Illegal B&Bs

Fulco (2018) states that many B&Bs in downtown Taipei City are not actually engaged in "home-sharing" but are really landlords who lease property to short-term tenants. These so-called B&B owners live elsewhere and rent apartments for a few nights at a time just like other B&Bs owners do. Many of these illegal B&B owners claim that not everyone is comfortable staying with or hosting a stranger in their home or in making a connection with an absent host (DeAeth, 2018). Ironically, despite its focus on close contact between host and guest, these "home-sharing" rentals are known as illegal B&Bs and are different from legal B&B operations. The Taiwan B&B Act clearly states that a B&B operation is an accommodation service operated as a side business in an owner-occupied residence. If the owner is absent from a property being used for short-term rentals, as is the case with most Airbnb rentals in Taiwan, then it's not really a B&B and is actually an illegal operation. Illegal B&Bs are subject to a maximum fine of NT$500,000 and must be shut down.

These illegal accommodation services are popular and booming in many metropolitan areas in Taiwan, especially in Taipei City, Taichung City, and Kaohsiung City. Beginning in 2013, Airbnb offered great services not only for legal B&Bs, but also for illegal B&Bs in Taiwan. Screening processes and listing qualifications on Airbnb are simple and easy for B&B owners to meet. Given sufficient demand in the metropolitan areas, property owners can earn more from illegal B&Bs than from long-term rentals; thus, there are more and more people willing to take a risk and investing illegal B&Bs in metropolitan areas and using Airbnb as a tool to sell their rental spaces. While these illegal B&Bs are meeting an increasing demand, they, on the other hands, often depict themselves as an insurgent embattled with an entrenched incumbent—the hotel industry. Truthfully though, traditional lodging businesses see illegal B&Bs and short-term rental providers as major competitors who are chipping away at their revenues.

Zoning Violations

According to the B&B Act, B&Bs regulations and their executions are administered and managed by county and city governments, but in recent years, illegal B&Bs have sprung up, mainly because local authorities do not actively regulate the B&B industry according the B&B Act and Zoning Code. Authorities have ignored its expansions in some unauthorized areas and have let it grow rapidly without any restrictions. The development of B&Bs and the tourism industry are closely related to each other, both bringing tourists and economic benefits to the locals. Many economically deficient towns rely on B&Bs and tourism activities to produce significant resources for the areas. Local government officials find it difficult to enforce the B&B Act and

Zoning Code that would negatively impact a major economic driving force to their areas.

The B&B Act and zoning regulations clearly state that B&Bs can only be set up in specific urban and rural areas, such as tourist destination cities, scenic areas, national parks, and aboriginal areas, but many illegal B&Bs violate zoning code by operating in unauthorized areas or not registering their buildings. Especially in aboriginal areas, tribal homes were often built before the regulation and law were announced, and tribal members have not registered their homes as permitted buildings. Therefore, many of these B&Bs fail to obtain a license for their operations and become "illegal B&Bs." Though most B&Bs want to operate legally, public opinion and environmental assessments are becoming more and more strict, making the chances of illegal B&Bs becoming legal almost impossible. Outdated regulations limit the development of the B&B industry, and the government should work on alternatives, otherwise it will need to strictly ban illegal B&Bs, and the problem will not be solved.

Avoid Paying Taxes

B&Bs in Taiwan initially arose from a concept of homeowners sharing unused spaces or rooms with guests and allowing visitors to experience local life and integrate themselves into the local cultural. It has become a profitable business model that has attracted many entrepreneurs to build B&Bs. The term "B&B" has been stretched, and many Taiwanese B&Bs have disregarded the original intent of sharing extra rooms with guests.

According to the B&B Act, the number of rooms in a B&B cannot exceed five before 2017 or eight after the revised B&B Act of 2017. Even special B&Bs cannot exceed 15 rooms. If the number of rooms exceeds the regulation, the owner of a B&B must apply for a hotel license; however, there are many B&Bs whose operations exceed allowed capacities by using "Nominal Accounts" to divide one operation into multiple operations that circumvent hotel fire codes, construction, and business income tax regulations.

The Taiwan B&B industry was introduced to the lodging market almost four decades ago, and it is now being driven in a new direction. Some B&B investors spend millions of dollars buying land and building luxurious B&Bs on remote sites in order to offer rooms for rates higher than a five-star hotel. Even this type of development can be billed as a side job for a family and not a tax producing business. According to the B&B Act, a B&B business with monthly revenues less than NT$ 200,000 can be exempted from business tax. In practice, many B&Bs generate revenues that exceed the minimum but are not reported which hurt legal B&B owners who report their sales honestly.

5.4 CHALLENGES IN TAIWAN B&BS

5.4.1 LONGER RETURN ON INVESTMENT

With the rapid development and growth of B&Bs in Taiwan, the new design of B&Bs is superior to what it was just 20 years ago. During the early growth stage, the return on investment for a B&B was slightly less than 5 years. Currently, the initial investment required for a modern B&B has increased because of decoration, design, and labor costs, and the average return on investment for a modern B&B is about 8–10 years, and B&Bs that invest in operation sites far from the cities can expect are turn on investment of more than 10 years. There are many Baroque, Greek-style, Mediterranean, and other European-style B&Bs with building costs of one or 200 million New Taiwan Dollars. Ironically, the supposition behind these luxurious B&Bs is to increase the chance of becoming an illegal B&B, but often the rate of return on investment is not as good as owners expect.

5.4.2 OVERDEVELOPMENT ISSUES

The lack of management manpower for competent authorities is another root cause of the chaos in the Taiwan B&B industry. One of the most popular B&B destinations—Chingjing, Nantou has been identified as an indulged B&B development area by Taiwan's Ministry of the Interior. In response, the local government has claimed that there are too many illegal B&Bs in the area, that they cannot all be banned at once, and that inventory and assessment cannot possibly be completed quickly and making the local governments' suspicion of a lack of management resources irresponsible.

The B&B Act and other regulations may limit single-player B&Bs in an area, but there is a B&B cluster effect in Chingjing, Nantou. Many current B&B developments are clustered and can be aggregated into a multifunctional community. A B&B community like this may include accommodations, restaurants, cafes, tea houses, farmer's markets, gift shops, etc. There are very few cases in which illegal B&Bs are punished by competent authorities or ordered to suspend business. Therefore, there are many risk takers that will claim to be B&Bs in order to utilize local life, culture, landscape, and natural resources to provide accommodations for travelers.

5.4.3 LACK OF MANAGEMENT COMPETENCY

Many B&B owners dream of returning to the basics and enjoying their lives as famers and enjoying the unique local life in rural areas. Some may quit a job to buy a historic home or fixer-upper in a rural area which may sound like a dream, but it is, in fact, not easy to run a B&B as a business that hosts and provides guests services on a steady basis. New B&B owners may not understand the amount of physical labor involved in owning a B&B. In addition to being the boss, the owner is also the housekeeper, maintenance worker, cleaning staff, chef, tour guide, driver, marketer, and business manager. There are few professional organizations where owners can find help in managing a B&B in Taiwan. One may rely on Social Media and OTA platforms to attract guests and run their reservations, but many hoteliers predict that many of these smaller businesses will be put on the market within the next 10 years.

5.4.4 OTAs IN TAIWAN

There are several OTAs that serve as a marketing tool for B&Bs in Taiwan, including illegal B&Bs. According to Huang (2008), B&Bs successfully generate income from vacancies by homeowners; that is, they produce surplus materials and generate economic benefits using existing resources. With a growing demand for B&Bs in metropolitan areas, owners rely heavily on OTAs as tools for their marketing and reservations.

Airbnb is an online marketplace where individual B&B operations that meet its qualifications can offer accommodation services in their own homes. Taiwan B&B owners see this platform not only as a means to sell their rooms, but also as a way to promote their business to overseas markets (Chen et al., 2013). In 2017, Airbnb stated that 1.3 million foreign travelers used Airbnb listings in Taiwan (Jao, 2018) and contributed a total of NT$16 billion to the Taiwan tourism industry in 2018 (Airbnb, 2019). The rise of Airbnb's emerging B&Bs has assuredly affected both the mode of travel and the increase in accommodation services in Taiwan.

However, Airbnb and Taiwanese regulators have been playing a cat-and-mouse game for years. The Taiwan B&B Act and lodging regulations limit short-stay accommodation rentals to legal hotels and inns, but many Taiwan properties listed on home-sharing platforms and Airbnb do not meet these requirements. The Taiwan government continues to hope that Airbnb will return to its roots as a home-sharing platform that allows travelers with an adventurous spirit to connect.

5.5 CONCLUSIONS

B&Bs feature a home-sharing lodging experience that showcases local geography, culture, natural resources, and allows the owner's creativity and attitude toward life and leisure offering guests to experience a different lifestyle. B&Bs have shared the market in Taiwan's lodging industry for about 40 years. As the B&B industry progressed, B&Bs in metropolitan areas became much different than those in rural areas. Suitable regulations and managements of B&B services are needed in order to addressed the issues raised here; however, Taiwan's government has not established regulations in regard to B&B development and management, and many illegal B&Bs continue to expand, raising many concerns. Although there will inevitably be more problems and issues as B&B development grows in Taiwan, the latest revisions to the B&B Act and support of local governments can help produce healthy growth and development for increasingly popular travel styles and lodging choices. To address discrepancies in zoning violations and operational scale in the B&B industry, the central and local governments should work together to resolve these unfavorable issues.

The trend toward information transmission and service functions via the Internet has played an increasingly important role in the B&B industry. Apps from mobile devices also serve as tools for B&B operations. Many management competencies are needed to sufficiently run a B&B, so area B&Bs may join together and develop powerful branding strategies and knowledge sharing Apps to better manage and operate B&B businesses. Liu and Cheng (2018) also stated that B&B owners' lifestyles, customers, market information, external knowledge, business expansion needs, policy, and familiar partners will be driving forces in growing sustainable B&Bs in Taiwan.

Mobile devices and apps provide convenient and easy access to multiple resources, and technology has boosted the home-sharing economy to a new stage. Yet, it is necessary to establish long-term efficiency and a sustainable transformation mode in order to resolve problems brought about by developments in the current B&B industry.

CASE STUDY

The Most Successful Rebuilding B&B Project—Taomi Village

The rebuilding of Taomi Village after the 921 Earthquake was an iconic rebuilding project that resulted in boosting the B&B industry. Taomi Village (Chinese:桃米里) is an urban village situated in Puli Township, Nantou County. Before 1999, this agricultural village was facing population decline due to Taiwan's rapid industrialization. In the event of September 21, 1999 earthquake, 70% of the small village was destroyed; 168 of the 369 houses were completely destroyed, and 60 houses were partially destroyed. While the natural disaster was devastating, the village was able to turn the crisis into an opportunity. They not only reconstructed the village, but they also rebranded it.

Ten years later, Taomi Village had established itself as an eco-village. From the efforts of volunteer groups, local residents, and other parties, this eco-village was reborn after the 921 Earthquake. Scientists and college students who volunteered or were invited to the village found that it was situated in a valley with an unprecedented diversity of wildlife. Taomi claims to be home to 23 different kinds of frogs, 60 species of dragonflies, and 72 varieties of birds (Wang and Cater, 2014). Local residents and research teams took further advantage of the unique natural resources to plan for bigger projects. Though they found that it wasn't always easy to convince investors and stakeholders that frogs and dragonflies could actually bring money into the village, many of the local residents were out of work after the earthquake, so it was an opportunity for them to engage and help with Rebuilding Taomi Village projects.

The New Homeland foundation created a work program under government contract, and instructors helped by providing new ideas. Local residents soon learned to utilize local produce in cooking and local natural resources to plan tours that attracted many tourists to Taomi Village. Local residents in Taomi Village also volunteered to construct wetland parks, educational and recreational areas, and they made sure the area was rebuilt in an ecologically friendly manner. As a result, more than 13,000 fireflies attract visitors from all around the country every April and May, and about 20 eco-friendly B&Bs have opened in Taomi, giving tourists the opportunity to get close to nature and to appreciate what the local residents built after the 921 Earthquake.

FIGURE 5.3 Taomi Agricultural Leisure Area Map.

Source: Hsi-Peng Tseng (https://www.taomiala.com/).

KEYWORDS

- **Taiwan**
- **B&B industry**
- **B&B development**
- **Minsu**
- **lodging market**
- **sharing concept**

REFERENCES

Airbnb. Community, Direct Economic Impact. https://press.airbnb.com/zh-tw/airbnb%E5%8E%BB%E5%B9%B4%E5%9C%A8%E5%8F%B0%E8%B2%A2%E7%8D%BB%E8%B6%85%E9%81%8E160%E5%84%84%E5%8F%B0%E5%B9%A3%E8%A7%80%E5%85%89%E8%B2%A1/ (accessed July 2, 2019).

Airbnb. Taiwan Airbnb. https://www.airbnb.com/a/Taiwan?af=43720035&c=.pi0.pk67447763257_344891810320_c_115729028352&sem_position=1t1&sem_target=kwd-115729028352&location_of_interest=&location_physical=9006279&gclid=Cj0KCQ-jwz8bsBRC6ARIsAEyNnvppqkyJLrlcO7Yv-WZwoEKA6LnKGqIfQF0Weaf1kydxdUbgis9NFCMaArtTEALw_wcB (accessed May 20, 2019).

Chen, L.; Lin, S.; Kuo, C. Rural Tourism: Marketing Strategies for the Bed and Breakfast Industry in Taiwan. *Int. J. Hosp. Manag.* **2013,** *32*(1), 278–286.

Chen, W.; Zhuang, B. Exploration of a Development and Sustainable Operation Strategy for Bed & Breakfasts in Mountain Areas- Case Study of the Nantou Chingjing Area. *Sloping Disaster Prev.* **2009,** *8*, 1–18.

Chiang, F.; Sun, C.; Lin, C. The Impact of Taiwan's WTO Entry on its Domestic Agriculture Sector, *The Applied Regional Science Conference*, Blackwell Publishing Ltd, 2004.

Chou, Y.; Ou-Yang, H.; Chen, K. Taiwan Tourism 2020: A Sustainable Tourism Development Strategy. *Taiwan Mod. Tour.* **2018,** *1*(1), 1–20.

DeAeth, D. Taiwan is Gunning for 'Unlicensed Rentals' with Airbnb in its Crosshairs. Taiwan News. https://www.taiwannews.com.tw/en/news/3456177 (accessed June 13, 2018).

Fulco, M. Is Airbnb Heading Into Regulatory Trouble in Taiwan? Economy. https://international.thenewslens.com/article/110475 (accessed Dec 19, 2018).

Hotch, R.; Glassman, C. *How to Start and Run Your Own Bed and Breakfast Inn*; Stackpole Books, 1992.

Hsiao, T.; Chuang, C. Independent Travelling Decision-Making on B&B Selection: Exploratory Analysis of Chinese Travelers to Taiwan. *Anatolia* **2015,** *26*(3), 408–420.

Hsiao, T.; Tsai, H.; Lockyer, T.; Sung, P.; Kuo, S. Factors for Development of Bed and Breakfast Management in Taiwan. *Anatolia* **2018,** *29*(3), 393–410.

Hsueh, S. L. Assessing the Effectiveness of Community-Promoted Environmental Protection Policy by Using a Delphi-fuzzy Method: A Case Study on Solar Power and Plain Afforestation in Taiwan, Renew. *Sust. Energ. Rev.* **2015,** *49*, 1286–1295.

Huang, H.; Perng, Y. Empirical Research on Key Factors in Environmental Protection Based Ecotourism of B&B. *EURASIA J. Math. Sci. Technol. Educ.* **2018,** *14*(1), 467–474.

Huang, L. Bed and Breakfast Industry Adopting E-Commerce Strategies in E-Service. *Serv. Ind. J.* **2008,** 633–648.

Jao, N. Airbnb's Latest Challenge is Coming from Taiwan. https://technode.com/2018/08/16/airbnb-taiwan/ (accessed Aug 16, 2018).

National Development Council, (2009). Vanguard for Excellence in Tourism from 2009 to 2014. https://www.ndc.gov.tw/News_Content.aspx?n=01B17A05A9374683&sms=32ADE0CD4006BBE5&s=CBE2AF4E776CBCD5.

Sharpley, R. Rural Tourism and the Challenge of Tourism Diversification: The Case of Cyprus. *Tour. Manag.* **2002,** *23*(3), 233–244.

Taipei Times. 921 Earthquake Reconstruction has Proceeded Well. http://www.taipeitimes.com/News/taiwan/archives/2004/09/22/2003203848 (accessed Sep 22, 2004).

Taiwan Tourism Bureau. Tourist Multiplier Program, (2002). https://admin.taiwan.net.tw/upload/contentFile/auser/b/doublep/double.htm?no=120.

Taiwan Tourism Bureau. Tour Taiwan Years from 2008 to 2009, (2008). https://admin.taiwan.net.tw/English/FileUploadListE003120.aspx?Cond=6dbdb38b-e558-47a1-b09a-a98468a8c0fd&appname=FileUploadCategoryListE003120.

Taiwan Tourism Bureau. B&B Industry Statistics, (2011, 2012, 2019b). https://admin.taiwan.net.tw/FileUploadCategoryListC003330.aspx?Pindex=5&CategoryID=ddeddb2a-dab1-40df-aef7-4e30cbdf35e2&appname=FileUploadCategoryListC003330.

Taiwan Tourism Bureau. Tourism Statistics (2019a). https://stat.taiwan.net.tw/

Taylor, M.; Taylor, R. *Start and Run a Bed & Breakfast*, 3rd ed.; International Self-Counsel Press: Bellingham, WA, 2004.

Wang, G. L.; Liu, C. C.; Tseng, S. R. An Evaluation of Taiwanese B and B Service Quality Using the IPA Model. *J. Bus. Res. Turk* **2012,** *4*(3), 20–27.

Wang, C.; Cater, C. *Tourism Crisis and Disaster Management in the Asia-Pacific*; Ritchie, B., Campiranon, K., Eds.; CPI Group Ltd: United Kingdom, 2014.

Wu, Q.; Zhang, Y.; Lin, Y.; Xie, M. The Bed & Breakfast Policy and System in Taiwan: On the Possibility of Sustainable Bed & Breakfast Operation Combined with Eco-Tourism. *Ch.* **2008,** *34*, 817–830.

Zheng, J. Feasibility Study of the Development of a Micro-Lodging Industry in Taiwan (Research Program of the Tourism Bureau, the Ministry of Transportation and Communications (MOTC), No.: MOTC-TBROC-99-007). Taipei: Tourism Bureau, Ministry of Transportation and Communications, 2010.

CHAPTER 6

DESTINATION IMAGES OF THAILAND: CURRENT AND FUTURE DEVELOPMENT

YODMANEE TEPANON[1], WANLANAI SAIPRASERT[2], and PIMTONG TAVITIYAMAN[3*]

[1]*Kasetsart University, Thailand*

[2]*Chiang Mai University, Thailand*

[3]*The Hong Kong Polytechnic University, Hong Kong*

[*]*Corresponding author. E-mail: pimtong.t@cpce-polyu.edu.hk*

ABSTRACT

This study aims to explain the overall positive and negative images of Thailand tourism from tourists' perspectives. In addition, this study is to describe the major impacts of tourism on the Thai society and present the current position of food/gastronomy and sex tourism in Thailand. Lastly, the managerial implications for Thailand tourism's future development are proposed.

6.1 INTRODUCTION

Thailand is a developing country that is popular for its plenty of positive and negative images. Delicious food, exotic natural sceneries, and a long historical culture comprise the positive image of the country. By contrast, the negative images of the country include traffic congestion, political instability, natural disasters, and sex trafficking.

This study explains the following key concepts. First, the overall destination images of Thailand with regard to push and pull factors and positive and negative images are investigated. Second, the major economic, sociocultural, and environmental impacts of the tourism industry to the Thai society are analyzed. Last, case studies on gastronomy and sex tourism, including their current positioning and future development, are discussed.

6.2 DESTINATION IMAGE

6.2.1 PUSH VS PULL FACTORS OF THAILAND AS A TOURISM DESTINATION

Push and pull motivations are important determinants in travel and tourism studies because these two factors directly influence a tourist's decision-making process. Push factors initiate the drive of an individual to travel, whereas pull factors particularly focus on the appeal of the destination. Several researchers have investigated push motivation in their studies. Crompton (1979) was the first to explore and classify the seven push motivations of pleasure tourists. These sociopsychological motives comprise escape, self-exploration, relaxation, prestige, regression, kinship relationships enhancement, and social interaction facilitation. The study also found two pull motivations or cultural motives, namely, novelty and education.

Sangpikul (2008) particularly applied the concept of push and pull motivations on Japanese senior travelers to Thailand and revealed three push factors, specifically, novelty and knowledge seeking, rest and relaxation, and ego enhancement. Four pull factors were also identified, namely, cultural and historical attraction, travel arrangements and facilities, shopping and leisure activities, and safety and cleanliness. In a later study, Yiamjanya and Wongleedee (2014) investigated international tourists and their decision-making process in selecting Thailand as their destination. The result corroborated that the top five travel motivations influencing them to select Thailand were to gain experience in a foreign land, eat Thai food, learn a new culture, relax in a foreign land, and learn new things.

The study further discussed the top five push-based motives to gain experience in a foreign land, to learn a new culture, to relax in a foreign land, to learn new things, and to explore their interest in the Thai culture. Moreover, the top five pull-based motives were Thai food, traditional markets, good weather, beaches and sand, and good services. Furthermore, Sastre and Phakdee-Auksorn (2017) investigated push and pull travel motivations and behavioral intentions among the sample of British outbound tourists to Phuket, Thailand. The result affirmed that British tourists' main push motivations were to have fun, to rest and relax, and to escape from daily routine and environment. By contrast, the main pull motivations were natural sceneries and landscapes, beaches, and hospitality and friendliness of people.

6.2.2 POSITIVE VS NEGATIVE IMAGE OF THAILAND

Thailand is geographically located in the heart of mainland Southeast Asia and is considered an important regional hub attributed to its well-developed infrastructure (Netherlands Embassy in Bangkok, 2017). The capital city Bangkok, with 20.5 million tourists, is listed number one among the top 20 most visited cities in the world. Interestingly, among the top 20, Phuket ranks 12th with 9.29 million visitors, and Pattaya ranks 18th with 8.76 million visitors (Mastercard Global Destination Cities Index, 2018). The majority of tourists of these three Thai cities were from China, Malaysia, and South Korea. The Ministry of Tourism and Sports in cooperation with the Tourism Authority of Thailand (TAT) developed a tactical plan for tourism from 2017 to 2022. The plan serves as a framework to promote a positive tourism image by expanding tourism in the segments where Thailand has a competitive advantage, that is, leisure, business, and medicine. In addition, the government purposes to stimulate positive images by developing new segments, sports, and gastronomy, to promote long-term growth in the country's tourism.

The first and most prominent positive image of Thailand is *leisure destination*. Aung, Nge and Hichitake (2015) studied international tourists' perceived destination image of Thailand. They found that the tourists' image of Thailand can be ranked as price, friendliness of people, shopping, cultural/historical sites, hotels/resorts, natural scenery, overall service quality, cuisine, convention/exhibition facilities, safety and security, ease of access to transportation, and climate. Moreover, the results contended that Thailand has been receiving considerable positive feedbacks for its cultural and historical places. Shopping in Thailand also exhibits remarkable value for money. Thailand receives leisure travelers and nonworkers, such as the retired, the unemployed, and students.

Second, Thailand positions toward the growing segment of *business destination* also known as "MICE." The Ministry of Tourism and Sports (2017) categorized this segment into four. M stands for meeting of the small groups of company employees or representatives from a particular industry; I stands for incentive trips given as rewards to top-performing employees; C denotes conventions, which are large meetings for exchanging ideas and information on specific industries and professions; and E stands for exhibitions that showcase of products or services from a certain industry. Thailand has a positive reputation in the MICE destination because of its strategic location, international connectivity, transportation system, basic infrastructure, and government support. In 2017, Thailand welcomed 1,047,959 international MICE visitors, thereby generating a revenue of 88.45 billion THB. The majority of MICE travelers were from East Asia, Europe, and South Asia. Five dynamic MICE cities in Thailand include Bangkok, Pattaya, Phuket, Chiang Mai, and Khon Kaen.

The third positive image of the country is *medical and health destination.* The national strategy focuses on becoming a regional health hub and building extensive tourism infrastructure specializing in cosmetic and gender transition surgery (Pocock and Phua, 2011). The Medical Tourism Index conducted by the International Healthcare Research Center ranked Thailand as the18th most popular medical tourism destination in 2016 (KPMG Phoomchai Audit, Ltd., 2018). The majority of medical tourists were from the Eastern Mediterranean and Southeast and South Asia, while the highest number of medical tourists comes from the United Arab Emirates, followed by Bangladesh and Myanmar (Noree et al., 2016).

Thailand also broadens its fourth image, *sport destination*, which is a new tourism market. Sports tourism is the tourism segment that appeals to many tourists. Traveling to a sporting event generates trips to the city where the sporting event transpires. Consequently travelers spend on accommodation, meals, and souvenirs (Worapinij et al., 2017). The TAT mentioned that sports, including golf, are already popular among tourists visiting the country. Other facilities where tourists can enjoy sporting activities such as boxing rings and Muay Thai training centers are also sought after. Furthermore, international sporting events that have transpired in the country, such as the World Weightlifting Championships, the World Touring Car Championships, the Asian Games, and the South East Asian Games, prove Thailand's destination image (TAT, 2016).

Thailand's final positive image is *gastronomy or food destination.* Gastronomy or food or culinary tourism refers to the enjoyment of prepared food and beverage and other food-related activities resulting in a great and memorable gastronomic experience. This tourism image also includes the consumption of local food, cuisine, meals, and eating style together with learning about the gastronomic tourist attraction's culture, lifestyle, and ethos/religion. Tourists can also engage in activities including cooking classes (Suanpang, 2015). Under this destination image, Thailand has gained the reputation as the legendary "kitchen of the world." In addition, Karim and Chi (2010) studied perceptions of the three food image dimensions, namely, food/cuisine, dining/restaurant, and food-related tourism activities in France, Italy, and Thailand. Results revealed that Thailand was perceived to exhibit the best dining experience among the three countries. Thailand offers various attractive restaurants with ease of access, reasonable prices, friendly service personnel, and menus in English.

Although Thailand occupied several positive tourism destination images, the country also suffered from certain negative tourism images. Several studies have highlighted the apparent negative destination image of Thailand. Rittichainuwat et al. (2001) claimed that negative images of Thailand are due to prostitution, AIDS, overpopulation, and traffic congestion. Rungsuwannarat et al. (2015) also mentioned similar negative images. In particular, their study found economic inequality, traffic jams, numerous night clubs, prostitution, heavy pollution, language barriers, and the AIDS problem as the negative destination images of Thailand. Moreover, Thailand's destination image reflects relatively low safety and security attributes (Aung et al., 2015; Tavitiyaman and Qu, 2013). Tourists' travel decisions are commonly influenced by the obtained information regarding a potential destination and perception of that information (Ingram et al., 2013). Thus, the negative images such as natural disaster, political instability, and sex trafficking and prostitution require immediate constructive corrections.

The first negative image is *natural disaster*. Thailand has encountered plenty of natural disasters, including the Indian Ocean Tsunami in 2004 and the Chao Phraya River floods in 2011. These catastrophes caused numerous deaths and huge economic damages. Nakasu (2017) noted that these disasters were considered international issues, given that over half of all deaths in Thailand caused by the Indian Ocean Tsunami are foreigners. In addition to natural disasters, *crimes and accidents* are included in Thailand's negative images. In contrast to hailing Bangkok as first among 20 of the world's most-visited cities, Thailand is one of the 20 most dangerous countries in the world for tourists. Travel and tourism competitiveness reports by the World Economic Forum stated that among the 136 countries surveyed, Thailand ranked 118th with a 4.0 score out of 7.0 for safety and security for tourists. The low ranking is due to high crime and violence rates and low police services' reliability (The Nation, 2017). The murder of two British tourists on Koh Tao islands in Southern Thailand in 2014, the Erawan Shrine bombing in Bangkok in 2015, and the boat capsizing in Phuket in 2018 prove the prevalence of crimes and violence in the country.

Second, tourism destinations can be negatively affected by *political instability*. The political crisis in Thailand and the information disseminated about it affect tourists' perceptions of the country. Thai tourism also faced a recession period from 2006 to 2010 due to the publicized riots. For instance, in 2008, anti-government protesters stormed into the terminal building of Suvarnabhumi airport. Consequently, scuffles broke out, and the airport was shut down. Furthermore, thousands of travelers were stuck as a result of political instability (Ingram et al., 2013). Unstable political situations can affect the country's image because tourists are more concerned about their safety and security than other destination attributes (Saechau et al., 2015).

Lastly, Thailand obtained negative branding on *sex trafficking and prostitution*. Henkel et al. (2006) investigated the perceptions of international visitors in Thailand and found that tourists' opinion on sex and erotic tourism lacks extreme importance when they consider Thailand as a tourist destination. However, 26.29% of the respondents listed sex tourism as one of their top three images of Thailand. Thailand is a source, destination, and transit country for men, women, and children subjected to forced labor and sex trafficking. Furthermore, sex trafficking remains a problem in Thailand's extensive cases involving prostitution and sex trade (Chanwanpen, 2018; U.S. Department of State, 2015).

6.3 MAJOR IMPACTS OF DESTINATION IMAGE TO THE THAI SOCIETY

The tourism industry and destination image of Thailand play important roles for generating revenues. The Thai government and tourism operators continuously promote hospitality and tourism products and services to attract local and international visitors. The TAT has formulated the Market Plan 2018, which aims to generate revenue and increase visitor arrivals, promote new ideas, and follow the development agenda of Thailand 4.0, and maintain a balance between the economy, society, and the environment (TAT Newsroom, 2017). Consequently, the image of Thailand influences local sustainable tourism in terms of economic, sociocultural, and environmental impacts. Considerable studies have summarized these impacts in detail.

6.3.1 ECONOMIC IMPACTS

The economic impact of tourism attempts to estimate changes in regional spending, output, income, and/or employment associated with tourist policies, events, facilities, or destinations (Tyrrell and Johnston, 2006). The World Travel and Tourism Council (2018) reported that the travel and tourism industry generated 21.2% (3.229 trillion Thai Baht [THB] or 95 billion USD) of the GDP in Thailand in 2017. Moreover, this value is predicted to increase to 28.2% (5.957 trillion THB or 175.2 billion USD) by 2028. In 2017, the tourism industry generated over 2.33 million jobs or 6.2% of the total employment rate. Furthermore, visitors generated over 2.02 billion THB (59.6 billion USD) or 19.2% of the total exports in 2017. By 2028, the amount of total visitor exports is expected to reach 4.01 billion THB (118.1 billion USD) or 24.1% of the total visitor exports.

Thailand's tourism industry consists of two segments, namely, domestic and international tourism. The number of domestic tourists, which are measured by the number of trips per population per year, stood at 138.8 million trips per person per year in 2015. Hence, a Thai tourist takes 1.57 trips per year on average, which is slightly above the global average of 1.5, and their length of stay is 2.61 days (Ministry of Tourism and Sports, 2017). Thailand's international tourism has been growing simultaneously with domestic tourism. In particular, the number of tourists who have visited the country has reached 20.5 million. These tourists' average length of stay is 4.7 days with $173 average spending per day (Mastercard Global Destination Cities Index, 2018). These figures present the potential growth and the subsequent positive economic impacts of the tourism industry in Thailand.

Similar trend represents international tourist markets. Table 6.1 explains the growth of Thailand tourism from the international tourists' perspective. In 2008, the number of international tourist arrivals was 14.58 million with an average of 9.51 days length of stays, thereby generating US$ 17,247 million tourism receipts. Until 2018, the number of international tourist arrivals had significantly increased by 38.17 million tourists with an average length of stays for 9.29 days, which yielded US$ 58,066 million tourism receipts. During the last decade, Thailand tourism has been attractive among international tourists, particularly from the East Asian region such as China and Malaysia (Ministry of Tourism and Sports, 2019). The projected number of international tourists has continuously grown with the forecast number of 41.1 million international tourists by 2019 (Bangkok Post, 2019).

A desirable destination image can encourage potential tourists to visit a certain destination, thereby generating high tourist expenditure (Arnegger and Herz, 2016). The supply chain of tourism business is characterized by the following: tourists visit the destination; revenue consumption (e.g., accommodation, catering expenses, and shopping) is generated during the trip; the demand for employment in tourism enterprises is directly and indirectly increased; the income of local families is improved; and local consumption is enhanced (Van Rooyen, 2018). For instance, sports tourism can stimulate the local economy, enhance the creation of jobs, and increase future tourist visitation (Cheung et al., 2016; Tyrrell and Johnston, 2006). Cheung et al. (2016) described the total expenditure of visitors during the East Asian Games, which involved shopping, food and beverages, entertainment, transportation, and accommodation. In addition, Poudel et al. (2016) explored stakeholders' perspectives of sustainable tourism development in economic measures. Positive economic impacts include the contribution to community development funds, increased incomes and new business development, new markets for local products, improved living standards, enhanced purchase of goods and services from local community, and new development in the local community.

TABLE 6.1 International Tourist Arrivals and Receipts for Thailand.*

Year	International tourist arrivals (million)	Tourism receipts ($US million)	Length of stay (days)
2008	14.58	17,247	9.51
2009	14.15	14,880	8.99
2010	15.94	18,706	9.12
2011	19.23	25,458	9.64
2012	22.35	31,657	10.02
2013	26.55	39,282	9.85
2014	24.81	36,108	9.83
2015	29.92	42,544	9.47
2016	32.53	46,274	9.56
2017	35.59	53,951	9.52
2018	38.17	58,066	9.29
2019	41.1#	N/A	N/A

*Ministry of Tourism & Sports (2019), #Bangkok Post (2019, Jan 28).

Many attractions in Thailand are economically attractive to tourists. For instance, many backpackers stay in Khao San Road, Bangkok because of the atmosphere, low costs, and variety of tourist services (i.e., restaurants, bars, and clubs) (Howard, 2005). Van Rooyen (2018) used Porter's five competitive forces to investigate the economic impact of Thai tourism. The high growth rate of tourism can entice investors and international hotel operators to expand their business in major destinations in Thailand.

By contrast, tourism can convey negative economic impacts to the destination and community. These impacts include an increase in the number of properties and land, as well as an increase in the prices of daily goods, food, services, and cost of living (Poudel et al., 2016). The appropriate distribution of income to all areas and stakeholders is challenging to a certain extent (McDowall and Choi, 2010). For instance, the economic impacts only benefit the large and international tourism operators but not the small and/or local tourism operators. These small and/or local tourism operators may feel taken advantage of and will thereby show lack of support to the tourism activities in the destinations. Gradually, many shopping malls, retail establishments, large hotels, and chain restaurants are established in major destinations to cater to tourists. These large-scale business establishments disrupt local and small business enterprises. These enterprises cannot survive in the long term with limited resources and a shortened supply chain.

6.3.2 SOCIOCULTURAL IMPACTS

Positive sociocultural impacts of tourism include knowledge enhancement about other cultures, facilitation of intercultural communication and understanding, incentive grants for the preservation of local culture, opportunities to participate in local cultural activities, and support for the restoration and maintenance of cultural and historic sites (Poudel et al., 2016). "Open to the New Shades of Thailand" is the proposed marketing plan to promote a deep understanding of unique Thai experiences. This plan can enhance the image of Thailand as a land of amazing diversity that is worth discovering (TAT Newsroom, 2017). Tourists can enjoy unforgettable experiences in Thailand. Howard (2005) corroborated that many tourists visited Khao San Road, a major backpacker tourist area in Bangkok, to experience different cultures, gain new experiences, and escape from their home and work. In addition, McDowall and Choi (2010) asserted that tourism can bring love and pride to local culture and support local traditions and culture. Thailand has a long history of culture, tradition, food, architecture, and festivals. Thus, the uniqueness of these elements can attract tourists to explore cultural offerings that are different from their country of origin.

However, negative sociocultural impacts include increased crime rates, traffic problems, crowded public places, and conflict between visitors and residents. Numerous illegal activities can be attributed to gambling, prostitution, sexual openness, drug use, and alcohol abuse. These activities can deteriorate the quality of life of the locals, disrupt traditional family values, and weaken the traditions of the native people (Poudel et al., 2016). Van Rooyen (2018) argued that tourists can affect local residents' lifestyles and contribute to the loss of native customs and traditions. Thai locals who live outside Khao San Road exhibit unfavorable attitudes toward tourists who stay in Khao San Road, whereas street vendors and Thai visitors have favorable attitudes toward the area (Howard, 2005). Other locals lack direct benefit from tourists, but they are affected by the traffic jam and crowds on Khao San Road. Hence, these locals fail to support tourists. By contrast, street vendors' income comes from tourists; hence, they prefer many tourists on Khao San Road.

The interrelation between sociocultural and economic impacts exists in tourism. The considerable demand for labor causes the influx of local and international manpower in a certain destination, thereby yielding sociocultural issues. Migrant workers may also be considered threats to the local community, particularly if they exhibit alcoholic tendencies, drug abuse, and aggressive behavior (Malam, 2003; Soontayatron, 2015). Cheap labor, sex trafficking, slums, and crimes are also challenges related to the influx of migrant workers (Soontayatron, 2015).

6.3.3 ENVIRONMENTAL IMPACTS

Tourism constitutes positive and negative impacts on the environment. Poudel et al. (2016) enumerated the positive environmental impacts, such as strong local environmental conservation efforts, promotion of incentives for the conservation of natural areas, and promotion of positive environmental ethics among parties. Many ecological tours and outdoor activities such as nature and wildlife tours, hiking and camping, and scuba diving and snorkeling are offered to international tourists in Thailand. Tourists will experience nature, which induce their strong desire to preserve the destinations.

From the environmental perspective, tourists tend to exhibit low obligation and responsibility toward the environment they visited (Miller et al., 2010). The deterioration of the natural environment; increased air, water, and noise pollution; increased environmental problems such as wastewater discharge, which produce long-term negative effects on the environment; and deforestation are negative environmental impacts (Poudel et al., 2016). Howard (2005) confirmed that noise and air pollutions develop on Khao San Road when plenty of tourists converge in the area. The local residents in Chiang Mai and Chiang Rai (North Thailand) perceive that the negative impacts of tourism are the damaged natural resources and pollution (Untong, 2007). The effects of water, air, and noise pollutions are in line with the study of McDowall and Choi (2010). Moreover, the future development of resorts and hotels would destroy the natural environment (McDowall and Choi, 2010).

Transportation infrastructure should be developed to facilitate tourists who visit Thailand. Enhanced public transportation can complement environmental protection efforts. Furthermore, new roads, railways, and infrastructure can affect the natural environment and local residents. Therefore, the feasibility plan should consider all the pros and cons of transportation development.

6.4 CASE OF FOOD TOURISM IN THAILAND

As previously mentioned, food and gastronomy tourism is considered as one of the main destination images of Thailand. The global hospitality industry is currently thriving given that a growing number of people spend money on food at restaurants or buying takeout (Business insider, 2018). Generation X, with ages ranging from 30 to 50, spends a higher average amount on food away from home than the Baby Boomers, the Silent Generation, and the Millennials (Business insider, 2018). Eating out serves several needs and motivations. People may dine at a restaurant for convenience, to escape from routine, and for social opportunities with friends and family. The excitement combined with expectation and satisfaction from restaurant experience can improve one's emotion (Epter, 2009). The gastronomy tourism trend expands beyond borders as a tourism phenomenon worldwide. Contrary to the ongoing globalization and the availability of global products and services, for example, McDonald's or Coca Cola, tourists look for an authentic taste from local food during their trip (Richards, 2004).

Gastronomy means the art or science of good eating (Gastronomy, 2019). Historically, the word *gastronomy* is rooted from the Greek term, *gastros*, which means stomach, and *gnomos*, which means knowledge or law (Kivela and Crotts, 2006). Consuming food and beverages during a trip has become an integral part in tourist experiences. Long (2013) describes that culinary tourism is focusing on food while visiting a destination. As a component of gastronomy, food is a fundamental component of life that embraces the manner of living, local beliefs, and culture and tradition. Moreover, such a fundamental component expresses how locals respond to their needs and how society has evolved through time (Moulin, 2000). The intertwining story of food and local environment attracts tourists' interest because people develop their senses toward more intangible experiences than tangible ones. Pine and Gilmore (1998) coined the term "experience economy," which presently motivates most industries as the result of consumers' value of real and extraordinary experiences offered in service, good, or commodity. Tasting the food itself and experiencing and learning the ingredients and the process of how the food is produced create a lasting memory for tourists. "We are what we eat" refers to tangible aspects of the food we consume and identifies the types of cuisine that tourists can experience on holidays (Richards, 2004). The importance of food in tourism extends more than tasting local cuisine. A food journey allows tourists to immerse themselves in another place, culture, or time. The UN World Tourism Organization (UNWTO, 2012) reported that international tourists spend roughly 20% of their budget on food when visiting other countries. Another study focusing on domestic food tourism confirms the vital role of food by discovering that approximately 50% of domestic tourists participate in food-related activities (Tourism Council of Thailand, 2013).

Thailand is considered one of the most famous food and gastronomy tourism destinations. Thailand, which offers a variety of food rooted from different regions in the country, processes the distinctive position that provides locals and tourists a scrumptious journey. Thai cuisine consists of a variety of local food influenced by weather, geography, and culture. From the mountains in the North, the opulent Central plains, the dry Northeastern plateau, and the rich, resourceful South, Thai food varies in ingredients, spices, cooking methods, and tastes. Interestingly, Thai food has merely been enjoyed by global diners for less than half a century ago (Sunanta, 2005). Notably, this cuisine has been ranked among the top mentioned cuisines, which merely falls behind French, Italian, and Chinese (Sunanta, 2005).

Nonetheless, Thai restaurants located in foreign countries tend to modify ingredients and tastes to cater the local taste in the neighborhood. This adaptation used to be favored by local residents. However, current consumers look for creative and authentic experiences when using products or services and thus prefer authentic cuisine that shows the original taste of unknown places. The TAT recognizes the significant potential of Thai cuisine. The agency has continuously promoted Thailand and Thai food from the 1994 "Amazing Thailand" slogan, which included all aspects of the Thainess to the "Thailand means 'good food' in any language" campaign attempt in 2019. This campaign aims to boost tourists' spending and enhance their gastronomic experience (TAT News, 2019). Moreover, this campaign focuses on a variety of food including street food and fine-dining dishes to increase understanding of Thai gastronomy and integrate tourism to the agricultural sector that comprises the majority of the Thai workforce. The determination is unwavering. Since 2017 TAT has teamed up with the prestigious Michelin Guide to publish the Michelin Guide Bangkok that aims to generate an approximate 10% increase in tourist spending on overall food (TTR weekly, 2017). Many restaurants are currently certified by Michelin in Thailand, with roughly 80% are located in the Bangkok metropolitan area (Table 6.2).

TABLE 6.2 Price Range and Certified Michelin Star Restaurants in Bangkok (Year 2019).

Stars given Price range	2 stars	1 star	Bib-gourmet	The Plate Michelin
More than USD 75	2	3		8
USD 50–75	2	7	2	4
USD 25–50		7	13	23
USD 6–25		5	57	48
Less than USD 6		1		35
Total	4	23	72	118

The list comprises of a combination of fine dining restaurants and street food stalls located throughout Thailand. Thailand has been awarded by CNN, one of the most popular global media sources, as the top-rated best street food in the world for two consecutive years (Bangkok Post, 2017). Promoting Thai food would generate revenues for locals given that the trend toward eating street food as a part of an exotic experience in faraway destination increases. From the supply chain perspective, the emphasis on promoting gastronomy tourism has a great potential to increase demand for local agricultural produces and products. Thus, gastronomy tourism is also beneficial to the Thai agricultural sector and, consequently, to the farmers who presently earn an approximately USD 1800 a year or USD 150 a month (Bangkok Post, 2018). In 2016, gastronomy tourism generated 329 billion THB or approximately 10 billion USD or 20% of the tourism income, following accommodation (29%) and shopping and souvenirs (24%) (Bangkok Post, 2017).

The rapid increase of the experience economy prompts tourists' expectations of authentic food and active participation during the cooking process. Passively dining local food may excite others, but restaurants and tourism-related food and beverage establishments should provide hands-on experience for visitors. Gastronomic experiences such as cooking classes, visiting vegetable farms, and picking out ingredients from the market amplify memorable experiences more than providing ready-to-eat dishes. Holistically creating involvement through food can connect people and their ability to connect all senses (Richards, 2015). The interaction between tourists and the hosts is apparently the reason for gastronomy experience. Satisfaction is increased when conversation and activities between two parties are present. This service escape can also enhance dining experience, and vice versa (Dixit, 2019). Thus, the overall satisfaction of gastronomy tourism ties to the cocreation concept that has been presently recognized.

6.5 CASE OF SEX TOURISM AND IMAGE OF THAILAND

Tourism professionals worldwide are well aware of the sex tourism phenomenon. Sex tourism is a silent but fast-growing industry, particularly in developing countries, due to its great ability to draw proceeds into countries. Certain researchers have looked into this phenomenon and attempted to scrutinize its perplexity (O'Connell Davidson, 1998; Oppermann, 1998, 1999; Ryan and Hall, 2000; Ryan and Kinder, 1996). However, sex tourism poses considerable challenges for all parties involved given its sensitive nature. Considerable attention has been controversially drawn to the illegal and cultural aspect when commercial sex operations are discussed.

Thailand has been globally known as one of the suppliers of sex tourism and has been viewed as a sex tourism destination for several decades. Thus far, the sex industry in Thailand may exceed the revenue from other main industries within the country (Mason, 1999). Historically, the global sex industry began in the early 1960s when assigned American military personnel arrived for the war in Vietnam. These service personnel enjoyed a 1-week rest and relaxation holidays in Thailand (Mason, 1999). The R&R or rest and relaxation proliferated Thailand's prostitution and sex tourism. The word about Thai ladies' submissive glamour has been out since then. The "good guys go to heaven, bad guys go to Pattaya (a district located in the Eastern coast of Thailand)" has been a slogan for tourists coming to Pattaya, thereby creating an unfavorable image of the location and the entire country. The number of tourists visiting the sex capital, Pattaya, was approximately 30% of all inbound tourists in 2017 (Asian Correspondent, 2018).

However, several tourism organizations in Thailand, mainly the TAT, have endeavored to communicate to the world that Thailand strongly opposes any form of sex tourism. Moreover, Thailand aims to turn the country's image into a "quality leisure destination" for all traveler groups (TAT News, 2018). Nonetheless, any agency can hardly correct this kind of image because the western world emphasizes the country's portrayal as one of the top sex tourism destinations while disregarding extended investigation on the causes. The mainstream literature, particularly newspapers and guidebooks, generally directs readers to the exotic narratives as they naturally elevate their curiosity. The country has attempted to present other tourism activities such as cultural tourism, ecotourism, long-stay tourism, wedding tourism, or medical tourism. However, a sensitive story such as sex tourism easily catches global attention (Boonchutima, 2009).

The struggle to change Thailand's image is persistent. The Branding Thailand project was initiated in 2001 to understand tourists' perception of Thailand as a destination. This project was carried out to determine Thailand's brand from consumers' perspectives of the three main industries, namely, cuisine, fashion, and tourism. This 1-year project was conducted using quantitative and qualitative methodology. Results validated that Thailand was viewed positively and negatively. The positive aspects included terms such as exotic, friendly people, beautiful, and good food. However, negative comments included terms such as sex/prostitute, poor/poverty, and pollution (Nuttavuthisit, 2007). The findings were fairly similar to the study of Henkel et al. (2006), which illustrated international tourists' image of Thailand relative to food, friendly people, beaches, nightlife, and entertainment. Approximately, 27% and 10% of the study respondents viewed sex tourism and erotic tourism as an image of Thailand, respectively. Nearly half of the international respondents perceived the sexual fun image of Thailand when nightlife and entertainment, which roughly accounted for 21%, were combined. Furthermore, local Thai tourists associated the country with terms such as beaches, food, nature, wildlife and parks, historic architecture, and friendly people. The sex tourism image type was mentioned by only roughly 9%. Therefore, different points of view are evident between locals and visitors. The obtained information resulted in several campaigns in the attempt to profile Thailand. The most effective slogan "Amazing Thailand" intends to communicate the human aspects of Thailand or "Thainess" to the global visitors (Thomsen, 2014). The country aims to rebrand its image by focusing on the further intangible aspects of tourist experiences.

As a tool for communication, the modern technology has presently enhanced user experience and, particularly, tourist experience, thereby influencing image building. Social media sites such as TripAdvisor or Facebook, which contain user-generated content, are considered credible and valid platforms (Hospitalitynet, 2015). Influencing other consumers with one's own opinions and experiences is effective (Živković et al., 2014). Travelers' image about the destination is automatically developed as they read through the comments and reviews. Thus, tourists' observations on social media gradually shape their perception of a destination. This changing environment inevitably poses challenges to Thai destination management agencies and the TAT to subdue the sex tourism image of Thailand.

6.6 CONCLUDING REMARK

The push motivation factors of Thailand as a tourist destination consist of relaxing, knowledge-seeking, and experiencing a new culture. Given that Thailand is renowned for its natural beaches, food, and friendly people, the pull motivation factors include natural landscapes, culinary, and Thai hospitality. The pull motivation typically engages with tourist destination attractiveness. Hence, a destination can be successful by creating a positive image according to the identified pull motivation, which can entice repeat visitors and referrals and recommendations.

The major stakeholders of Thailand tourism should continue generating accurate destination image to yield positive impacts. In addition, the negative impacts of tourism on the economy, society, culture, and the environment should be minimized. Certain implications are as follows. The government, together with the public should develop proper tourism policies and plans to achieve sustainable development. Local residents should also be involved in tourism planning and policy making. Moreover, serious efforts to minimize negative impacts are recommended to benefit the locals (McDowall and Choi, 2010). Emotional attachment toward tourist destinations should also be established to enhance the likelihood of sustainable tourism and environmental and social protection (Chubchuwong et al., 2015). Local tour guides and tourist information officers should be trained to provide accurate information regarding the uniqueness of each destination to tourists. Tour programs with local community activities can be created for different tourist segments. When tourists gain a full understanding of the culture and the natural environment in Thailand, their attitudes and behavior toward natural and cultural protection could be established.

KEYWORDS

- **destination image of Thailand**
- **major impacts to the Thai society**
- **food tourism**
- **sex tourism**

REFERENCES

Arnegger, J.; Herz, M. Economic and Destination Image Impacts of Mega-Events in Emerging Tourist Destinations. *J. Destination Mark. Manag.* **2016,** *5*, 76–85.

Asian Correspondent. Sex Tourism in Pattaya: Is the Party Over?. https://asiancorrespondent.com/2018/01/sex-tourism-pattaya-party/ (accessed December 18, 2018).

Aung, H. M.; Nge, E. T.; Hichitake, Y. Destination Image of Thailand and Singapore. *Apheit J.* **2015,** *4*(2), 59–48.

Bangkok Post. Food the Focus of New Travel Campaign. https://www.bangkokpost.com/business/tourism-and-transport/1343819/food-the-focus-of-new-travel-campaign (accessed Jan 29, 2019).

Bangkok Post. Survey Finds 40% of Farmers Live Under Poverty Line. https://www.bangkokpost.com/news/general/1475901/survey-finds-40-of-farmers-live-under-poverty-line (accessed Jan 21, 2019).

Bangkok Post. Record 38.27m Tourists in 2018: 41m Expected in 2019. https://www.bangkokpost.com/business/1619182/record-38-27m-tourists-in-2018-41m-expected-in-2019 (accessed Jan 28, 2019).

Boonchutima, S. Resistance to Change: Thailand's Image as a Sex Tourist Destination. *Asian Congress Media Commun. J.* **2009,** *1*, 62–72.

Business Insider. How Much the Average American Millennial, Gen Xer, and Baby Boomer Spends Each Year Eating Out. https://www.businessinsider.com/average-restaurant-takeout-spending-generation-united-states-2018-9 (accessed Jan 23, 2019).

Chanwanpen, K. Human Trafficking Remains a Serious Issue in Thailand. *The Nation* **2018,** http://www.nationmultimedia.com/detail/national/30351252.

Cheung, S. Y.; Mak, J. Y.; Dixon, A. W. Elite Active Sport Tourist: Economic Impacts and Perceptions of Destination Image. *Event Manag* **2016,** *20*, 99–108.

Chubchuwong, M.; Beise-Zee, R.; Speece, M. W. The Effect of Nature-Based Tourism, Destination Attachment and Property Ownership on Environmental-Friendliness of Visitors: A Study in Thailand. *Asia Pac. J. Tour. Res.* **2015,** *20*(6), 656–679.

Crompton, J. Motivations for Pleasure Vacation. *Ann. Tour. Res.* **1979,** *6*(4), 408–424.

Dixit, S. K. *The Routledge Handbook of Gastronomic Tourism,* Eds.; Routledge: New York, 2019.

Epter, A. *Eating Out in Modern American Society: Why Do People Make the Choice to Eat Outside the Home?.* University of Vermont, Graduate College Dissertations and Theses. Paper 77, 2009.

Gastronomy. Marriam-Webster.com. https://www.merriam-webster.com/dictionary/gastronomy 2019.

Henkel, R.; Henkel, P.; Agrusa, W.; Agrusa, J.; Tanner, J. Thailand as a Tourist Destination: Perceptions of International Visitors and Thai Residents. *Asia Pac. J. Tour. Res.* **2006,** *11*(3), 269–287.

Hospitalitynet. The Impact of Social Media on the Tourism Industry. https://www.hospitalitynet.org/news/4071855.html (accessed Feb 6, 2019).

Howard, R. W. Khaosan Road: An Evolving Backpacker Tourist Enclave Being Partially Reclaimed by the Locals. *Int. J. Tour. Res.* **2005,** *7*, 357–374.

Ingram, H.; Taberi, S.; Watthanakhomprathip, W. The Impact of Political Instability on Tourism: Case of Thailand. *Worldw. Hosp. Tour. Themes.* **2013,** *5*(1), 92–103.

Karim, S. A.; Chi, C. G. Q.. Culinary Tourism as a Destination Attraction: An Empirical Examination of Destinations' Food Image. *J. Hosp. Mark. Manag.* **2010,** *19*, 531–555.

Kivela, J.; Crotts, J. C. Tourism and Gastronomy: Gastronomy's Influence on How Tourists Experience and Destination. *J. Hosp. Tour. Res.* **2006,** *30*(3), 354–377.

KPMG Phoomchai Audit, Ltd. *Industry Focus.* https://assets.kpmg/content/dam/kpmg/th/pdf/2018/03/th-medical-tourism-industry-focus-secured.pdf (accessed Jan 10th, 2019).

Long, L. M. Culinary Tourism. In *Encyclopedia of Food and Agricultural Ethics*; Thompson, P., Kaplan, D., Eds.; Springer: Dordrecht, 2013.

Malam, L., (2003). Performing Masculinity on the Thai Beach Scene (Working Paper no.8). Australia: Gender Relations Centre.

Mason, A. Tourism and the Sex Trade Industry in Southeast Asia. *Totem Univ. West. Ont. J. Anthropol.* **1999,** *7*, 1.

Mastercard Global Destination Cities Index. *Big Cities, Big Business: Bangkok, London and Paris Lead the Way in Mastercard's 2018 Global Destination Cities Index, Press Releases.* https://newsroom.mastercard.com/press-releases/big-cities-big-business-bangkok-london-and-paris-lead-the-way-in-mastercards-2018-global-destination-cities-index/ (accessed Jan 10, 2019).

McDowall, S.; Choi, Y. A. Comparative Analysis of Thailand Residents' Perception of Tourism's Impacts. *J. Qual. Assur. Hosp. Tour.* **2010,** *11*, 36–55.

Miller, G.; Rathouse, K.; Scarles, C.; Holmes, K.; Tribe, J. Public Understanding of Sustainable Tourism. *Ann. Tour. Res.* **2010,** *37*(3), 627–645.

Ministry of Tourism & Sports. *Tourism Statistics.* https://www.mots.go.th/more_news.php?cid=411 2019.

Ministry of Tourism and Sports. *The Second National Tourism Development Plan (2017–2021).* Bangkok: Office of Permanent Secretary: Bangkok, Thailand, 2017.

Moulin, C. Gastronomy and Tourism: Must Every Tourist Wish be Our Command?. *The Tour. Rev.* **2000,** *55*(1), 19–24.

Nakasu, T. *Natural Disasters and Disaster Management in Thailand: Status, Risks, and Trends.* 13th International Conference on Thai Studies: Globalized Thailand? Connectivity, Conflict and Conundrums of Thai Studies, Chiangmai: 15–18 July, 2017.

Netherlands Embassy in Bangkok. *Tourism Industry in Thailand.* www.rvo.nl/sites/ default/files/2017/06/factsheet-toerisme-in-thailand.pdf (accessed Jan 10, 2019).

Noree, T.; Henefeld, J.; Smith, R. Medical Tourism in Thailand: A Cross-Sectional Study. *Bull. World Health Organ.* **2016,** *94*, 30–36.

Nuttavuthisit, K. Branding Thailand: Correcting the Negative Image of Sex Tourism. *Place Branding Public Dipl.* **2007,** *3*(1), 21–30.

O' Connell Davidson, J. *Prostitution, Power and Freedom.* T. J. International: Great Britain, 1998.

Oppermann, M. Sex Tourism. *Ann. Tour. Res.* **1998,** *26*(2), 251–266.

Oppermann, M. *Sex Tourism and Prostitution: Aspects of Leisure, Recreation, and Work.* Cognizant Communication: New York, 1999.

Pine, J.; Gilmore, J. H. Welcome to the Experience Economy. *Harv. Bus. Rev.* **1998,** *July–August*, 97–105.

Pocock, N. S.; Phua, K. H. Medical Tourism and Policy Implications for Health Systems: A Conceptual Framework from a Comparative Study of Thailand, Singapore and Malaysia. *Glob. Health* **2011,** *7*(12), 1–12.

Poudel, S.; Nyaupane, G. P.; Budruk, M. Stakeholders' Perspectives of Sustainable Tourism Development: A New Approach to Measure Outcomes. *J. Travel Res.* **2016,** *55*(4), 465–480.

Richards, G. Gastronomy: An Essential Ingredient in Tourism Production and Consumption?. In *Tourism and Gastronomy*; Hjalafer, A., Richards, G., Eds.; Routledge: New York, 2004.

Richards, G. Evolving Gastronomic Experiences: From Food to Foodies to Foodscapes. Pre-Publication Version Gastronomic Experiences: From Foodies to Foodscapes. *J. Gastron. Tour.* **2015,** *1*, 5–18.

Rittichainuwat, B. N.; Qu, H.; Brown, T. J. Thailand's International Travel Image. *Cornel. Hosp. Quart.* **2001,** *42*(2), 82–88.

Rungsuwannarat, C.; Noraphat, N.; Michiels, T.; Fujiwa, D.; Lin, F. A Comparative Study of Destination Image between Thailand and Indonesia. *Aphiet J.* **2015,** *4*(2), 5–26.

Ryan, C.; Hall, M. C. *Sex Tourism: Marginal People and Liminalities*. London: Routledge, 2000.

Ryan, C.; Kinder, R. Sex, Tourism and Sex Tourism: Fulfilling Similar Needs? *Tour. Manag.* **1996,** *17*(7), 507–518.

Saechau, E.; Nge, E. T. N.; Phrombun, T.; Htun, H. M.; Mouton, K. International Tourists' Perception toward Thailand and Myanmar. *Aphiet J.* **2015,** *4*(2), 37–55.

Sangpikul, A. Travel Motivations of Japanese Senior Travelers to Thailand. *Int. J. Tour. Res.* **2008,** *10*, 81–94.

Sastre, R. P.; Phakdee-Auksorn, P. Examining Tourists' Push and Pull Travel Motivations and Behavioral Intentions: The Case of British Outbound Tourists to Phuket, Thailand. *J. Qual. Assur. Hosp. Tour.* **2017,** *18*(4), 437–464.

Soontayatron, S. Residents; Perceptions Towards Sociocultural Impacts Induced by Internal Migration. *Asia Pac. J. Tour. Res.* **2015,** *20*(11), 1234–1245.

Suanpang, P. Gastronomic Tourism Strategy Implication for Thailand. *Int. J. Adv. Res.* **2015,** *3*(7), 581–589.

Sunanta, S. The Globalization of Thai Cuisine. *The Canadian Council of Southeast Asian Studies Conference Proceedings*, Toronto, Oct. 14–16, 2005. http://citeseerx.ist.psu.edu/viewdoc/download?doi=10.1.1.404.1934&rep=rep1&type=pdf (accessed Jan 30, 2019).

TAT News. *TAT Reaffirms that 'We Stay On The Right Course And Do Not Promote Sex Tourism'.* https://www.tatnews.org/2018/02/tat-reaffirms-stay-right-course-not-promote-sex-tourism/ (accessed Jan 15, 2019).

TAT News. *Gastronomy Tourism: Thailand Means "Good Food" in Any Language*. https://www.tatnews.org/2019/02/gastronomy-tourism-thailand-means-good-food-in-any-language/ (accessed March 1, 2019).

Tavitiyaman, P.; Qu, H. Destination Image and Behavioral Intention of Travelers to Thailand: The Moderating Effect of Perceived Risk. *J. Travel Tour. Mark.* **2013,** *30*(3), 169–185.

The Nation. *Thailand "Among Top 20 Most Dangerous Countries" to Visit. June 9th*. https://www.nationmultimedia.com/news/national/30317678 (accessed Jan 12, 2019).

Thomsen, S. N. *Tourist Agency Launches New Slogan for Thailand: Amazing Thailand. It Begins with the People*. https://scandasia.com/tourist-agency-launches-new-slogan-thailand-amazing-thailand-begins-people/ (accessed Jan 13, 2019).

Tourism Authority of Thailand. *Thailand Aims at Becoming Asia's Destination for Sport Tourism*. https://www.tatnews.org/2016/06/thailand-aims-at-becoming-asias-ultimate-destination-for-sports-tourism/ (accessed Feb 12, 2019).

Tourism Authority of Thailand Newsroom. *TAT's Marketing Plan 2018 to Heighten Thailand as a Preferred Destination*. https://www.tatnews.org/2017/07/tats-marketing-plan-2018-to-heighten-thailand-as-a-preferred-destination/ (accessed July 17, 2017).

Tourism Council of Thailand. *The 2012 Survey of Thai Culinary Travelers*. http://service.nso.go.th/nso/nsopublish/citizen/news/news_travel.jsp, 2013.

TTRweekly. *Food Tourism to Star in TAT Promotions*. https://www.ttrweekly.com/site/2017/06/food-tourism-to-star-in-tat-promotions/ (accessed Feb 15, 2019).

Tyrrell, T.; Johnson, R. The Economic Impacts of Tourism: A Special Issue. *J. Travel Res.* **2006,** *45*(1), 3–7.

Untong, A. Local Residents' Attitudes Towards Tourism: A Case Study in Chiang Mai and Chiang Rai, Thailand. In *Mekong Tourism: Blessing for all?*; Kaosa-Ard, M., Ed.; pp 125–136. Social Research Institute: Chiang Mai, 2007.

UNWTO. *Global Report on Food Tourism*; Detectivegrafico: Spain, 2012.

U.S. Department of State. *Trafficking in Persons Report*. https://www.state.gov/ documents/organization/243562.pdf, 2015.

Van Rooyen, J. The Economic Impact of Geopolitical Unrest on Thailand's Tourism Industry. *J. Bus. Econ. Dev.* **2018,** *3*(2), 30–42.

Worapinij, P.; Jantarakolica, K.; Kanittinsuttitong, N. *Sports Tourism Management Model and the Development of Sport Tourism in Thailand*. The 22nd International Society for Business Innovation and Technology Management Conference. Bangkok: August 31–September 2, 2017.

World Travel & Tourism Council. Travel & Tourism Economic Impact 2018 Thailand. https://www.wttc.org/-/media/files/reports/economic-impact-research/countries-2018/thailand2018.pdf (accessed Jan 12, 2019).

Yiamjanya, S.; Wongleedee, K. International Tourists' Travel Motivation by Push-Pull Factors and the Decision Making for Selecting Thailand as Destination Choice. *Int. J. Social Behav. Educ. Econ. Bus. Ind. Eng.* **2014,** *8*(5), 1348–1353.

Živković, R.; Gajić, J.; Brdar, I. The Impact of Social Media on Tourism. *The 11th Annual International Religious Tourism and Pilgrimage (IRTP) Conference*, 2014; pp 758–761.

CHAPTER 7

INNOVATION EXERTIONS IN THE THAI HOTEL INDUSTRY

NATE-TRA DHEVABANCHACHAI[1*], HENRY TSAI[2], and KAEWTA MUANGASAME[1]

[1]*Tourism and Hospitality Management Division, Mahidol University International College, Thailand*

[2]*School of Hotel and Tourism Management, The Hong Kong Polytechnic University, Hong Kong*

[*]*Corresponding author. E-mail: natetra.dhe@mahidol.ac.th*

ABSTRACT

The purpose of this study is to understand the innovation determination in the hotel industry in Thailand. The investigation focuses on independent hotels in Bangkok. The study also intends to cover innovation implementation approach. Managerial guidelines and framework for the innovation implementation within the hotel industry and a direction for future development, particularly for independent hotels, are discussed.

7.1 INTRODUCTION

The Asian continent is well known for its unique service, people, and culture. Thailand has gained popularity for its beautiful beaches, history, culture, culinary delights, tourist attractions, and shopping. Thus, Thailand is one of the most well-liked destinations in Asia (Henkel et al., 2006). The top locations are Bangkok, Pattaya, Chiang Mai, and Phuket, where many five-star international chains and independent units serve the market. The number of tourists visiting Thailand has consistently been increasing from 10.08 million in 2003 to 38.28 million in 2018 (Vanhaleweyk, 2019).

Lunkam (2017) explains that the number of accommodations in Thailand has increased from approximately 300,000 rooms in 2000 to over 600,000 rooms in 2016, where 29% are located in Southern Thailand, 11% in Northeastern Thailand, 15% in Eastern Thailand, 6% in Western Thailand, 14% in Northern Thailand, 4% in Central Thailand, and 21% in Bangkok.

According to Agoda.com, approximately 171,931 Thai hotels (international chain and independent hotels) are listed, and almost 1 million rooms operate across the country (Sritama, 2018), especially small hotels that are booming in Thailand. Presently, 45% of hotels in Thailand are international chains, whereas 55% are local brands or independent hotels. Hence, hotel operators need to differentiate themselves to increase revenue, profit, market share, sales volume, and to lower costs (Gridffin & Page, 1993; Gallouj, 1998; Coombs & Miles, 2000; King, 2000; Chapman et al., 2002). For 4 years in a row, Bangkok has been recognized as the world's best city. In 2019, Bangkok has been ranked as the top destination for global travelers and welcomed 22.7 million international visitors (Talty, 2019). By 2022, first-class international chains will lead the market in Bangkok, increasing up to 27.2% of revenues (Kasemsuk, 2019). All major players in the market that are independent properties must work hard to be unique and innovative to sustain in a competitive market.

Rueckert et al. (1985) and Kotler et al. (2002) state that chain operations will continue to strengthen their position as market leaders through innovation. Storey (1994) also states that international chain hotels need to overcome the challenges of innovation more than independent hotels. Independent hotels tend to face endless uncertainty in the market. They often have fewer hierarchical obstacles, resulting in a less-structured approach to innovation. Thus, independent firms can be more flexible to changing conditions, and therefore should have the ability to respond to customer needs/wants better. However, with the less-structured approach, independent hotels may lack a proper and systematic approach toward every facet of an innovation project. A large number of SME businesses in Thailand are found lacking proper accounting systems, business plans, and have outdated technological support and poor innovation (OSMRJ, 2008).

Thailand has more local and independent hotels than international chain hotels, and the former continues to grow. Thus, this research intends to obtain information from executives, such as owners or management team, of independent hotels about their innovation determinations and implementation approach. Obtaining data on independent hotels' innovation efforts will be beneficial because they are major players in this industry. Moreover, given that implementation approach to innovation is crucial to organizational success, this study will therefore use the model of Hansen and Birkinshaw's (2007) "Innovation Value Chain: An Integrated Flow" as a basis during the interview with respondent hotels. The answers to these questions will be interesting to know.

- What innovation activity is (activities are) currently being implemented in the property?
- How is the innovative idea generated?
- What is the approach toward the innovation's implementation?

The research is expected to serve as a good source for viewing the innovation activities and the corresponding approach of independent hotels in Bangkok, Thailand. Specifically, this study aims to explore innovation activities and determination of independent hotels and the approach/process of innovation implementation of independent hotels on the basis of Hansen and Birkinshaw's (2007) Innovation Value Chain model.

7.2 INNOVATION AND ITS TYPOLOGIES

Schumpeter (1974) is one of the many researchers who developed innovation theory. He defines innovation as "new ways of doing things, or better, unique combinations of the factors of production," and identifies it as a core practice of business (McGuire, 1996). Therefore, organizations must view innovation as an opportunity to create new things or to make changes for producing a different product or service (Drucker, 1985). Rogers (1983) defines innovation as knowledge, an idea, product, or object that is perceived as new by the adopters. An idea does not need to be new as measured by space or time since it was first used or discovered. If an idea appears new to an adopter, then it is an innovation. Moreover, innovations are akin to inventions, as they have similar processes that bring them into implementation and commercialization (Hjalager, 1994). Lancaster and Taylor (1988) refer to innovation as an idea in practice, which is regarded as original and novel by stakeholders of a community.

Ottenbacher (2006) adds that innovation ranges from totally new-to-the-world services with an entirely new market to minor adjustments of existing services. Innovation is the process of using an idea to make work or a task function better and solve problems. Thus, innovation is implementing a new idea, product, or process to the stage of acceptance and then to the stage of change and adaptation (Hall & William, 2008). Although ample literature has focused on product innovation, consideration has only been recently given to service innovation (Vila et al., 2012; Enz, 2011; Verma et al., 2008), such as constant improvements via technology, investment in employee performance, or management efforts for enhancing customer experience (Verma et al., 2008). Vargo and Lusch (2004) add that service innovation involves the improvement and development of sets of different skills, attitudes, and capabilities in providing better value of services/products to customers. **Product innovation** refers to new knowledge, product, service, or technology improvements that result from a shift in customers' needs and wants and degree of global competition (Gunday et al., 2011), for example, the in-room "plug & play" internet system. Booz-Allen and Hamilton (1982) define **service innovation** as something new to the world of services or new service lines that are added into existing services, such as improvements to existing services, repositioning, or reduction of cost. **Process innovation** is intended to increase the quality of products or their delivery method, and it is objectively implemented to reduce costs and to significantly improve production (Fagerberg and Verspagen, 2002). **Organizational innovation** is related to the strategies that an organization uses to improve their business. It also looks at improving workplace satisfaction to increase labor productivity (Oslo Manual, 2005). **Marketing innovation** is a set of new methods of marketing, such as packaging design, or new ways of promotion and pricing. It can also refer to detecting new markets (Lashley, 2008).

7.3 INNOVATION ACTIVITIES OF INDEPENDENT HOTELS

The term "independent hotel" is used to represent small- and medium-sized hotels that are not a domestic chain hotel and that do not have any international brand affiliation. Independent hotels are normally SMEs owned and managed by families or business investors (Beaver & Prince 2004). The common defining criteria of SMEs are the number of employees and fixed asset level. A small enterprise has < 50 employees or possesses fixed assets of not more than 50 million baht, whereas a medium enterprise employs 51–200 employees or has fixed assets worth between 50 and 200 million baht. SME businesses play a vital role in Thailand's economy and its societal growth. SMEs in the service sector were also steadily increasing from 2007 to 2012 (Charoenrat and Harvie, 2017). However, small businesses lack economies of scale, and are less likely to raise sufficient profit margin to invest in product development or any creativity project, including innovative ideas (Peters and Weiermai, 2002).

Pikketmaat and Peters (2006) published findings from surveys of 107 SME hotel businesses on innovations in alpine tourism destinations (Table 7.1). The highest percentage of innovation activities is in the area of architecture/design (57.14%), and the rest are all <10%. Though efforts on innovation have been made, improvements can still be achieved as each organization's innovation is influenced by various factors, including culture (de Brentani, 1999).

TABLE 7.1 Innovation Activities and Degree in Functional Area.

Functional area	% of innovation activities
Information technology	9.07%
Gastronomy	9.30%
Operational procedures	3.17%
Quality assurance	2.04%
Wellness	1.36%
Product bundling	5.22%
HRM	4.76%
Animation	4.08%
Marketing	1.13%
Architecture/design	57.14%
Total	100.00%

Source: Pikketmaat and Peters (2006).

7.4 INNOVATION IMPLEMENTATION THEORIES

This study employs the model of Innovation Value Chain: An Integrated Flow by Hansen and Birkinshaw (2007). The flow covers the entire process of innovation implementation (see Exhibit 7.1). It explains the transformation of materials into finished goods, and it can serve as a guideline on effective innovation implementation.

Idea generation: Executives understand that innovation commences with good ideas, but where do these thoughts and concepts come from? First, executives spontaneously seek "flash ideas" from their working groups or business units for creativity, and they usually find they have a moderately good sense of what is possible. Executives must allow the opportunity for team members to brainstorm and feedback to generate the best ideas. **Idea conversion:** Creating great ideas is one thing, but implementing ideas is another. New concepts will not be created without strong screening and capitalizing mechanisms. If an idea gets screened and funded but weakens because an organization is preoccupied, then this idea will not lead to innovative ideas coming to fruition. **Idea diffusion:** Ideas that have been selected, funded, and developed still need buy-in not only from guests. Gaining the support of relevant stakeholders within the company has greater significance, because their support will facilitate the adoption of new practices or products across locations, units, locations, channels, and customer groups.

This study aims to explore innovation activities, determination, and approach of independent hotels based on Hankin and Birkinshaw's (2007) Innovation Value Chain: an Integrated Flow (see Exhibit 7.1).

7.5 INNOVATION CRITERIA MATCHING

This study presents the results of this research work that are derived from the qualitative methodology of an in-depth interview with the owner/department heads of five independent hotels in Bangkok. A letter code has been given for each respondent from A to E, which are "hotel A to hotel E." The codes are to be utilized throughout the research paper.

The researcher commenced the selection process by choosing independent properties in Bangkok via hotel website. Eight independent hotels accurately correspond to the innovation criteria list, and five hotels accepted to participate in the research (Table 7.2).

The results from four hotels (A, B, D, and E) are found with product innovation. Hotel C exhibits service innovation, which is their pet-friendly policy. Hotels A and D offer minor product innovations, such as free in-room mini bar; complimentary tea, coffee, and snack at the check-in area; and a pillow menu selection. Hotel E offers in-room touch pad, which is a technological innovation, whereas hotel B has been constructed with a unique interior design that depicts Thai literature.

7.6 CURRENT INNOVATION IMPLEMENTATIONS: INDEPENDENT HOTELS

The following section discusses the efforts of independent hotels on innovation activities.

EXHIBIT 7.1 Innovation Value Chain: an Integrated Flow.

	IDEA GENERATION			**CONVERSION**		**DIFFUSION**
	IN-HOUSE Creation within a unit	**CROSS-POLLINATION Collaboration across units**	**EXTERNAL Collaboration with parties outside the firm**	**SELECTION Screening and initial funding**	**DEVELOPMENT Movement from idea to first result**	**SPREAD Dissemination across the organization**
KEY QUESTIONS	Do people in our unit create good ideas on their own?	Do we create good ideas by working across the company?	Do we source enough good ideas from outside the firm?	Are we good at screening and funding new ideas?	Are we good at turning ideas into viable products, businesses and best practices?	Are we good at diffusing developed ideas across the company?
KEY PERFORMANCE INDICATORS	Number of high-quality ideas generated within a unit	Number of high-quality ideas generated across units	Number of high-quality ideas generated from outside the firm	Percentage of all ideas generated that end up being selected and funded	Percentage of funded ideas that lead to revenues; number of months to first sale	Percentage of penetration in desired markets, channels, customer groups; number of months to full diffusion

Source: Hansen & Birkinshaw (2007)

TABLE 7.2 Matching Innovation Criteria List for Independent Hotels.

Cat.	Criteri	Researchers	Matching criteria with the respondent hotels, 4-5 star properties in Bangkok	Innovation activity	Keywords from criteria
1.	It is a new way of doing things, or better, **unique combinations** of the factors of production.	Schumpeter (1974)	Hotel B	Traditional Thai literature as a décor concept	Unique combination
2.	It is a **new thing** or to **make changes** to produce a **different product** or service.	Drucker (1985)	Hotel A Hotel D	Free mini-bar/pillow menu (A) Sky lobby with 24-h free of charge drink and snack at the lobby for in-house guests (D)	Make changes for different product/ service (A) New thing for different product/service (D)
3.	It is a **new-to-the-world service** with an entirely new market, to the minor **adjustments of existing services.**	Ottenbacher (2006)	Hotel C	Pets allowed in the city hotel	New-to-Thailand service, adjustments in existing services
4.	It is new idea delivered with a new service model through constant **improvements via technology,** investment in employee performance, or management efforts to enhance customer experiences.	Verma et al. (2008)	Hotel E	In-room touchpad	Improvements via technology

7.6.1 INNOVATION IMPLEMENTATION OF INDEPENDENT HOTELS

Five respondents of hotels A to E were cooperative, and the results (Table 7.3) show that the innovation activities exhibited efforts of the owners. Each activity, such as the free mini-bar, pillow menu, in-room touch pad, and 24-h coffee and tea at the lobby, could easily be conducted by a staff with low degree of guest interaction. Even with the Thai literature decor of hotel B, such information was not explained to guests by the staff while checking-in. This observation shows that most respondents establish a small scale of product innovation, which requires less complexity in operation.

Hotel C's service innovation of allowing pets displays a brief information of the policy on the hotel website. No details, such as type, size, or any limitation of the pet, including fees, are indicated. No insightful motivation about this service innovation was expressed during interview, and each reply was brief.

7.6.2 TYPES OF INNOVATION ACTIVITIES IN INDEPENDENT HOTELS

Owners of independent hotels in Bangkok are willing to establish innovation activities, though not very systematically. Most independent hotels pursue product innovation, and one respondent establishes service innovation. Many researchers have explained that firms can engage in many forms of innovation. Sundbo and Gallouj (1998) and Oslo Manual (2005) describe different types of innovation, such as marketing innovation, process innovation, and organizational innovation, such as human sources innovation apart from product and service innovations. Innovation also includes information technology or other management innovations (Drucker, 1985; Hjalager, 2010). Independent hotels could develop more innovation if owners buy in and have knowledge/motivation. Hua and Batra (2015) emphasized that five-star hotels in Bangkok implement technological innovations to provide a more convenient stay for guests. J.W. Marriott has a worldwide mobile application that guests can use to view real-time occupancy, make instant bookings, and confirm a reservation. The Peninsula offers online newspapers for their in-house guests. Furthermore, most guests in Bangkok are leisure and business travelers, which make establishing technological innovations especially by independent hotels highly possible (Damanpour et al., 2009). Anthonisz (2014) and Rayna and Striukova (2009) state that in the hotel

TABLE 7.3 Innovation Activity by the Independent Hotels.

Hotel	Type of innovation	Innovation activities	Descriptions of innovation activities
Hotel A	Product	Free mini-bar and pillow menu	A free mini bar is replenished on a daily basis with soft drinks, juice water, and snacks. The pillow menu is made from wood nicely placed on the guest's bed.
Hotel B	Product	Traditional Thai literature as décor concept with pillow menu	The hotel has a clear concept, which is decorated with the face of Ravana, one of the great characters of the Ramakien in traditional Thai literature. Each room is painted with Ravana. Guests are inspired to learn more about Thai literature by staying in an ambiance of Thai traditional literature. The pillow menu is simply made from paper and is placed on the guest's pillow.
Hotel C	Service	Pet is allowed	Modern hotels in Bangkok rarely allow pets within the vicinity. The overall size of the building and the guest room is rather small. Pets are allowed in all areas: lobby, food and beverage outlets, and guest rooms. No written policy regarding this feature is available. Guests are fully responsible for their pet.
Hotel D	Product	Sky lobby with 24-h, free-of-charge drinks and snacks for in-house guests	All in-house guests are eligible to enjoy tea/coffee/snacks, which are welcoming and accommodating. The snack are local products, such as dried fruits, Thai snacks, cookies, various fruit juices, fresh coffee, and a wide selection of tea. The overall presentation is excellent. This service is rare even in 5-star international chains or any 5-star independent hotel.
Hotel E	Product	In-room touchpad	Most rooms feature a one-of-a-kind touchpad control that lets guests control the lighting, in-room temperature, curtains, TV–audio system, and other room amenities.

industry, it is difficult to survive without using technology. Independent hotels prefer product innovation and do not capitalize in high technology, but once they invest, they will claim it as an innovation.

7.6.3 INNOVATION EFFORTS OF INDEPENDENT HOTELS

Independent hotels prefer simple product innovation. This fact is supported by Peter and Weiermai (2002); they state that small business lack economies of scale and are less likely to raise sufficient profit margin to invest in large-scale innovation. Moreover, Chae (2012) and Wooder and Baker (2011) state that service innovation is crucial and it can create a competitive advantage. However, only one respondent established service innovation by implementing a pet-friendly policy. Thus, the innovation activities of independent hotels are focused on providing a good image, such as Thai literature interior design, convenience and choices for customers, a free mini-bar, a pillow menu, 24-h free coffee, and in-room touchpad. These activities require minimal human interaction. Furthermore, Pikketmaat and Peters (2006) find that innovation activities in SME hotels are simple and uncomplicated.

Hua and Batra (2015) indicate that hotels in Bangkok, especially five-star hotels, often utilize interior and exterior designs to catch up with innovation, and these innovations have improved customer satisfaction and business performance. Thus, for independent hotels, using certain interior designs can be considered innovative. Therefore, this finding reflects that independent hotels can work hard on the delivery of innovation, such as highlighting the Thai literature décor to guests while they are checking in. Hackett and Melia (2012) point out that current hotel clientele seeks for unique services, new experiences, new forms of entertainment, and warm hospitality. Hence, the research finding shows that independent hotels in Bangkok have considerable work in this regard.

7.7 IDEA GENERATION, CONVERSION, AND DIFFUSION

7.7.1 APPROACH TO INNOVATION IMPLEMENTATION

Hansen and Birkinshaw (2007)'s innovation value chain: an integrated flow model describes that to successfully implement innovation activity, executives need to be able to link and collaborate each stage of idea generation, idea conversion, and idea diffusion, and not focus solely on any one part. By doing so, the strengths and weakness at each stage can be identified. Innovation implementation begins from idea generation and jumps right to idea diffusion. A significant amount of work needs to be done in each of the three categories of idea generation, conversion, and diffusion (Exhibit 7.2).

7.7.2 IDEA GENERATION

All five independent hotels' idea generation stages are at the owner level. After that, the owner relates the innovative idea to the head, and then the head will communicate to their staff. Sutthijakra (2011) stated that multinational companies have a certain degree of tension because they need to remain standardized as set by the corporate office, yet local adaptation is needed to meet consumers' requirements. Similarly, the owners who participated in this survey can involve their head team and staff member at the idea generation stage.

7.7.3 IDEA CONVERSION

No idea conversation step takes place for all respondents, and only top-down strategy is implemented. The idea conversion stage refers to screening an idea—selection and movement from idea to first result—development (Hansen & Birkinsahw, 2007). An innovative idea needs a certain extent of system and mechanism for the most effective way of innovation implementation. All independent hotels have been concluded lacking a systematic approach to idea generation and conversion.

7.7.4 IDEA DIFFUSION

Roger (2003) states that in a large organization, diffusion is not programmed, but necessitates a step-by-step implementation by the management and employees. This view is supported by Hansen and Birkinshaw (2007). Kostova and Roth (2002) and Blind (2004) state that multinational companies establish standards as essential and written instructions/procedures to direct the actions of properties worldwide in their operation. The study shows that all respondents are lacking the idea diffusion stage as well. This study affirms that independent hotels in Bangkok could use Hansen and Birkinshaw (2007)'s model for improving their innovation implementation approach (see Exhibit 7.2).

EXHIBIT 7.2 Innovation Implementation Flow of Independent Hotels.

	Idea Generation			Conversion		Diffusion	
	Stage 1			Stage 2		Stage 3	
	IN-HOUSE Creation within a unit	**CROSS-POLLINATION Collaboration across unit**	**EXTERNAL Collaboration with parties outside the firm**	**Selection Screening and initial funding**	**Development Movement from idea to first result**	**Trial Trialing the idea in the operation**	**Spread Dissemination across the organization**
Independent Hotel	**By Owner**	**None**	**None**	**Uncertain**	**Uncertain**	**None**	**By Property with verbal cues**
Key Questions	Do people in our unit create good ideas on their own?	Do we create good ideas by working across the company?	Do we source enough good ideas from outside the firm?	Are we good at screening and funding new ideas?	Are we good at turning ideas into viable products, businesses and best practices?	How is trial process managed? What is the duration of trail period? Does every property need to have a trial period?	Are we good at diffusing developed ideas across the company?
Key Performance Indicators	Number of high quality ideas generated within a unit	Number of high quality ideas generated across units.	Number of high quality ideas generated from outside the firm.	Percentage of all ideas generated that end up being selected and funded.	Percentage of funded ideas that lead to revenues; number of months to first sale.	Result and feedback from operators (employee), customer and the management.	Percentage of penetration in desired markets, channels, customer groups, number of months to full diffusion.

7.8 CONCLUSION

The research results have urged the researcher to propose the model that owners can use along with the guideline of Hansen and Birkinshaw (2007) (Fig. 7.1). The model starts from a strong vision of the owner in involving the team in innovation implementation. Sulistiyani and Harwiki (2015) add that the innovation capacities of SME business can be enhanced by establishing knowledge sharing among owners and employees to enable the "best idea generations" and most effective "idea conversation" stages.

Once idea generation and conversion stages are approved, the trial period is to be embarked upon along with the systematic training approach, which is the stage of "idea diffusion." During the trial period, feedback from staff, management, and customers are gathered. Once the trial period is executed, the next stage is the real launch. The launch of the innovation activity should go along with evaluating support for continuous improvement. Therefore, using the model in Figure 7.1 can help owners to involve their team members in the implementation process for the best outcome.

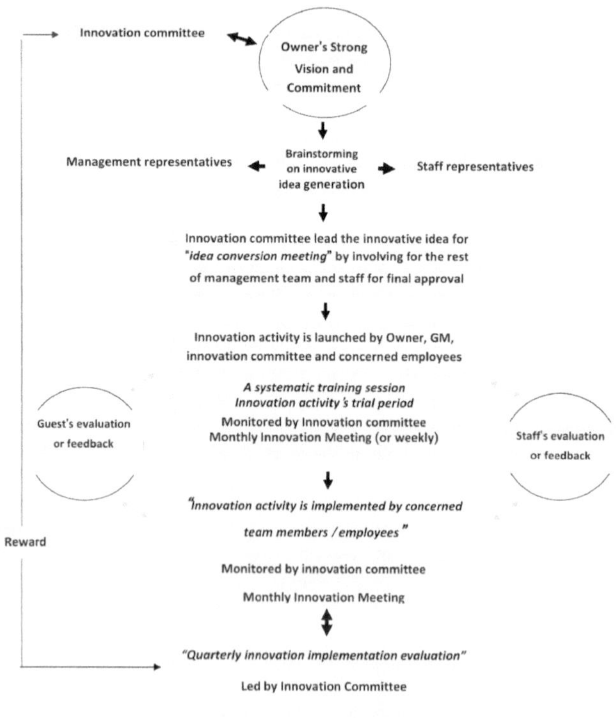

FIGURE 7.1 Innovation approach model for independent hotels.

KEYWORDS

- **innovation**
- **innovation activities**
- **innovation implementation**
- **independent hotel**
- **Thailand**

REFERENCES

Anthonisz, A. Assessing the Future of Housekeeping Operations in Dubai's Five-Star Hotel Industry–Room for Innovation *Worldwide Hospital. Tourism Themes.* **2014,** *6*(4), 352–361.

Blind, K. The Role of Standards in the Service Sector. In *The Economics of Standards: Theory, Evidence, Policy*; Blind, K., Ed.; Cheltenham, UK: Edward Elgar, 2004; pp. 162–180.

Boon-itt, S.; Chomvong, V. (2010). In Analyzing Service Quality in Thai Hotel Industry: An Application of the SERVQUAL Model, Proceedings of the 2010 IEEE IEEM, Thailand: Thammasat Business School, pp.395–399..

Chae, B. K. An evolutionary framework for service innovation: insights of complexity theory for service science. *Int. J. Prod. Econ.* **2012,** *135*, 813–822.

Chapman, R. L.; Soosay, S.; Kandampully, J. Innovation in logistic Services and the New Business Model: A conceptual Framework. *Manag. Service Qual.* **2002,** *12*(6), 358–371.

Charoenrat, T.; Harvie, C. Thailand's SMEs in Economic Integration. *J. Southeast Asian Econ..* **2017,** *34*(1), 148–174.

Coombs, R; Miles, I. Innovation, Measurement and Services: The New Problematique. In *Innovation Systems in the Service Economy,* Metcalfe, J. S., Miles, I., Eds.; Boston,: Kluwer Academic Publ., 2000; pp. 85–103.

Creswell, J.W. (2008). *Selection of a research design.* Retrieved from http://www.sagepub.com/upm-data/22780_Chapter_1.pdf

Damanpour, F.; Walker, R. M.; Avellaneda, C. N. Combinative effects of innovation types of organizational performance: a longitudinal study of service organizations. *J. Manag. Studies* **2009,** *46*(4), 650–675.

Denzin, N. K. *The research act,* Aldine: Chicago, 1978.

Drucker, P. *Innovation and Entrepreneurship: Practice and Principles*, Heinemann: London, 1985.

Erickson, F. Qualitative Methods in Research on Teaching. In *Handbook of Research on Teaching*; Wittrock, M. C., Ed.; Macmillan: New York,1986, pp 119–161.

Fagerberg, J.; Verspagen, B. Technology-Gaps, Innovation-Diffusion And Transformation: An Evoluntionary Interpretation. *Res. Policy* **2002,** 31:1291-1304.

Gallouj, F. Innovating In Reverse: Services and The Reverse Product Cycle. *European J. Innov. Manag.* **1998,** *1*(3), 123–138.

Günday, G.; Ulusoy, G.; Kılıç, K.; Alpkan, L. Effects Of Innovation Types On Firm Performance. *Int. J. Prod. Econ.* **2011,** *133*(2), 662–676.

Griffin, A. PDMA Research On New Product Development Practices: Updating Trends and Benchmarking Best Practices. *J. Prod. Innov. Manag.* **1997,** *14*(6), 429–458.

Hall, C. M.; Williams, A. M. *Tourism and Innovation.* Routledge: London, 2008.

Hackett, G.; Melia, D. The Hotel as the Holiday/Stay Destination: Trends and innovations. Presented at THRIC Conference, Belfast, Ireland, 2012.

Hansen, M.; Birkinshaw, J. The sophisticated Innovator. *Harvard Business Review*, pp. 121–130, 2007.

Henkel, R.; Henkel, P.; Agrusa, W.; Agrusa, J.; Tanner, J. Thailand as a Tourist Destination: Perceptions of International Visitors and Thai Residents. *Asian Pacific J. Tourism Res.* **2006,** *11, 269-287.*

Hennink, M.; Hutter, I.; Bailey, A. *Qualitative Research Methods,* Sage: Los Angeles, 2011

Hjalager, A. M. (1994). Dynamic Innovation in the Tourism Industry. *Progr. Tourism Hosp. Manag.* **1994,** *6*, 197–224.

Hua, N.; Batra, A. The New Trends and Innovation in Selected Five-Star Hotels in Bangkok, Thailand. *Int. Res. E-J. Business Econ.* **2015,** *1*(2), 42–53.

Kasemsuk, N. 2019, *A luxe rescue*. www.bangkokpost.com/business/1766624/a-luxe-Rescue. (accessed Oct 7, 2019).

King, W. R. Measuring police innovation: issues and measurement policing. *Int. J. Police Strat. Manag.* **2000,** *23*(3), 303–317.

Kotler, P.; Bowen, J.; Markens, J. *Marketing for Hospitality and Tourism*, Prentice Hall: Upper Saddle River, 2002.

Kostova, T.; Roth, K. Adoption of an Organizational Practice by Subsidiaries of Multinational Corporations: Institutional and Relational Effects. *Acad. Manag. J.* **2002,** 45, 215–233.

Lancaster, G.A.; Taylor, C.T. A Study of Diffusion Of Innovation in Respect of The High Speed Train. *Eur. J. Market.* **1988,** *22*(3), 21–27.

Lashley, C. Studying Hospitality: Insights from Social Sciences. *Scandinavian J. Hospital. Tourism* **2008,** *8*(1), 69–84.

Lunkarm, P. (2017). *Krungsri Research.* Thailand Industry Outlook 2018–2020.

Maxwell, J. A.; Loomis, D. Mixed Methods Design: An Alternative Approach, In *Handbook of Mixed Methods in Social and Behavioral Research*; Tashakkori, C. and Teddlie, C., Eds.; Sage: Thousand Oaks, 2002; pp 241–271.

McGuire, J.W. *Theories of Business Behaviour*, Prentice-Hall, Inc: Englewoods, 1996.

Miles, M. B.; Huberman, A. M. *Qualitative Data Analysis: An Expanded Sourcebook,* Sage: Thousand Oaks, 1984.

OCED. Oslo Manual: Guidelines for Collecting and Interpreting Innovation Data, 3rd Ed, Paris, 2005.

OSMRJ Small & Medium Enterprise Development Policies in Thailand. Tokyo: Organization for Small & Medium Enterprises and Regional Innovation Japan, 2008.

Ottenbacher, M.; Shaw, V.; Lockwood, A. An Investigation of the factors Affecting Innovation Performance in Chain and Independent Hotels. *J. Quality Assurance Hospital. Tourism* 2006, *6*(3-4), 113–128.

Peters, M.; Weiermair, K. Innovationen und Innovationsverhalten im Tourismus, In *Schweizer Jahrbuch für Tourismus 2001/2002;* Bieger, T., Ch. Laesser, Eds.; St. Gallen, 2002; pp 157–178.

Pikkemaat, B.; Peters, M. Towards the Measurement of Innovation—A Pilot Study in the Small and Medium Sized Hotel Industry. *J. Quality Assur. Hospital. Tourism* **2006,** *6*:3-4, 89–112.

Rayna, T.; Striukova, L. Luxury Without Guilt: Service Innovation in the All-Inclusive Hotel Industry. *Service Business* **2009,** *3*(4), 359–372.

Rhee, J.; Park, T.; Lee, D.H. Drivers of Innovativeness and Performance forInnovative SMEs in South Korea: Mediation of Learning Orientation. *Tech-Novation* **2010,** *30*(1), 65–75.

Rogers, E.M. *Diffusion of Innovations,* 3rd ed.; The Free Press: New York, 1983.

Rueckert, R.W.; Walker, O.C.; Roering, K.J. The Organization of Marketing Activities: A Contingency Theory of Structure and Performance. *J. Market.* **1985,** *49*(1), 13–25.

Schumpeter, A. *Capitalism, Socialism and Democracy*, 2nd ed.; Harper & Row Publishers: New York, 1947.

Storey, D. J. *Understanding the Small Business Sector*, Routledge: London, 1994.

Strauss, A.; Corbin, J. *Basics of Qualitative Research: Grounded Theory Procedures and Techniques.* Sage Publications: Thousand Oaks, 1998.

Sritama, S. (2018). Why Are so Many Hotels for Sale? https://www.bangkokpost.com/business/1590886/why-are-so-many-hotels-for-sale-.

Sulistiyani, R.; Harwiki, W. How SMEs Build Innovation Capacity Based On Knowledge Sharing Behavior? Phenomenological Approach. *Procedia–Social Behav. Sci.* **2015,** *219* (2016) 741–747.

Sundbo, J.; Gallouj, F.*Innovation in Services*, Tourism Satellite Accounting Research World; World Travel and Tourism Council: London, 1998.

Sutthijakra, S. Managing Service Subsidiaries Through an Innovation Perspective: A Case of Standard Interpretation in Multinational Hotels. *Service Indus. J.* **2011,** *31*(4), 545–558.

Talty, A. 2019 Bangkok Is the Most Visited City In The World...Again. https://www.forbes.com/sites/alexandratalty/2019/09/04/bangkok-is-the-most-visited-city-in-the-world-again/#209b2c185f1b

Vargo, S.; Lusch, R. Evolving to a New Dominant Logic for Marketing. *J. Market.* **2004,** *68*(1), 1–17.

Vanhalweyk, G. 2019 Tourism Statistics Thailand 2000-2019. www.thaiwebsites.com/tourism.asp. (accessed Oct 7, 2019.

Verma, R.; Anderson, C.; Dixon, M.; Enz, C.; Thompson, G.; Victorino, L. In *Key Elements in Service Innovation: Insights for The Hospitality Industry*, Cornell University Roundtable Proceedings, No. 1 (November), Ithaca, NY: Center for Hospitality Research Report, Cornell University.

Vila, M.; Enz, C.; Costa, G. Innovative Practices in the Spanish Hotel Industry. *Cornell Hospitality Quarterly,* **2012,** *51*(1), 75–85.

Wooder, S.; Baker, S. Extracting Key Lessons in Service Innovation. *J. Prod. Innov. Manag.* **2011,** *29*(1), 13–20.

CHAPTER 8

THE DEVELOPMENT OF THE VIETNAMESE PARK SYSTEM: MANAGEMENT MODEL AND CARRYING CAPACITY

TUAN PHONG LY[1*] and THI HONG HAI NGUYEN[2]

[1]*Macao Institute for Tourism Studies, Macao, China*

[2]*University of Greenwich, United Kingdom*

[*]*Corresponding author. E-mail: jack@iftm.edu.mo*

ABSTRACT

With beautiful scenic and rich biodiversity, national parks (NPs) have been included in the national strategy for tourism development, especially ecotourism in Vietnam. Its management, however, has shown weaknesses and hence merits attention in the light of sustainable development. Utilizing proper management model and implementing carrying capacity are prerequisites for NP striving to meet the triple mandates of its tourism business management (i.e., recreation, conservation, and economic pursue). This chapter provides an overview of Vietnamese NP system, focusing on its development history as well as the current state of tourism operations. It reviews the transition of Vietnamese park management models in different periods. At the same time, it investigates the current application of carrying capacity management in different parks. Findings show that Vietnamese park system continues to transform into different management models, from parastatal management to public and for-profit management. There are three different approaches to the implementation of carrying capacity in Vietnamese parks, including (1) a laisser-faire approach; (2) a relaxed approach; and (3) a strict approach. At such, this chapter analyzes the rationales leading to current development trend in Vietnamese park system.

It contributes a theoretical explanation of the choice of park management model and carrying capacity application in Vietnamese context. Finally, this chapter discusses its practical implications to tourism management in NPs in Vietnam as well as other developing countries.

8.1 INTRODUCTION

Vietnam's tourism industry has started from the 60s of the last century, yet it has really grown and started to play an important role in the economy since the 1990s (Nguyen, 2011). Since then, its tourism industry has achieved remarkable results with over 15 million international arrivals and 80 million domestic visitors in 2018 (VNAT, 2019). Heritage sites and national parks (NPs) are among the best tourist attractions in Vietnam. With beautiful scenic and rich biodiversity, NPs have been included in the national strategy for tourism development, especially ecotourism in Vietnam. Its management, however, has shown weaknesses and hence merits attention in the light of sustainable development. Utilizing proper management model and implementing carrying capacity are prerequisites for NP striving to meet the triple mandates of its tourism business management, including recreation, conservation, and economic pursue (Eagles and Hallo, 2013; Ly and Nguyen, 2017; Ly and Xiao, 2016a). Thus, this chapter provides an overview of Vietnamese NP system, focusing on its development history as well as the current state of tourism operations. It also reviews the transition of Vietnamese park management models in different periods and discusses the current application of carrying capacity management in various NPs in the country.

8.2 THE DEVELOPMENT OF VIETNAM TOURISM AND HOSPITALITY

Home to diverse natural and cultural attractions, Vietnam provides appeal for a wide range of travelers. Vietnam is rising in its popularity as a tourist destination in South-East Asia. In 2018, the number of international arrivals reached over 15 million people, representing an increase of 19.9% compared to the previous year (VNAT, 2019). The direct contribution of travel and tourism to GDP was VND294,660bn (USD12,965.5mn), 5.9% of total GDP in 2017, and forecasts to rise by 6.0% pa, from 2018 to 2028 (World Travel and Tourism Council, 2018). Vietnam is considered to be one of the ASEAN's most attractive markets for foreign investors with a steadily increasing GDP and booming foreign direct investment, a pattern that is forecasted to remain stable in the upcoming years (EVBN, 2018). According to Vietnam Investment Review (2018), Vietnam has experienced unprecedented growth in the hospitality sector in recent years, with an average occupancy rate of nearly 60%. At some popular tourist destinations like Ho Chi Minh, Ha Noi, Hoi An, Da Nang, Hue, Nha Trang, and Phu Quoc, the number of accommodation facilities has surged in the last few years.

It takes a long way for Vietnam to develop its tourism and hospitality reaching the current stage. The development of Vietnam's tourism can be divided into three key periods (Truong, 2012). The first period (1960–1975), tourism was served as a political tool (Thanh, 2005). During that time, tourism was only meant for Government's guests and other special invites and only took place in the North of Vietnam. The second period (1976–1990), tourism begun to develop and became an economic sector (Cooper, 2000; Thanh, 2005). Since the 1990s, tourism has been considered as a vital tool of economic growth and poverty alleviation (The Government of Vietnam, 2005). It was credited that the success of Vietnam's economy in general and its tourism and hospitality industry in particular, was due to a significant transfer from state monopoly to joint-venture businesses, named as *doi moi* (renovation) policy (Cooper, 2000). The reform in 1986 started to introduce "open door" policy that encouraged a shift from a socialist-oriented to a market-driven economy. This transfer was most obvious in the hospitality sector. Before 1986, most state-owned hotels were styled and equipped with standard facilities and unskilled staff (Mok and Lam, 1997). After the law on foreign investment was applied in 1987, joint-venture hotels have started to emerge. In general, the *doi moi* policy has called for the decentralization of Vietnam's tourism. As in a dynamic and complex setting, conventional management structures and roles based on a centralized and hierarchical authority are no longer believed to be adequate to achieve the sustainable development of tourism industry. This transformation also made a great impact on the management model in Vietnamese NPs.

8.3 VIETNAMESE NATIONAL PARKS: THE TRIPLE MANDATES OF PARK MANAGEMENT AND THE CHANGE UNDER *DOI MOI* POLICY

While the tourism industry shows no signs of decrease, the increasing concerns for the environment are likely to spur tourists' desires to visit impressive landscapes in NPs and other protected areas (Beunen et al., 2008;

Mowforth and Munt, 2003). Nature-based tourism or ecotourism is a large and growing global industry, partially dependent upon the attributes of the natural environment, and especially occurring within NPs (Eagles, 2002; Huang, 2008; Ly and Xiao, 2016a). In the emerging destination like Vietnam with its rich natural resources, nature-based tourism or ecotourism have also experienced tremendous growth (Ly and Nguyen, 2017; Suntikul et al., 2010; The Government of Vietnam, 2011). In response, the Vietnamese government has acknowledged ecotourism to be one of the country's key tourism products, and many resources and policies are allocated to the development of NPs (Suntikul, 2010; The Government of Vietnam, 2010).

The increasing popularity of NP brings both opportunities and challenges. The opportunity is to provide proper recreational level, while the inherent challenge is to conserve park resources for the enjoyment of future generations (Plummer, 2009). Unlike developed countries with its recereation and conservation mandates on park management (Pigram and Jenkins, 2006), Ma et al. (2009) argue that developing countries need to consider one more mandate, the *economic pursue*. Sustaining the triple mandate appears to be most challenging in developing countries as economic value may overshadow ecological conservation (Pigram and Jenkins, 2006). As a consequence, park management has found it difficult to balance visitor recreational use, with conservational and economic objectives.

The *doi moi* policy has called for the decentralization toward Vietnamese parks. In a dynamic and multifaceted milieu, traditional centralized management structures are no longer believed to be adequate to attain the triple mandates of park management (Abrams et al., 2003; Lockwood, 2010). Uniformity is no longer the core in order to serve the best interest of the tourism industry or its stakeholders. The Vietnamese parks continually search for and apply new tourism policies for adaptive management. The decentralization of Vietnamese parks have experienced two main periods.

The *first* one referred to *deconcentration* or *administrative decentralization* (Ribot, 2002), while central government has been hanging over part of its governance power to provinces, without the joining of any private sector in park management system (The Government of Vietnam, 2006). There are two types of NPs: (1) the *cross-provincial* parks, under the governance of the Ministry of Agriculture and Rural Development (MARD) and (2) the *within-provincial* parks, under the administration of every Provincial People's Committee (PPC). Among 33 NPs in Vietnam, four of them are under the governance of the MARD while the rest belong to the latter category. The Vietnamese park system is considered as a lack of management experiences in tourism development, especially after the decentralization of the power to the lower levels of the government, such as those parks managed by PPCs. Hence, the latter has even less management experience than the former (Crewell and Maclaren, 2000; Ly and Xiao, 2016a; Phan et al., 2002; Suntikul et al., 2010).

Under this circumstance, Vietnamese parks have entered their *second* period of decentralization, referred as *delegation* (Ribot, 2002), with the joining of private sector into park tourism management (The Government of Vietnam, 2010). Delegation means the transfer of management power to semi-independent political units, which means that a park is operating by a government-owned company and includes the participation of private sector (Larson, 2002). In 2006, the Vietnamese Special-Use Forest (SUF) policy introduced a coexisting management, entailing the distribution of power between the public and private sectors (The Government of Vietnam, 2006). Then, it was updated and redeveloped in 2007, 2010, and 2011, respectively (MARD, 2007, 2011; The Government of Vietnam, 2010). Accordingly, the National Park Management Board (NPMB) is designated as the forest owner with rights to manage tourism-related activities within a park under the following modes. First, the NPMB and its public (*state management*) tourism institution can directly manage tourism activities. Second, the NPMB can lease out forests to the private sector for operating tourism and recreation activities (*private management*). Third, public and private sector bodies can engage in joint management and association of tourism activities (*joint ventures*).

8.4 THE TRANSITION OF VIETNAMESE PARK MANAGEMENT MODELS

Since the SUF tourism management policy was introduced in 2006, the Vietnamese park system has witnessed a change from a *parastatal management model*, which can be defined as a corporation owned or fully controlled by the government, to a *public and for-profit model*, which includes other bodies apart from government agencies in park tourism management (Eagles, 2009, More, 2005). Many approaches are used to distinguish park management models (Buckley, 2002; Eagles, 2009; Glover and Burton, 1998; More, 2005). Eagles (2009) recommends analyses of management models by investigating their three aspects: (1) ownership of the land and related resources, (2) source of income, and (3) types of management bodies. Table 8.1 showed the differences between two different park management models in Vietnam at the moment. This development is a core step toward decentralization in the SUF system from deconcentration to delegation periods (Ribot, 2002).

TABLE 8.1 Differences Between Parastatal Model and Public and For-Profit Model.

Model functional aspects	Parastatal Model in deconcentration period	Public and for-profit model in delegation period
1. Ownership of land-related resources	Government agency	Government agency
2. Income sources	User fees and charges	User fees and charges; forest-renting fee
3. Management bodies	Parastatal corporation owned or wholly controlled by the government	Combinations of any management modes (state management, private management, and joint ventures)

Since the announcement of the coexisting/public and for-profit model (Eagles, 2009) in 2006, its first and only application has been found in Phong Nha-Ke Bang National Park (PNKB NP) in Quang Binh Province, Vietnam in 2010 (Ly and Xiao, 2016a, 2016b). Such public–private concession, which determines the best alternative for fulfilling the triple mandate of park management are not novel to park tourism development (Su and Xiao, 2009). This section explains how the Vietnamese park system has ended up choosing the coexisting model for new park management with evidence from PNKB NP (Fig. 8.1).

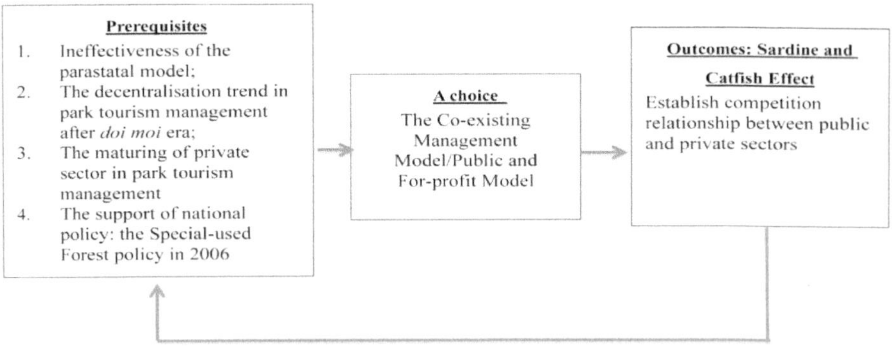

FIGURE 8.1 The choice of Vietnamese Park Management Model.

Source: Ly and Xiao (2016a).

There are four prerequisites leading to the choice of Vietnamese park system. The first is the ineffectiveness of the parastatal model in reaching the triple mandate of park management. Suntikul (2010) and Suntikul et al. (2010) reported that the centralized governance of public sector showed inability to reach park management sustainability, especially in the conservation aspect. Therefore, the Vietnamese park system keeps an open attitude toward park management model application. The second prerequisite relates to the decentralization trend of park management in the last decades in general (Abrams et al., 2003; Lockwood, 2010) and after the *doi moi* era of Vietnam in particular. The decentralization trend of park management is matching with political milieu of Vietnam in promoting the polycentric regime. The country has recognized that a multicomponent economic model is better than the monopoly state-management model (Ly and Xiao, 2016b). At the same time, the positive rise of private sector in Vietnamese tourism industry also brings the basis for coexisting model development. The central government has witnessed good economic efficiency in private-management units within tourism businesses. Moreover, some private sectors have showed interests to park tourism management and requested for investment in park tourism businesses. Under this circumstance, the last important prerequisite is the announcement of the SUF policy in 2006, which officially allowed concurrent management of tourism sites between public and private sectors within one park.

The core strategy of the new model is the coexistence of public and private sectors in establishing *coopetation* (i.e., cooperation and competition) for park management. Given the involvement of private management in park tourism businesses, the new model brings about changes in state management of NPs. In the case of PNKB NP, it is a within-provincial park governed by Provincial People's Committee of Quang Binh (PPC of QB). However, daily park operation and management issues are direct responsibilities of the PNKB's NPMB, which is under the authority of PPC of QB. The PNKB's NPMB governs a tourism management unit called the Phong Nha Tourism Centre (PNTC) to oversee tourism activities in the park under the state-management typology. Truong Thinh Group is a local company in Quang Binh Province. In 2010, the PPC of QB has allowed the group to invest, operate, and manage the Paradise Cave within the park under a 50-year forest renting contract. This cooperative form can be considered as a format of the private management, as shown in Figure 8.2 below.

The major contribution of the new model is the addition of the *competitive element* to park tourism management, which cannot be found in any previous models. The integration of private management under the group has brought some positive impacts to park tourism management. The strong private sector (*a catfish*) starts to keep the public sectors (*the sardines*) more active to avoid being eliminated by the free market. The Vietnamese government seems to have made a right choice to keep the catfish in the same tanks with the sardines for the improvement of park management effectiveness.

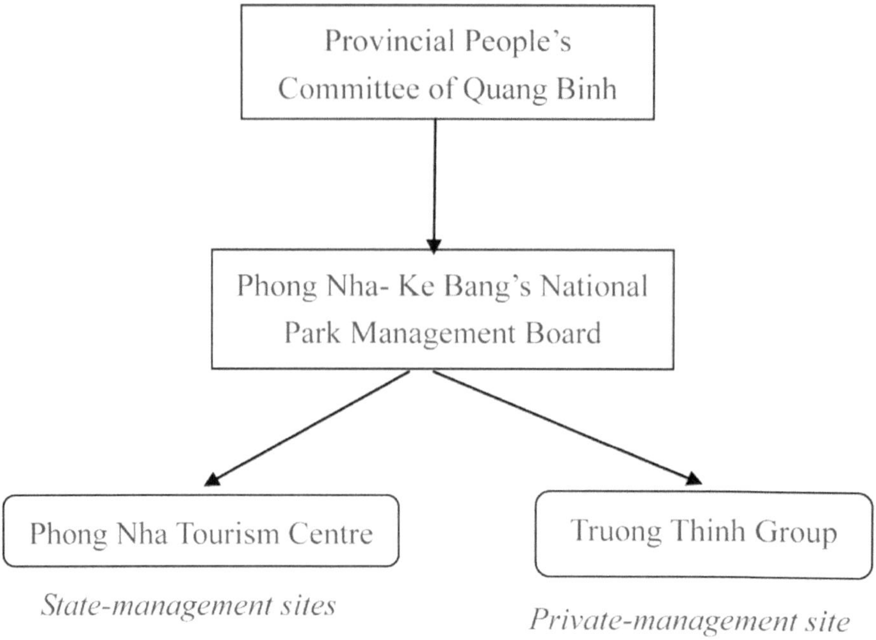

FIGURE 8.2 The coexisting management model of Phong Nha-Ke Bang National Park.

8.5 THE APPLICATION OF CARRYING CAPACITY MANAGEMENT IN VIETNAMESE PARKS

Apart from using the proper park management model, implementing carrying capacity is another important condition for parks to meet the triple mandates of park management. The World Tourism Organisation defines carrying capacity as "the level of visitors' use an area can accommodate" (Buckley, 1999, p 706). The literature on carrying capacity has blossomed since the 1970s and has been developed extensively in recent decades (Manning et al., 2000; Prato, 2009). However, its application to parks is quite limited and has sometimes resulted in failure, even in developed countries (Manning, 2001). Lindberg et al. (1997) suggested that carrying capacity remains a good idea to propose, but a political nightmare to implement. The main difficulty lies in determining how much impact is too much, or how much impact should be allowed before park management should intervene into park tourism capacity (i.e., *the intervention point*) (Manning, 2001). Park management can decide on the subjective/objective extent of intervention point in park capacity management, which is ranged from extremely subjective to extremely objective, as shown in the Figure 8.3 below. Literature has proved that both extremes are either outdated or difficult to practice in reality. What park management can do is keeping visitor capacity within these two extremes and revise the management by objectives (MBO) approach with proper management objectives, indicators, and standards for sustainability which suitable for each park context (Manning, 2001).

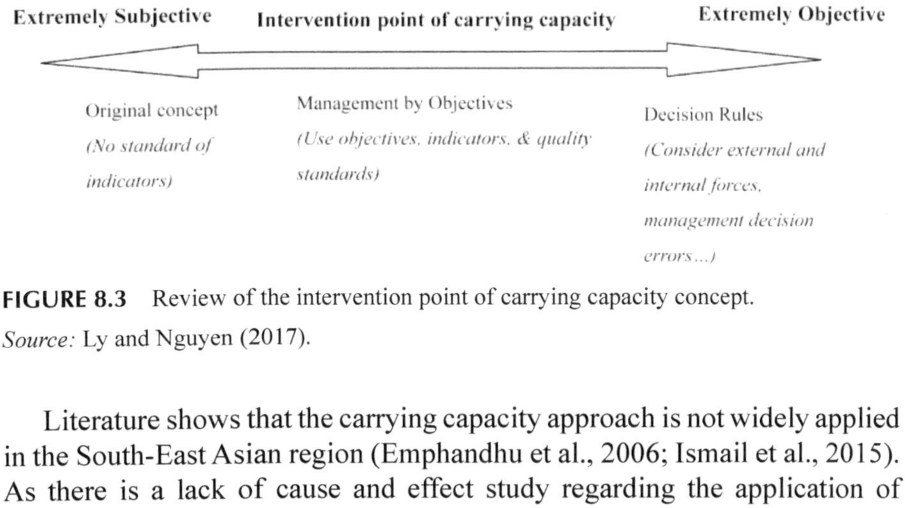

FIGURE 8.3 Review of the intervention point of carrying capacity concept.
Source: Ly and Nguyen (2017).

Literature shows that the carrying capacity approach is not widely applied in the South-East Asian region (Emphandhu et al., 2006; Ismail et al., 2015). As there is a lack of cause and effect study regarding the application of carrying capacity management in a whole NP system within one country. Therefore, it is not easy to stimulate park management to implement this controversial concept (Posavac, 2011). This section investigates the current situation of carrying capacity approaches applied to the Vietnamese park system and assesses its application process for further development.

The concept of carrying capacity has been mentioned in Vietnamese park system management in 2007. MARD (2007, p 5) has required that authorities and individuals who are operating ecotourism/nature-based tourism activities for NPs must "based on physical, ecological, landscape, and social impact assessment of ecotourism activities on parks and protected areas, to regulate the maximum visitors per day/per site." This policy, however, includes no follow-up guidelines for further implementation of the concept. The application of carrying capacity concept therefore opens for the interpretation of park managers in the country. Therefore, it is normal to witness that within the same country, the intervention point of park managers toward visitor capacity are different. In the study of Vietnam park system by Ly and Nguyen (2017), three approaches to the implementation of carrying capacity were found, including (1) *laissez-faire* approach, (2) *relaxed* approach, and (3) *strict* approach.

Among these approaches, the first one, that is, *laissez-faire approach*, is found to be the most popular in the Vietnamese parks, applying in Ba Vi, Cuc Phuong, Nam Cat Tien, Bach Ma NPs. This approach does not have any regulations regarding carrying capacity management. NPMBs provide no guidelines and communications to tourists after entering the park, even during peak visiting days. When overcrowding happened, the NPMBs make decisions on a case-by-case basis to address the issues. For example, in February 2015 there were approximately 5000 people coming to Ba Vi NP at sunset due to rare snowy weather in the region. The park manager decided to stop selling tickets and close the entrance gate to protect visitors from any unseen danger. This intervention was highly subjective, passive, and made on a case-by-case basis. This approach indicates the typical management style and attitude of top-down management culture in Vietnam. Park management will not take any action while there is no official guideline from the central government. The biggest disadvantage of this passive governance is that it is difficult to reach triple mandates of park management.

The second approach is the *relaxed approach*, which is being applied at PNKB NP. As abovementioned, this park has adopted the most up-to-date coexisting management model, which involves a collaboration of public and private sectors. Both sectors claimed that the park did not regularly apply any carrying capacity tools for managing tourist arrivals. The NPMB only interferes in rare cases when they "feel" overcrowding in peak visiting days, or when tourists visit some signature tours inside the strictly protected area of a park. For example, at the time of this study taken place, Hang En Cave was limited for 16 people to visit as a time, or Son Doong Cave was 10 people for each visit. However, a management issue was found, the NPMB were not aware of how the intervention point was determined. All numbers were decided and guided by the consultant institution, the British Caving Association in this case. The park managers actually did not actually know how to calculate those numbers by themselves.

Park managers provided further explanation regarding the fact that not many parks had been applied carrying capacity in their operation management. Although national policies declare that parks should only offer ecotourism (MARD, 2007; The Government of Vietnam, 2010), the main source of tourists visiting parks is in form of mass tourism. At the same time, each park management board is put under pressure from the Central Financial Department to meet annual targets of tax revenues and number of tourists. Therefore, it is reasonable that most parks are focused on meeting these targets, rather than limited tourist capacity. This situation demonstrates Pigram and Jenkins (2006)'s argument that economic considerations might be surpassed ecological preservation.

The *strict approach* is suggested to be the ideal method to obtain the triple mandates of park management, as the intervention point is rather objective. This approach is taken by Con Dao NP at the moment. The decision on the intervention point is considered the reactions of the ecosystem toward park management decisions (Prato, 2009). Although this park is managed under the parastatal model, the public sector has actively extended and applied the government's policy on carrying capacity in all park tourism activities and sites, including nice land sites and eight underwater snorkeling sites (Con Dao National Park Management Board, 2016). The initiative of Con Dao NP suggested that other parks in Vietnam could take more action toward implementation of carrying capacity policy. The problem seems to lie in the will to change of other NPMBs.

8.6 CONCLUSION

The Vietnamese park system continues to transform into different management models in order to find a balance for the triple mandates of park management. Since the introduction of SUF policy in 2006, the Vietnamese parks have witnessed the existence of two management models simultaneously, which are the *parastatal model* and the *coexisting model*. Most of Vietnamese parks are still applied the parastatal model to govern park tourism activities now, only PNKB NP is applied the new model. This new model in the Vietnamese context is one version of the public and for-profit model (Eagles, 2009). The core and innovative strategy of the new model is the addition of the competitive element to park tourism management between public and private sectors. To some extent, people witnessed some benefits of the new model toward park management development in the case of PNKB NP (Ly and Xiao, 2016a, 2016b). Although Vietnamese national policy is calling for decentralization strategy in park management, it is hard to change people's top-down management culture within a short time. It explained why the majority park management boards are applied the old model, while holding a hesitation attitude toward the application of the new model. The public–private cooperation participation is a promised model; however, it is rarely applied in Vietnamese park system. Therefore, the researchers are calling for more application of the coexisting management model to witness the full functions of it.

This chapter also provides a nationwide picture of the current implementation of carrying capacity situation in Vietnamese parks. Three approaches have been adopted to manage carrying capacity, including (1) a *laissez-faire* approach, means zero regulation from local authorities, (2) a *relaxed* approach, means irregular polices for different cases and for different tourist sites, and (3) a *strict* approach, means preestablished rules and norms for all tourist sites. On the one hand, NP management boards in Vietnam seems to hold an alternative attitude toward carrying capacity application, as there is not big overcrowd situation being reported that make severe consequence to park's nature and landscape. On the other hand, it is indicated that park management model has no direct influence on the application of carrying capacity and that the public and for-profit model is not always a promise for best practice in carrying capacity implementation. While the parastatal model seems not always support (e.g., Con Dao NP) or concern (e.g., Ba Vi and Cuc Phuong NPs) this issue. The extent to which the carrying capacity tool is applied depends on the willingness to change of Vietnamese park management boards. Moreover, this chapter also indicates that economic constraints make it harder to implement carrying capacity management in the developing countries. This information on park management model and particular focus on carrying capacity of Vietnamese NPs is useful reference for other developing countries, especially those in South-East Asia. However, it might be useful to remind that analytical generalization in alignment with country context should be keep in mind when making park management decision.

KEYWORDS

- **national park**
- **park management model**
- **carrying capacity**
- **sustainable tourism**
- **Vietnam**

REFERENCES

Abrams, P.; Borrini-Feyerabend, G.; Gardner, J.; Heylings, P. *Evaluating Governance: A Handbook to Accompany a Participatory Process for a Protected Area*. Parks Canada: Ottawa, 2003.

Beunen, R.; Regnerus, H. D.; Jaarsma, C. F. Gateways as a Means of Visitor Management in National Parks and Protected Areas. *Tour. Manag.* **2008,** *29*(1), 138–145.

Buckley, R. An Ecological Perspective on Carrying Capacity. *Ann. Tour. Res.* **1999,** *26*(3), 705–708.

Buckley, R. Public and Private Partnerships Between Tourism and Protected Areas: The Australian Situation. *J. Tour. Studies* **2002,** *13*, 26–38.

Con Dao National Park Management Board. *Quyết định ban hành Quy định về sức chứa của môi trường tại các tuyến, điểm tham quan, du lịch sinh thái trong phạm vi quản lý của Vườn quốc gia Côn Đảo* [Decision regarding Regulations on environmental carrying capacity management for all tours and ecotourism sites within Con Dao National Park]. Ba Ria-Vung Tau, Vietnam: Con Dao National Park Management Board, 2016.

Cooper, M. Tourism in Vietnam: doi moi and the Realities of Tourism in the 1990s. In *Tourism in South and Southeast Asia: Issues and Cases*; Hall, C. M., Page, S., Eds.; Butterworth-Heinemann: Oxford, 2000; pp 167–177.

Creswell, C.; Maclaren, F. Tourism and National Parks in Emerging Tourism Countries. In *Tourism and National Parks: Issues and Implications*; Boyd, S., Butler, R., Eds.; Wiley: Chichester, 2000, pp 281–299.

Eagles, P. F. J. Governance of Recreation and Tourism Partnerships in Parks and Protected Areas. *J. Sustain. Tour.* **2009,** *17*(2), 231–248.

Eagles, P. F. J.; Hallo, J. C. Parks and Protected Areas in Canada and the United States. In *Introduction to Recreation and Leisure*, 2nd ed.; Kinetics, H., Ed.; Human Kinetics: Champaign, IL, 2013; pp 93–126.

Eagles, P. F. J.; McCool, S. *Tourism in National Parks and Protected Areas: Planning and Management*; CABI: New York, NY, 2002.

Emphandhu, D.; Yemin, T.; Pattanakiat, S.; Tantasirin, C.; Ruschano, R.; Chettamart, S.; Nasa, M. *Recreation Carrying Capacity Analysis at Khao Leam Ya Mu Ko Samed National Park, Thailand*. Paper Presented at the Third International Conference on Monitoring and Management of Visitor Flows in Recreational and Protected Areas, Rapperswil, Switzerland, September 2006.

EVBN, 2018. *Vietnam Hospitality Report: Hospitality Sector (Edition 2018)*. https://evbn.org/vietnam-hospitality-report/

Glover, T.; Burton, T. A Model of Alternative Forms of Public Leisure Services Delivery. In *Leisure Management: Issues and Applications*; Collins, M., Cooper, I., Eds.; London: CABI, 1998; pp 139–155.

Huang, X.中國國家公園：自然遊憩資源管理比較研究 [The Comparison Research of Nature-Based Recreation Resource Management Between Canada and China's National Parks]. China Travel and Tourism Press: Beijing, 2008.

Ismail, M. M.; Jaafar, M.; Mohamad, D. Social Carrying Capacity of Ecotourism National Park: Case Study of National Park Kual Tahan, Malaysia. *J. Malays. Inst. Plan.* **2015,** *8*, 1–10.

Larson, A. Natural Resources and Decentralization in Nicaragua: Are Local Governments up to the Job? *World Develop.* **2002,** *30*(1), 17–31.

Lindberg, K.; McCool, S.; Stankey, G. Rethinking Carrying Capacity. *Ann. Tour. Res.* **1997,** *24*(2), 461–465.

Lockwood, M. Good Governance for Terrestrial Protected Areas: A Framework, Principles, and Performance Outcomes. *J. Environ. Manag.* **2010,** *91*, 754–766.

Ly, T. P.; Nguyen, T. H. H. Application of Carrying Capacity Management in Vietnamese National Parks. *Asia Pac. J. Tour. Res.* **2017,** *22*(10), 1005–1020.

Ly, T. P.; Xiao, H. The Choice of Park Management Model: A Case Study of Phong Nha-Ke Bang National Park in Vietnam. *Tour. Manag. Perspect.* **2016a,** *17*, 1–15.

Ly, T. P.; Xiao, H. An Innovative Model of Park Governance: Evidence from Vietnam. *J. Ecotour.* **2016b,** *15*(2), 99–121.

Ma, X. L.; Ryan, C.; Bao, J. G. Chinese National Parks: Differences, Resource Use and Tourism Product Portfolios. *Tour. Manag.* **2009,** *30*(1), 21–30.

Manning, R. E. Visitor Experience and Resource Protection: A Framework for Managing the Carrying Capacity of National Parks. *J. Park Recreat. Admin.* **2001,** *19*(1), 93–108.

Manning, R. E.; Valliere, W.; Minteer, B.; Wang, B.; Jacobi, C. Crowding in Parks and Outdoor Recreation: A Theoretical, Empirical, and Managerial Analysis. *J. Park Recreat. Admin.* **2000,** *18*(4), 57–72.

MARD. *Ban hành quy chế Số 104/2007/QĐ-BNN quản lý các hoạt động du lịch sinh thái tại các vườn quốc gia, khu bảo tồn thiên nhiên* [Regulations No.104/2007/QĐ-BNN on management of ecotourism activities in national park and nature reserves]. Hanoi: Author, 2007.

MARD. *Thông tư Số 78/2011/TT-BNNPTNT quy định chi tiết thi hành Nghị định Số 117/2010/NĐ-CP ngày 24/12/2010 của Chính Phủ về tổ chức và quản lý hệ thống rừng đặc dụng* [Circular No. 78/2011/TT-BNNPTNT on guide the implementation of the Decree No. 117/2010/NĐ-CP on organization and management of special use forest system]. Hanoi: Author, 2011.

Mok, C.; Lam, T. Hotel and Tourism Development in Vietnam. *J. Travel Tour. Mark.* **1997,** *7*(1), 85–91.

More, T. From Public to Private: Five Concepts of Park Management and their Consequences. *George Wright Forum* **2005,** *22*(2), 12–20.

Mowforth, M.; Munt, I. *Tourism and Sustainability: New Tourism in the Third World*, 2nd ed.; Taylor & Francis: London, 2003.

Nguyen, T. H. H. Hue Cultural Heritage Tourism: Heritage Tourists and Length of Stay (Master Dissertation). Breda University of Applied Sciences, Breda, the Netherlands, 2011.

Phan, N.; Quan, T.; Le, K. *Ecotourism in Vietnam: Potential and Reality*, 2002. https://kyotoreview.org/issue-2-disaster-and-rehabilitation/ecotourism-in-vietnam-potential-and-reality/

Pigram, J.; Jenkins, J. *Outdoor Recreation Management,* 2nd ed.; Routledge: Abingdon, 2006.

Plummer, R. *Outdoor Recreation: An Introduction*. New York, NY: Routledge, 2009.

Posavac, E. *Programme Evaluation: Methods and Case Studies*, 8th ed.; Prentice Hall: Upper Saddle River, NJ, 2011.

Prato, T. Fuzzy Adaptive Management of Social and Ecological Carrying Capacities for Protected Areas. *J. Environ. Manag.* **2009,** *90*(8), 2551–2557.

Ribot, J. *Democratic Decentralization of Natural Resources: Institutionalising Popular Participation*; World Resource Institute: Washington DC, 2002.

Su, D.; Xiao, H. The Governance of Nature-Based Tourism in China: Issues and Research Perspectives. *J. China Tour. Res.* **2009,** *5*, 318–338.

Suntikul, W. Tourism and Political Transition in Reform-Era Vietnam. In *Tourism and Political Change;* Butler, R., Suntikul, W., Eds.;. Goodfellow: Oxford, 2010; pp 133–143.

Suntikul, W.; Butler, R.; Airey, D. Implications of Political Change on National Park Operations: doi moi and Tourism to Vietnam's National Parks. *J. Ecotour.* **2010,** *9*(3), 201–218.
The Government of Vietnam. *Chiến lược quốc gia về phát triển du lịch năm 2020, tầm nhìn 2030* [National Strategy for Tourism Development to 2020, Vision 2030]. Hanoi: Author, 2011.
The Government of Vietnam. *Luật du lịch* [Law on Tourism]. Hanoi: Author, 2005.
The Government of Vietnam. *Nghị định Số 23/2006/NĐ-CP về việc thực hiện các quy định của pháp luật về bảo vệ và phát triển rừng* [Decree No. 23/2006/NĐ-CP on the implementation of the law on forest protection and development]. Hanoi: Author, 2006.
The Government of Vietnam. *Nghị định số 117/2010/NĐ-CP về tổ chức và quản ly hệ thống rừng đặc dụng* [Decree No. 117/2010/NĐ-CP on organisation and management of the Special-use Forest system]. Hanoi: Author, 2010.
Truong. V. D. Tourism Policy Development in Vietnam: A Pro-Poor Perspective. *J. Policy Res. Tour. Leis. Events* **2012,** *5*(1), 28–45.
Vietnam Investment Review, 2018. *Changing Face of Vietnam Hospitality*. https://www.vir.com.vn/changing-face-of-vietnam-hospitality-59073.html.
VNAT, 2019. *International Visitors to Vietnam in December and 12 Months of 2018*. http://vietnamtourism.gov.vn/english/index.php/items/13551.
World Travel and Tourism Council, 2018. *Travel and Tourism Economic Impact 2018 Vietnam*. https://www.wttc.org/-/media/files/reports/economic-impact-research/countries-2018/vietnam2018.pdf.

CHAPTER 9

SUSTAINABLE TOURISM AND COMMUNITY DEVELOPMENT IN MALAYSIA

NEETHIAHNANTHAN ARI RAGAVAN and
PUVANESWARAN KUNASEKARAN*

Faculty of Social Sciences and Leisure Management,
Taylor's University, Malaysia

**Corresponding author.*
E-mail: puvaneswaran.kunasekaran@taylors.edu.my

ABSTRACT

This chapter explores tourism and community development in Malaysia toward sustainable tourism practice. Tourism development is closely aligned with the aim to eradicate poverty, rural community development, reduce the disparity between the rich and poor, leading to national cohesion and cultural commodification. Community-based tourism (CBT) is a type of rural tourism usually managed by communities and intended to deliver wider community benefit. Currently, studies on sustainable tourism are heavily dictated by economic and environmental aspects, ignoring the significance of sociocultural perspective. The local people are strongly attached to their culture and they believe that tourism helps them not only to protect their culture but also to learn foreign culture from tourists. "Outsiders" through various developmental initiatives have been always guiding the local communities to engage in tourism business effectively. However, the generally used top-down approach has resulted in forced involvement of the community in tourism development in their areas. Thus, the issue of sustainable tourism practice is still questionable when the planning and implementation of development projects are mainly influenced by the outsiders. This chapter specifically presents the importance of community-based initiatives toward sustainable tourism in Malaysia. This unique contribution will be a significant addition to the body of knowledge in the sustainable tourism aspect. In a nutshell, CBT can be considered as a desired form of tourism toward achieving sustainable development goals (SDGs).

9.1 TOURISM INDUSTRY

According to Presenza et al. (2019) tourism is one of the largest and fastest growing industries in the world and it is a social phenomenon of major importance. Due to this positive expansion, tourism has achieved a position, where it can bring various advantages to the people involved and the particular country where it is to the people involved and the particular country where it is developed. According to Theobald (2005), the tourism industry expanded massively in the 1960s, has grown tremendously by uplifting the economy and community in the system. Kayat (2011) stated that global tourism began to grow exponentially since the 1950s due to stimulation by interrelated "push" factors, which originated mostly from individual and social needs and demand, as well as the "pull" factors stimulated by tourist destinations worldwide.

The development of tourism in the contemporary era was elevated by the significant growth in tourist arrivals internationally. Kunasekaran (2013) noted that with growing economic activity in the emerging markets, the tourism industry provides an important opportunity for developing countries to move up the value chain toward the production of innovative tourism products and quoted UN Secretary-General Ban Ki-moon (2011) who stated "at a time of profound global economic uncertainty, tourism's ability to generate socio-economic opportunities and help reduce the gap between rich and poor is more important than ever."

Kayat (2011) cited Mason (2003) who suggested five major reasons for the growth of travel at the end of the 19th century; improved standard of living due to a rise in industrial output associated with the Industrial Revolution, cheaper and more accessible travel due improvements in transportation, the introduction of annual holidays, changing perceptions of the environment, and an increasing desire to travel due to education. Despite the "shocks" that the industry went through in the past 30 years due to exogenous factors such as war, terrorism, recession, economic downturn, SARS, Bird flu, and Tsunami, it has shown to be strong and resilient. Tourism is an important tool for countries' economic development and growth, and is a catalyst for creating employment and poverty alleviation.

As global travel increases every year, tourism development has become an important developmental agenda for countries throughout the world. The main reason for this tourism development is that tourism is seen as a business that provides economic assistance to a community. She also highlighted that, however, tourism has the potential to generate harmful effects that are costly to the community. She stressed rightly that as such, concerted determinations must be made on how these developments should look like, as well as to see who should benefit the most, who should benefit the least, and who and what should change.

Kayat (2011) concluded that in aiming for growth in tourism, which is always measured by tourist arrivals or tourist spending from year to year, many nations worldwide concentrate their resources and efforts in developing tourism as an important economic sector for their countries. She reiterated that tourism should also be developed as a means to enhance the living standards of the nations' population and to enrich the lives of both the hosts and the guests. According to Fridgen (1990), tourism development is the long-term process of preparing for the arrival of tourists. It involves planning, building, and managing the attractions, transportations, accommodations, services, and facilities that serve the tourist, while taking into consideration all of the factors that may influence travel to a specific destination.

Thus, if managed strategically, tourism development efforts usually result in the growth of the number of tourists to a specifically developed destination, unless something that is beyond the control of the management takes place that actually hinders the tourists from visiting that destination, as what had happened during the recent cases of natural disasters or terrorist attacks. Tourists are defined as "people who travel to, and stay in, places outside their usual environment for not more than one consecutive year, for leisure, business and other purposes not related to the exercise of an activity remunerated, from within the place visited" (World Tourism Organisation, 1995). There are many different purposes and motivations for people to travel. According to Richardson and Fluker (2004), the primary reasons for travel may be vacation and leisure, visiting friends and relatives (VFR), or business. Whatever is the purpose or motivations of the tourists, the travel to a destination directly or indirectly affects the host community as they are the ones who eventually provide hospitality to the tourists and the tourists experience a different atmosphere from their place of origin.

9.2 TOURISM INDUSTRY IN MALAYSIA

Situated in the Southeast Asia region, Malaysia is bordered with Singapore, Thailand, Brunei, and Indonesia. In 2018, Malaysia has a population of 32.6 million (Department of Statistics Malaysia, 2019). The country has different ethics group from different background, beliefs, languages, and culture. Before Independence Day, tourism was not a notable type of industry in Malaysia (Musa, 2000; Marzuki, 2010). Malaysia once highly depended on the primary commodities; however, the global economic regression and slashed of oil price forced the government to find an alternative source of income (Habibi et al., 2009). Thus, tourism was recognized as a tool to improve the economy of the country.

The tourism industry is the primary contributor to the services sector, constituting approximately 15% of Malaysia GDP or RM201.4 billion in 2017 (Ministry of Finance Malaysia, 2018). Tourism is also the second main contributor of foreign exchange income (Mosbah and Al Khuja, 2014), anticipating foreign tourists to contribute at least RM100 billion to Malaysian economy by 2020. Tourism delivers major benefits to the local people and is a tool to grow employment opportunities and offer better income distribution.

Malaysia is a famous tourist destination in the Asia Pacific region (Habibiet al., 2009). Based on the Travel and Tourism Competitiveness Index published by the World Economic Forum (2017), Malaysia ranked 26th place as the most competitive tourism destination. It is also a global top 10 medical tourism (Malaysia Healthcare Travel Council, 2018). Table 9.1 summarizes some global achievements and recognition.

Malaysia has been experiencing tourism difficulty over the past 5 years when the two most airline tragedies involved Malaysian Airlines Flight MH17 and the MH370 flight which occurred in 2014 (Yang et al., 2018). This has asserted heavy impacted on Malaysia's tourism industry particularly the inbound Chinese tourists. It was reported that over 60% of the booking from Chinese were cancelled (Bizcommunity.com, 2019). Meanwhile, the security and safety issue has called an alert for the tourism industry (Ayob and Masroni, 2014). Moreover, haze in Malaysia has caused Chinese tourists to divert their vacation to other destinations (Malaysiakini, 2019). In addition to this, high competition from neighboring countries such as Vietnam and Thailand had further impacted the tourism industry (Jaafar, 2018). In order to gain the interest of foreign tourists, the Malaysian government has allocated RM100 million for promotional and marketing campaign (Ministry of Finance Malaysia, 2018).

TABLE 9.1 Selected Global Achievements and Recognitions.

Title	Year	By
1. 4th World's Top Retirement Haven	2015	InternationalLiving.com
2. Medical Travel Destination of the Year	2015	International Medical Travel Journal (IMTJ) Medical Travel Awards 2015
3. 11th Most Visited Country in the World 2014	2015	UNWTO
4. Asia's Best Golf Destination	2014, 2015	Creative and Innovative In Culture And Tourism Sector
5. Kuala Lumpur—2nd World's Best Top Shopping Cities	2015	Muslim Travel Shopping Index (MTSI) 2015
6. World's Travel Award Asia & Australasia 2016	2015, 2016	World Travel Awards.com
7. Lonely Planet's Top 10 Cities For Best In Travel	2017	Lonely Planet
8. Kuala Lumpur—World's 5th Top Shopping Destination 2016	2016	Expedia UK
9. No.1 World's Top Muslim-Friendly Destination	2015–2017	Mastercard-Cresent Rating Global Muslim Travel Index
10. Kuala Lumpur; Ranked as 10th City in terms of international tourist arrivals by Top 100 City Destinations Ranking 2017 Edition	2017	Euromonitor International

Source: Adapted from Ministry of Tourism, Arts, and Culture Malaysia (2019).

Table 9.2 shows that a total of 25.83 million tourists visited Malaysia in 2018 compared to 25.95 million in the year 2017, a slightly dropped by 0.46% and the tourist receipts also reduced by 2.4%. However, the Malaysian government is optimistic about the future of tourism.

TABLE 9.2 Tourist Arrivals and Receipts to Malaysia from 2009–2018.

Year	Arrivals	Receipts
2018	25.83 million	84.1 billion
2017	25.94 million	82.1 billion
2016	26.76 million	82.1 billion
2015	25.72 million	69.1 billion
2014	27.44 million	72.0 billion
2013	25.72 million	65.4 billion
2012	25.03 million	60.6 billion
2011	24.71 million	58.3 billion
2010	24.58 million	56.5 billion
2009	23.65 million	53.4 billion

Source: Tourism Malaysia (2019).

9.3 TOURISM DEVELOPMENT IN MALAYSIA

Tourism development in Malaysia is closely aligned with the state's objectives which are to eradicate poverty, rural community development, reduce the disparity between the rich and poor, leading to the cohesion of the Malaysian culture and ultimately, the promotion of national unity. Goeldner et al. (2000) recommended tourism to gain competitiveness in tourism development; the government agencies must work with relevant stakeholders to achieve sustainability.

Generally, it can be claimed that the strategies put forth by the government to achieve competitiveness and sustainability is linked by the need for a management information system specifically for tourism in order to support policy formulation, strategic planning, routine decision-making, and overall performance evaluations. The competitiveness of a tourism destination can be defined as its capacity to compete and grow profitably in the tourism business. Sustainability, on the other hand, relates to the ability of a destination to maintain the sociocultural and environmental attributes while competing in the business platform.

According to Kayat (2011), the development of tourism in Malaysia depends on several factors which are mainly the "pull" factors and government interventions in the form of policies and strategies for the destination. Kayat cited Page (2003), who described "pull" factors as those factors within the destination zone that attract tourists to travel from one zone to another zone. The pull factors that influence tourist arrivals to Malaysia includes among others the geographical proximity between originating zone and destination zone, accessibility, attractions, cultural relationships, infrastructure includes among others the geographical proximity between originating zone and destination zone, accessibility, availability of attractions, cultural links, availability off infrastructure and services within the destinations, affordability, peace and stability, positive market image, and pro-tourism policies (Kayat, 2011).

Malays, Chinese, Indians, and various indigenous people of Sabah and Sarawak live and work alongside harmoniously in a multiethnic society. This is also reflected in the Malaysia, Truly Asia promotional campaign by Tourism Malaysia—"To know Malaysia is to love Malaysia. A bubbling, bustling melting pot of races and religions where Malays, Indians, Chinese and many other ethnic groups live together in peace and harmony" (Tourism Malaysia, 2012). According to Khan et al. (2019), the Malaysian government has strongly supported and encouraged longer staying tourists by promoting "Malaysia My Second Home" (MM2H) campaign. As for the MM2H campaign, the selling point for the target market is very much zoomed on retirement and the associated health care or fitness especially with Malaysia being marketed as a health care hub within the region. What more with the National Heart Institute extending special deals of US$6,000–US$7,000 for a regular bypass heart surgery at the Institute (Malaysia Healthcare Association, 2017). In the 1970s, Malaysia's economy was well sustained by good prices for export commodities such as tin, rubber, oil palm, and timber. However, the economy suffered in the early 1980s due to the plummeting prices of these commodities which are highly influenced by the fluctuations in the world economy. Kayat (2011) noted that such instabilities coupled with budget deficits and balance of payment deficiencies had made the policy makers recognize the importance of service industries such as tourism.

Kayat rightly pointed out, that the hosting of the Pacific Asia Travel Association (PATA) Conference in 1972 by the Malaysian government was an impetus for tourism development. With that, the government established a Tourism Development Corporation under the Ministry of Trade and Industry in the same year. On May 20th, 1987, the Ministry of Tourism and Culture was established which evolved with a new identity in 2002—The Ministry of Tourism or in short MOTOUR.

9.4 TOURISM PRODUCTS IN MALAYSIA

Attractions form the very basis for travel to a destination. Attractions are the tourism products which form the important element that attracts tourists to visit a destination. A product is anything that can be presented to a potential group of buyers for attention, procurement, use, or consumption that can satisfy a want or a need. It includes physical objects, services, places, organizations, and ideas (Saidhani and Sudiarditha, 2019). Therefore, a tourism product in the Malaysian context is any attraction that includes physical objects such as *batik, songket,* and local handicrafts, services such as accommodation, transport, food, and tour guide, scenic places like Cameron Highlands, historic places like Malacca, establishments like the Bario community in Sarawak and niche areas like community-based tourism (CBT). Planned events like Formula 1 racing and Le Tour de Langkawi Cycle Racing could also be included as tourism products. All of these are packaged and catered to tourists to accelerate the fulfillment of their travel intentions.

Tourism in Malaysia can easily fall under any one or a combination of several categories as follows:

- Nature-based—ecotourism, jungle trekking, scuba diving, and snorkeling.
- Adventure-based—mountain climbing, cave exploration, and kayaking.
- Culture-based—culinary tourism, music festival, and indigenous tourism.
- Recreational-based—parasailing, jet skiing, golf tourism, and horse riding.
- Health-based—spa tourism, medical treatment, and physiotherapy.
- Heritage-based—urban tourism, educational tourism
- Community-based—CBT tourism and indigenous tourism
- Religious-based—rituals, festivals, visiting temples, and places of worship
- Rural-based—CBT, agro-tourism, and cottage industries.

9.5 COMMUNITY-BASED TOURISM IN MALAYSIA

In Malaysia, communities involve in tourism sector in several ways. Most of the communities are using the concept of CBT to the process of community development. This section reviews and analyzes the previous tourism development and the implementation of CBT in the community in Malaysia. Table 9.3 summarized the CBT studies in Malaysia. It appears that most of the CBT's in Malaysia focus on homestay program and conducted in rural areas (Harris, 2009; Yusof et al., 2013; Amran and Haniza, 2012). For instance, Othman et al. (2013) studied the operation of the homestay initiative in the state of Kedah focusing on CBT.

CBT have been promoted in a community as a result of the government's policy of using tourism to maximize the positive economic impacts (Ibrahim, 2008). A study conducted in Redang, Mabul, and Perhentian Island reported that the local communities in those islands enjoyed the benefits brought by tourism (Daldeniz and Hampton, 2013). However, language, seasonality, and cost of training are some of the main barriers from stopping local community in tourism. Harris (2009) conducted a case study how pro-poor CBT integrated into community well-being in Bario, Sarawak, Malaysia. Meanwhile, Davison et al. (2005) studied the role of e-commerce for CBT in Bario, Sarawak, Malaysia. A rural ICT project, e-Bario, was raised to promote CBT tourism (Harris, 2009). As a result, the social and economic life of the community has improved.

TABLE 9.3 Summary of the CBT Studies in Malaysia.

Researcher	Study site	Subject matter
Harris (2009)	Bario, Sarawak	Community development
Kayat (2002)	Kampung Relau, Kedah	Homestay
Kayat and Nur Farrah (2002)	Throughout Malaysia	Homestay
Jamal et al. (2011)	Banghuris, Selangor; Tanjung Piai, Johor; and Pelegong, Negeri Sembilan	Homestay
Ibrahim and Razzaq (2010)	Malaysia	Homestay
Razzaq et al. (2011)	Muar, Johor	Homestay
Parveen et at. (2016)	20 homestay owners from three villages in Sarawak	Homestay
Rosazman and Velan (2014)	Sabah	Homestay
Othman et al. (2013)	Kedah	Homestay
Amran and Haniza (2012)	Kinabatangan, Sabah	Homestay
Azizan et al. (2012)	Langkawi Island	Public participation
Munikrishnan et al. (2015)	Lower Kinabatangan, Sabah	Cooperatives
Goh (2015)	Lower Kinabatangan, Sabah	Homestay
Shazali et al., 2015; Shazali et al. (2017)	Bau, Sarawak	Indigenous community tourism
Hitchner et al. (2009)	Kelabit Highlands, Sarawak	Ecotourism

Source: Self developed.

In many rural areas, Malaysian government recognized community empowerment in cooperatives is vital to improve the socioeconomic benefits (Munikrishnan et al., 2015; Nor Haniza and Amran, 2013). One of the good successful CBT story in Malaysia is the establishment of community-based cooperative (KOPEL) in Lower Kinabatangan, Sabah (Munikrishnan et al., 2015). Local communities were found actively participate in KOPEL's tourism activities. It was reported that 14% of the total population are the member of cooperative. They addressed four important criteria in successful CBT such as (1) planning and empowerment, (2) awareness and training, (3) collaboration with other stakeholders, and (4) impacts of rural tourism. KOPEL actively work together with many government agencies, local and international non-governmental organizations (NGOs) and private organizations. The collaboration allows KOPEL to get training, funds to protect wildlife, and forest restoration. As a result, the income generated from KOPEL increased from RM38, 000 (USD12, 000) in 2000 to RM1.6million

(USD500, 000) in 2012. Thus, government encourages local community to participate and involve in the rural tourism to gain economic benefits (Yusof et al., 2013).

Nevertheless, some researchers include the indigenous local communities into CBT (Majin et al., 2016; Kunasekaran et al., 2013; Shazali et al., 2015). For instance, an indigenous community study conducted by Majin et al. (2016) in Carey Island revealed that tourism has impacted on the community life in term of economically, socially, culturally, and environmentally. However, the community feels anxious the lack of natural resources will cause the degradation of their culture and tourism (Kunasekaran et al., 2013). On the other hand, the unique culture is used by Bidayuh community to promote tourism (Shazali et al., 2015; Shazali et al., 2017). In the absence of development plan by the government, Bidayuh community is still willing to participate in tourism and to share their culture to outsiders.

9.6 TOWARD SUSTAINABLE TOURISM IN MALAYSIA

Sustainable tourism is no longer a buzz word as many governments in developing countries encourage sustainable tourism even though sustainable tourism was arguably lack of practical and challenges (Gossling, 2018). Buckley (2012) opined that tourism is not close to sustainability. For example, Higgins-Desbiolles (2018) critiqued the sustainability of tourism and mentioned that tourism has "problem" because the growth of development of tourism is not in line with sustainable goals. Failing to incorporate well-planning for a tourist destination will result in serious negative consequences (Teh and Cabanban, 2007). In Cambodia, sustainable tourism remains questionable (Carter et al., 2015). As Fodness et al. (1993) contented, "it is necessary to go beyond traditional tourism problem-solving paradigms and approaches to achieve reliable replicable results." Font and McCabe (2017) stressed that tourism products require a more innovative and sustainable strategy so that it can be marketed.

Despite the economic benefits of tourism, it could also cause adverse impacts such as the destruction of flora and fauna, impact natural resources such as water and soil. In addition, the implementation of sustainable tourism is always a challenge and difficult (Muangasame and McKercher, 2015). Apart from that, tourism can also threaten the communities from excessive development, carrying capacity, and manipulation of local culture. Tourism also heavily consumes the resources it depends on. In contrast, sustainable tourism is planned to benefit the local community, appreciate their culture, protect natural resources, channel the economic benefits directly to the local community, and create awareness to both tourists and local community about the significance of conservation.

Sustainable tourism practice in Malaysia mainly focuses on the community well-being perspectives as compared economics and environmental pillars (Kunasekaran et al., 2017; Bhuiyan, 2019). Four success factors of CBT such as dynamic leadership style and organizational relationship, the establishment of a tourism cooperative, corporations with government agencies, and tourism industry stakeholders and a resilient pledge to biodiversity conservation were highlighted in Miso Walai homestay in Kinabatangan, Sabah (Amran and Haniza, 2012). In a study in Langkawi Islands, Malaysia, inadequate information, an ineffective approach, residents' attitudes, and exclusion from participation processes were found barriers in public participation in decision-making processes (Azizan et al. 2012). They agreed that public participation in decision-making processes is viewed as imperative for successful tourism planning. Similarly, a study carried out in Tioman Island revealed that local community's welfare was neglected and this certainly affect the local communities' participation in tourism (Ho et al., 2017).

Similarly, Nor Haniza and Amran (2013) studied how a tourism cooperative in Kinabatangan River, Sabah creates sociocultural and environmental benefits to its members. The Model of Economically Sustainable Community Tourism (MESCOT) is perceived as a successful business model that adopted in CBT. This business model has been proven to grow the community business ventures through cooperative. On the other hand, Goh (2015) conducted a study in Kinabatangan, Sabah to identify the impacts of CBT. The researcher concluded that CBT are able to reduce the poverty rate, enhance local community participation, and protect the nature environment. The researcher also highlighted that cooperating with different stakeholders at different levels are important.

9.7 CONCLUSION

Sustainable Development Goals 2030 (SDG 17) heavily stresses the importance of community development as compared to economic sustainability. This is evident by several goals which are relevant to enhance community well-being:

- Goal 1: End poverty in all its forms.
- Goal 2: Zero Hunger.

- Goal 3: Health.
- Goal 4: Education.
- Goal 5: Gender equality and women's empowerment.
- Goal 6: Water and sanitation.
- Goal 7: Decent work and economic growth.
- Goal 8: Reduced inequalities.
- Goal 9: Global partnerships.

By giving specific attention to the significant issues determining sustainable tourism in terms of local community's involvement, this aspect could be addressed. Usually, the opinion of the local people is often overlooked by the governmental agencies. This scenario leads to dissatisfaction of the local people because they feel that they are not valued and empowered. Although the communities are included in preliminary developmental plans, the ultimate decision of developmental projects, however, will be taken by the government or outsider funding agencies. If this situation prolongs, rural communities will feel isolated with any tourism developmental programs structured by the government. Therefore, the relevant authorities should not simply consult the community to satisfy procedures, on a token basis. They should allow and value the community's viewpoints and prepare a platform of trial and error. By doing this, the community can be self-reliant and move toward community capacity building in long run.

In addition, the approach used by the Tourism Authority of Thailand (2003), by making successful ecotourism operators motivate other villagers into being involved in tourism, brought major development in Ban Mae Kampong in Chiang Rai, can be adopted in Malaysia. This scenario shows that the villagers tended to be involved in tourism when they were convinced by fellow villagers, rather than outsiders like government agencies. If this strategy is implemented by Malaysian authorities, rural people with a negative perception, in terms of economic benefits, awareness, and entrepreneurial knowledge, may have a better perception of tourism. This effort will also provide employment opportunities for the local people to be involved in tourism activities. In terms of environmental sustainability, indigenous tourism initiatives should be encouraged for preservation of community resources (Timothy, 1999) by targeting all the community members regardless of their state of involvement in the industry. Coetzee (2003) also suggested that the carrying capacity problem of tourism should be addressed to avoid environmental impacts.

Dernoi (1981) stressed that rural tourism will help to keep the next generation of villagers in the villages, instead of moving to cities to find other jobs. Negative perception on sociocultural impacts can be reduced if tourism premises are developed in the rural areas to provide better job opportunities, such as tour guides, chefs, or event planners. These new and varied attractive jobs will make the next generation of the rural people stay on, rather than finding jobs in urban areas. In terms of facilities, road and land development, long-term financial support by the government is essential. Sharpley (2001) identified that short-term development often involves; very high investment costs but low yields, low demand, a lack of critical skills and the dominance of mass tourism operators, as main obstacles. Therefore, long-term financial support is important, if tourism is to play an efficient rural revitalizing role.

In a nutshell, a major effort that needs to be taken especially by governmental agencies such as Tourism Malaysia and *Kraftangan Malaysia* under the Ministry of Culture Arts and Tourism is to create greater visibility of the existence of this indigenous community. The uncertain tourist arrivals have been the biggest difficulty for the community to involve in tourism. Without a strong and consistent demand, operating tourism is a high risk business. At the moment, the arrival of international tourists depends on several tour guides who have established network with the villagers. Thus, the relevant agencies should study and identify the target market which has interest in CBT. Then, direct promotion to these target groups may draw higher tourists' arrival to these areas. This initiative will lead the local communities guided toward sustainable tourism practice.

KEYWORDS

- **tourism development**
- **community tourism**
- **sustainability**
- **Malaysia**

REFERENCES

Amran, H.; Haniza, M. N. Critical Success Factors of Community Based Ecotourism: Case Study of Miso Walai Homestay, Kinabatangan, Sabah. *Malays. For.* **2012,** *75*(1), 27–40.

Ayob, N. M.; Masroni, T. Issues of Safety and Security: New Challenging to Malaysia Tourism Industry. *SHS Web Conf.* **2014,** *12*, 1–10.

Azizan, M.; Hay, I.; James, J. Public Participation Shortcomings in Tourism Planning: The Case of the Langkawi Islands, Malaysia. *J. Sustain. Tour.* **2012,** *20*(4), 585–602.

Bhuiyan, M. A. H. Homestay for Community Based Tourism Development at Kampung Jelawang in Kelantan, Malaysia. *Soc. Sci. J.* **2019,** *3,* 73–81.

Buckley, R. Tourism, Conservation and the Aichi Targets. *Parks*. **2012,** *18*(2), 12–19.

Carter, R. W.; Thok, S.; O'Rourke, V.; Pearce, T. Sustainable Tourism and its Use as a Development Strategy in Cambodia: A Systematic Literature Review. *J. Sustain. Tour*. **2015,** *23*(5), 797–818.

Coetzee, K.; Graff, J.; Hendricks, F.; Wood, G.; Development: Theory, Policy and Practice in South Africa. *Commun. Dev. J.* **2012,** *32*(1), 122–140.

Daldeniz, B.; Hampton, M. P. Dive Tourism and Local Communities: Active Participation or Subject to Impacts? Case Studies from Malaysia. *Int. J. Tour. Res.* **2013,** *15*(5), 507–520.

Davison, R. M.; Harris, R. W.; Vogel, D. R.; *E-Commerce for Community-Based Tourism in Developing Countries.* **1995,** 9th Pacific Asia Conference on Information Systems, Bangkok.

Department of Statistics Malaysia Official Portal. https://www.dosm.gov.my/v1/index.php?r=column/cthemeByCat&cat (accessed Oct 15, 2019).

Dernoi, L. A. Farm Tourism in Europe. *Tour. Manag.* **1983,** *12*(3), 155–166.

Higgins-Desbiolles, F. Sustainable Tourism: Sustaining Tourism or Something More?. *Tour. Manag Perspec.* **2017,** *25*, 157–160.

Fodness, D.; Pitegoff, B. E.; Truly Sautter, E. From Customer to Competitor: Consumer Cooption in the Service Sector. *J. Serv. Mark.* **1993,** *7*(3), 18–25.

Font, X.; McCabe, S. Sustainability and Marketing in Tourism: Its Contexts, Paradoxes, Approaches, Challenges and Potential. *J. Sustain. Tour*. **2017,** *25*(7), 869–883.

Fridgen, J. D. *Dimensions of Tourism,* Vol. 12; Educational Institute of the American Hotel & Motel Association; 1990; pp 43–57.

Goh, H. C. Nature and Community-Based tourism (CBT) for Poverty Alleviation: A Case Study of Lower Kinabatangan, East Malaysia. *Geogr. Malays. J. Soc. Space* **2015,** *11*(3), 42–52.

McIntosh, R. W.; Goeldner, C. R.; Ritchie, J. B. *Tourism: Principles, Practices, Philosophies*; John Wiley and Sons, 2000.

Gössling, S. Tourism, Tourist Learning and Sustainability: An Exploratory Discussion of Complexities, Problems and Opportunities. *J. Sustain. Tour*. **2018,** *26*(2), 292–306.

Habibi, F.; Rahim, K. A.; Ramchandran, S.; Chin, L. Dynamic Model for International Tourism Demand for Malaysia: Panel Data Evidence. *Int. Res. J. Finance Econ.* **2009,** *33*(1), 208–217.

Harris, R. W. Tourism in Bario, Sarawak, Malaysia: A Case Study of Pro-Poor Community-Based Tourism Integrated Into Community Development. *Asia Pac. J. Tour. Res.* **2009,** *14* (2), 125–135.

Hitchner, S. L.; Apu, F. L.; Tarawe, L.; GalihSinah, N. A.; Supang, Yesaya, E. Community-Based Transboundary Ecotourism in the Heart of Borneo: A Case Study of the Kelabit Highlands of Malaysia and the Kerayan Highlands of Indonesia. *J. Ecotour.* **2009,** *8*(2), 193–213.

Ho, J. A.; Chia, K. W.; Ng, S. I.; Sridar, R. Problems and Stakeholder Responsibilities in Island Tourism: The Case of Tioman Island in Malaysia. *J. Hosp. Tour. Res.* **2017,** *7*(1), 21–31.

Ibrahim, Y. *Pembangunan Pelancongan Dan Perubahan Komuniti*; Kuala Lumpur: Dewan Bahasa dan Pustaka, 2008.

Ibrahim, Y.; Razzaq, A. R. A. Homestay Program and Rural Community Development in Malaysia. *J. Ritsumeikan Soc. Sci. Humanit.* **2010,** *2*, 7–24.

Kayat, K. Exploring Factors Influencing Individual Participation in Community-Based Tourism: The Case of Kampungrelau Homestay Program, Malaysia. *Asia Pac. J. Tour. Res.* **2011,** *7*(2), 19–27.

Khan, M. J.; Misnan, S. H.; Ismail, H. N. International migration, second home programmes development and Malaysia experience. *International Journal of Built Environment and Sustainability.* **2019,** *6(1-2),* 63–74.

Kunasekaran, P.; Gill, S. S.; Talib, A. T.; Redzuan, M. R. Culture as an indigenous tourism product of MahMeri community in Malaysia. *Life Science Journal.* **2013,** *10*(3), 1600-1604.

Kunasekaran, P.; Gill, S.; Ramachandran, S.; Shuib, A.; Baum, T.; Herman Mohammad Afandi, S. Measuring sustainable indigenous tourism indicators: a case of MahMeri ethnic group in Carey Island, Malaysia. *Sustainability,* **2017,** *9(7),* 1256.

Majin, J. J.; Azman, I.; Jailani, F. Support for future tourism development, quality of life and perceived impacts of tourism industry within indigenous community: MahMeri of Carey Island, Malaysia. In M. A. Abdullah, W. K. Yahya, N. Ramli, S. R. Mohamed & B. E. Ahmad (Eds.), *Regional Conference on Science, Technology and Social. Business and Social Sciences.* **2016,** (pp. 1063-1073). Singapore: Springer Singapore.

Malaysia Healthcare Travel Council. Top 10 Medical Tourism Destinations in the World. **2018,** Retrieved from https://www.mhtc.org.my/top-10-medical-tourism-destinations-in-the-world/

Malaysiakini. Thousands of China tourists cancel holidays in M'sia. **2019,** [online] Available at: https://www.malaysiakini.com/news/492930 [Accessed 25 Oct. 2019].

Marzuki, A. Tourism development in Malaysia. A review on federal government policies. *Theoretical and Empirical Researches in Urban Management,* **2010,** *5*(8 (17), 85–97.

Muangasame, K.; McKercher, B. The challenge of implementing sustainable tourism policy: a 360-degree assessment of Thailand's "7 Greens sustainable tourism policy". *Journal of Sustainable Tourism.* **2015,** *23*(4), 497–516.

Ministry of Finance Malaysia, Budget 2019. **2018,** Retrieved from http://www.treasury.gov.my/pdf/budget/speech/bs19.pdf

Ministry of Tourism, Arts and Culture Malaysia. Achievements and Recognitions. **2019,** Retrieved from http://www.motac.gov.my/en/profile/achievements-and-recognitions.

Mosbah, A.; Al Khuja, M. S. A. A review of tourism development in Malaysia. *European Journal of Business and Management.* **2014,** 6(5), 1–9.

Munikrishnan, U. T., Rajaratnam, S. D., Mura, P., & Nair, V. Local residents' participation in rural tourism: The case of KOPEL in Kinabatangan, Sabah, Malaysia. *Asia Pac. J. Innov. Hosp. Tour.* **2015,** *8,* 181.

Musa, G. Tourism in Malaysia. In *Tourism in South and Southeast Asia: Issues and Cases*; Hall, C. M., Page, S., Eds.; Butterworth-Heinemann: Oxford, 2008.

Nor Haniza, M.; Amran, H. Tourism Cooperative for Scaling Up Community-Based Tourism. *Worldw. Hosp. Tour. Themes.* **2013**, *5*(4), 315–328.

Othman, F.; Sazali, F.; Mohamed, B. Rural and Community Based Tourism Development in Malaysia: Prospects for Homestays as a Social Economy Enterprise. *Team J. Hosp. Tour.* **2013,** *10*(1), 65–76.

Parveen, K.; Arif, J.; Norizan, B. A. O. The Impact of Community Based Tourism on Community Development in Sarawak. *J. Borneo Kalimantan.* **2016,** *2*(1).

Presenza, A.; Messeni Petruzzelli, A.; Natalicchio, A. Business Model Innovation for Sustainability. Highlights from the Tourism and Hospitality Industry. *Sustainability* **2019,** *11*(1), 212.

Kunasekaran, P., Sarjit, S. G., Talib, A., Marof, R. Culture as an Indigenous Tourism Product of Mah Meri Community in Malaysia. *Life Sci. J.* **2013,** *10*(3), 1600–1604.

Richardson, J.; Fluker, M. *Understanding and Managing Tourism.* Pearson Hospitality Press: Australia, 2004.

Rosazman, H.; Velan, K. Sustainable Community-Based Tourism (CBT) Through Homestay Programmed in Sabah, East Malaysia. *Proceeding of the Social Sciences Research ICSSR*, 2014, pp 41–61.

Saidani, B.; Sudiarditha, I. K. R. Marketing Mix-7Ps: The Effect on Customer Satisfaction. *JurnalPendidikanEkonomi Dan Bisnis (JPEB)*. **2019,** *7(1),* 72–86.

Salamiah, J.; Nor'Ain, O.; NikMaheran, M. N. Tourist Perceived Value in a Community-Based Homestay Visit: An Investigation into the Functional and Experiential Aspect of Value. *J. Vacat. Mark.* **2011,** *17*(1), 5–15.

Shazali, J.; Ahmad, S.; Sridar, R.; Syamsul, H.; Kunasekaran, P. Bidayuh Community of Malaysia: Presenting Culture and Nature as a Package for Sustainable Development of Indigenous Tourism. In *Balancing Development and Sustainability in Tourism Destinations: Proceedings of the Tourism Outlook Conference 2017*; Saufi, A., Andilolo, I. R., Othman, N., Lew, A. A., Eds.; Springer Singapore: Singapore; pp 89–96.

Sharpley, R. Rural Tourism and the Challenge of Tourism Diversification: The Case of Cyprus. *Tour. Manag.* **2006,** *23*(3), 233–244.

Shazali, J.; Sridar, R.; Ahmad, S.; Syamsul, H. Participation of the Bidayuh Community in Tourism Developmental Initiatives in Bau, Malaysia. *Life Sci. J.* **2015,** *12*(2), 46–48.

Teh, L.; Cabanban, A. S. Planning for Sustainable Tourism in Southern Pulau Banggi: An Assessment of Biophysical Conditions and their Implications for Future Tourism Development. *J. Environ. Manag.* **2007,** *85*(4), 999–1008.

Timothy, D. J. Participatory Planning, a View of Tourism in Indonesia. *Ann. Tour. Res.* **1999,** *26*(2), 371–391.

Theobald, W. F. *Global Tourism*, 3rd ed.; Butterworth-Heinemann/Elsevier: New York, 2005.

Tourism Malaysia, *Tourism Malaysia: A bit of culture*. 2012. [Online] http://www.tourism.gov.my/en/master/web-page/experiences (accessed Sep 2, 2012).

World Tourism Organization. *UNWTO Technical Manual: Collection of Tourism Expenditure Statistics*, Madrid, Spain, 1995.

Yang, L.; Tjiptono, F.; Poon, W. C. Will You Fly with this Airline in the Future? An Empirical Study of Airline Avoidance After Accidents. *J. Travel Tour. Mark.* **2018,** *35*(9), 1145–1159.

Yusof, Y.; Muda, M. S.; Amin, W. A.; Ibrahim, Y. Rural Tourism in Malaysia: A Homestay Program. *China USA Bus. Rev.* **2013,** *12*(3), 300–306.

CHAPTER 10

DEVELOPMENT AND INNOVATIONS OF THE HOTEL AND AIR TRANSPORT INDUSTRY IN SINGAPORE

ZHAOPING (GEORGE) LIU* and ELIVER CHEUK KI LIN

Singapore Institute of Technology, Singapore

**Corresponding author. E-mail: zhaoping.liu@SingaporeTech.edu.sg*

ABSTRACT

Singapore is a popular tourism destination. Its international visitor arrivals (IVA) and tourism receipts (TR) have achieved record highs in the past 3 years. The objectives of this chapter are to describe the development of Singapore's hotel industry and air transport industry and summarize innovations in those two industries. Two case studies such as Open Hotels Weekend and Hotel Innovation Challenge Series have been used to explain government inspired innovations in Singapore's hotel industry; while Changi Airport's many innovations and initiatives were cited to justify why Singapore ranks very high on many top city lists.

10.1 INTRODUCTION

Centrally located in Southeast Asia, Singapore is a very popular tourism destination. It ranked 5th on the list of Mastercard Global Destination Cities 2018 and 4th for Overnight International Visitor Spending (Master Card, 2019). Singapore ranked 8th for international tourist arrivals in the Asia Pacific region and 26th in the world. International tourism revenue accounted for 4.1% of Singapore's national GDP in 2017 (Budget Direct, 2019).

In Singapore, the tourism sector has been achieving record highs in international visitor arrivals (IVA) and tourism receipts (TR) in the past 3 years (2016–2018). In 2018, IVA increased 6.2% and reached 18.5 million (STB, 2018a). The top 15 source markets of Singapore are China (IVA of 3,418,000), Indonesia (3,021,000), India (1,442,000), Malaysia (1,254,000), Australia (1,107,000), Japan (830,000), Philippines (778,000), the United States (643,000), South Korea (629,000), Vietnam (592,000), The United Kingdom (589,000), Thailand (546,000), Hong Kong SAR (473,000), Taiwan (423,000), and Germany (357,000) (STB, 2019a). In 2018, TR increased 0.5% to reach S$26.9 billion. The five major components of TR are shopping (20%); accommodation (21%); food and beverage (10%); sightseeing, entertainment, and gaming (22%); and other TR components (28%) (STB, 2019a).

This chapter is composed of two major sections. Section one is on hotel industry and section two is on civil aviation industry. Section one includes the history and current development of the hotel industry in Singapore, hotel administration in Singapore, and two case studies of innovations in Singapore Hotel Industry. Section two includes Singapore civil aviation industry in 100 years and Changi Airport's innovations.

10.2 DEVELOPMENT AND INNOVATIONS OF SINGAPORE'S HOTEL INDUSTRY

The objective of this section is to review the history of Singapore hotel industry, describe its current status, and examine the public administration system relevant to hotel industry. Two recent events, the Open Hotel Weekends and Hotel Innovation Challenge Series, were used to showcase government inspired innovations in Singapore hotel industry.

10.2.1 HISTORY AND CURRENT DEVELOPMENT OF THE HOTEL INDUSTRY IN SINGAPORE

From London Hotel to Marina Bay Sands: A History Across Three Centuries

The history of Singapore hotels can date back to 1831, when a blacksmith by trade named Stephen Hallpike opened a boarding house at High Street in Singapore. London Hotel, the first real hotel in Singapore, was founded by Gaston Dutronquoy in 1839 (SHA, 2011). Even London Hotel was named the best hotel in Singapore at that time, many guests were not satisfied with the establishment (e.g., as described by Charles Walter Kinloch in 1852). In December 1855, a one night stay at the London Hotel would cost about $3.

In 1887, the Sarkies brothers (Tigran Sarkies and Martin Sarkies) opened the Raffles Hotel, a hotel that has become synonymous with Singapore (Liu, 2006). Innovation was in Raffles hotel's DNA since day one. For example, in 1915, a bartender at Raffles named Ngiam Tong Boon created the signature cocktail—"The Singapore Sling." Another innovation is the well-known Tiffin with Curry dishes (SHA, 2011). As one of the few 19th century hotels in the world, Raffles was considered more of a legend than a hotel (Sun, 2011). Being re-gazetted as a National Monument in 1995, Raffles' hotel played three roles of a landmark, a historical hotel, and a social venue (Henderson, 2001).

In early 1890s, when Singapore became world's seventh largest port due to the huge demand in rubber and tin, the demand for high-quality hotels also surged. New luxury hotels such as The Sea View were opened to address this market (SHA, 2011). In 1929, the Menasseh brothers opened the Goodwood Park Hotel. It was converted from a property seized from Germans in Singapore during the First World War. During the World War II, when Singapore was occupied by the Japanese, hotel operations were significantly affected. Raffles Hotel was called Syonan Ryokan and the Goodwood Park Hotel was occupied by the Japanese military as their headquarters (Goodwood Park Hotel, 2019).

When tourism was selected as one of the most important industries to boom its economy after Singapore became independent in 1965, many new hotels such as the Mandarin Hotel (1971) and the Shangri-La (1971) were built to serve tourists from Europe and America. The first international hotel chain (The Singapura InterContinental) arrived in Singapore in 1968 followed by others like Hilton and Hyatt (SHA, 2011).

The economic recession in early 1980s had a negative impact on the tourism industry. Hotel rates and occupancies were affected enormously, including many new hotels opened during the period (e.g., The Amara Hotel opened in 1986). When tourism industry bounced back in late 1980s and early 1990s, people saw a booming of the hotel industry. More than 20 new hotels opened in the 1990s including those of international hotel chains (e.g., Four Seasons, Marriot, and Ritz-Carlton) and local budget hotel chains (e.g., Hotel 81). The openings of Suntech Singapore International Convention and Exhibition Center in 1994 and Singapore Expo in 1999 brought more business travelers to Singapore, which helped to improve the performance of the hotel industry.

In 2001, Fullerton Hotel opened for business. This heritage hotel was converted from the former General Post building (Fullerton Hotel, 2019). In the 21st century, the most significant events in Singapore's hotel industry are probably the opening of Resorts World Sentosa and Marina Bay Sands (MBS) in 2010. They generate billions of spending on an annual basis. At the same time, the two integrated resorts (IRs) attracted incremental visitors to Singapore and didn't take away guests from current hotels (LVS, 2018). Other iconic events include the opening of Capella hotel (2009) and the Valley Wing of Shangri-La hotel, both targeting luxury guests.

Singapore Hotel Industry (2010–2018)

In this decade, Singapore has seen a steady growth of hotel supply (Fig. 10.1). In 2010, there were 282 hotels in Singapore. In 2017, it grew to 422 (STB Hotel, 2011–2018). The total number of rooms increased from 47,312 to 68,444 (Table 10.1) (STB, 2018b).

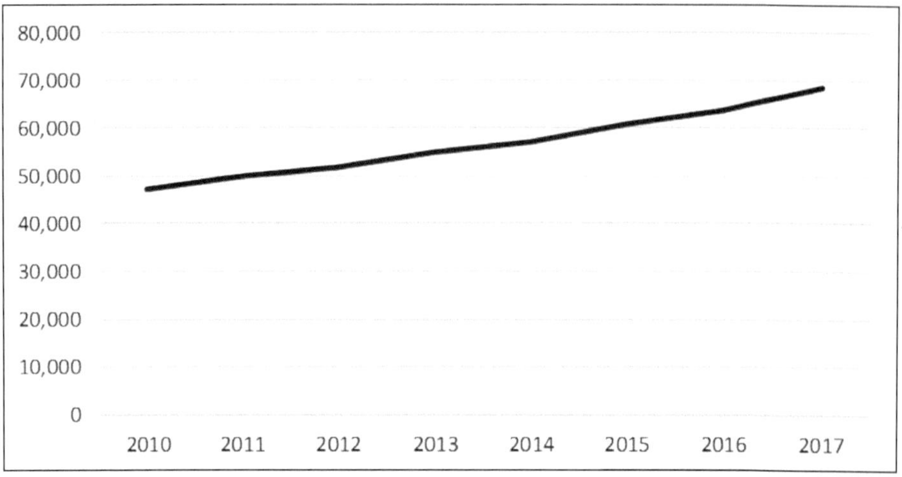

FIGURE 10.1 Hotel room stock in Singapore (2010–2017).

The total hotel room revenue also had significant growth during this period. In 2010, the opening of the two IRs (Marina Bay Sands and Resort Sentosa) pushed the room revenue of Singapore hotels to increase for more than 40% over the previous year. The impact extended to 2011 with a year-on-year growth of 26% (STB Hotel, 2012). In 2018, the Gazetted hotel room revenue grew 7.5% year-on-year and reached S$4.0 billion (STB hotel,

2019). Gazetted hotels are hotels that have been declared to be tourist hotels under the Singapore Tourism (Cess Collection) Act (STB hotel, 2019).

The occupancy rate was stable in this decade, ranging between 84% and 87%. During this period, the average room rate (ARR) started at $205.5 in 2010, reached peak between 2012 and 2014 (about $240) and dropped below $220 in 2017 and 2018 (Table 10.2).

TABLE 10.1 Numbers of Hotels and Rooms and Total Room Revenue (Singapore, 2010–2017).

Year	Number of hotels	Rooms stock	Room stock % change	Total room revenue ($m)	Year-on-year % change
2010	282	47,312	+10.8	2302.3	+42.8
2011	304	49,719	+5.1	2901.3	+26.0
2012	338	51,579	+3.7	3184.3	+9.8
2013	373	55,018	+6.7	3313.6	+4.1
2014	391	57,050	+3.7	3547.2	+7.0
2015	398	60,841	+6.6	3506.8	-1.1
2016	413	63,706	+4.7	3559.1	+1.5
2017	422	68,444	+7.4	3697.2	+3.9

Note: The currency in this chapter is in Singapore Dollars unless otherwise stated.

TABLE 10.2 Singapore Hotels—AOR, ARR, and RevPar (2010–2018).

Year	Standard average occupancy rate (AOR) (%)	Standard average room rate (ARR) ($)	Revenue per available room (RevPar) ($)	Year-on-year % change
2010	84.4	205.5	173.5	+22.6
2011	85.3	233.3	199.0	+14.7
2012	85.7	243.5	208.8	+4.9
2013	85.1	239.6	203.9	-2.4
2014	84.2	242.5	204.2	+0.2
2015	84.0	232.3	195.0	-4.5
2016	83.1	223.1	185.4	-4.9
2017	84.5	215.5	182.2	-1.7
2018	86.2	218.6	188.5	+2.3

The RevPar peaked in 2012 at $208.8 and then started to slide in the next few years. It bounced back to $188.5 in 2018 but was still $20 short of what in 2012. The year-on-year percentage change of RevPar is very dynamic (Fig. 10.2).

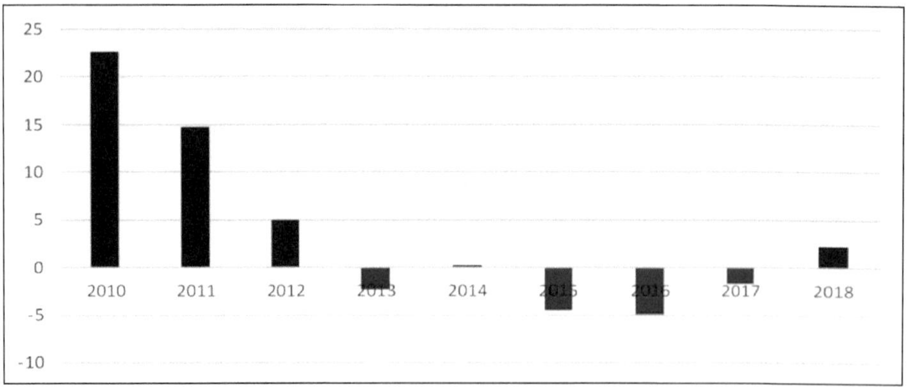

FIGURE 10.2 Singapore hotels: RevPar year-on-year percentage change (2010–2018).

Using a combination of factors such as ARRs, location, and product characteristics, Singapore Tourism Board (STB) has developed a unique hotel tiering system to categorize the different hotels. The four tiers of hotels are: "Luxury—includes hotels in the luxury segment and are predominantly in prime locations and/or in historical buildings; upscale—includes hotels in the upscale segment and are generally in prime locations or hotels with boutique positioning in prime or distinctive location; mid-tier—includes hotels in the mid-tier segment and are primarily located in prime commercial zones or immediately outlying areas; economy—includes hotels in the budget segment and are generally located in outlying areas" (STB hotel, 2019). The AOR, ARR, and RevPar data of different hotel tiers in the past 5 years (2014–2018) were summarized in Table 10.3. Both luxury and upscale tiers enjoyed high occupancy rates, while the AOR of the economy sector is the lowest.

10.2.2 HOTEL ADMINISTRATION IN SINGAPORE

Singapore hotels are mainly regulated by Singapore Tourism (Cess Collect) Act, Hotels Act, and Hotels Licensing Regulations. Since Singapore Tourism Act is generally applied to the broader tourism industry, our discussion focuses on the latter two.

TABLE 10.3 Singapore Hotel Tiers—AOR, ARR, and RevPar (2014–2018).

		2014	2015	2016	2017	2018
Luxury	Average occupancy rate (AOR)	87.0	85.7	85.1	83.3	87.3
	Average room rate (ARR)	457.4	446.5	448.0	443.6	449.0
	Revenue per available room (RevPar)	398.0	382.6	381.2	369.7	392.1
Upscale	Average occupancy rate (AOR)	87.0	86.0	84.7	85.6	86.1
	Average room rate (ARR)	266.8	264.8	261.5	257.8	263.0
	Revenue per available room (RevPar)	232.2	227.6	221.4	220.8	226.4
Mid-Tier	Average occupancy rate (AOR)	84.3	84.7	84.6	85.7	87.2
	Average room rate (ARR)	184.9	175.7	169.6	167.0	168.3
	Revenue per available room (RevPar)	156.0	148.9	143.4	143.2	146.8
Economy	Average occupancy rate (AOR)	78.0	78.5	77.2	82.2	83.9
	Average room rate (ARR)	109.6	104.6	101.2	103.2	105.4
	Revenue per available room (RevPar)	85.5	82.1	78.1	84.8	88.3

Hotels Act (Singapore)

In Singapore, a hotel is defined as "a boardinghouse, lodging house, guesthouse, and any building or premises not being a public institution and containing not less than four rooms or cubicles in which persons are harbored or lodged for hire or reward of any kind and where any domestic service is provided by the owner, lessee, tenant, occupier, or manager for the person so harbored or lodged" (Hotels Act, 1999). A hotel has to be registered before being used as a hotel. Hotels Act was first commenced on Oct 12, 1954 and has been revised multiple times. The current version was based on the 1985 edition. Its operation date is Aug 1, 1999.

Hotels Licensing Regulations (Singapore)

Hotels Licensing Regulations regulates the licensing of Singapore Hotels. The Hotels Licensing Board is consisted of one chairman and four other members (Hotel Licensing regulations, 2017). The current chairman is Mr Toh Yung Cheong, also the Chairman of State Courts. Four other members come from Urban Redevelopment Authority, STB, Singapore Police Force and Singapore Civil Defence Force. All questions before the board shall be decided by a majority of votes.

The classification of hotels by Hotel Licensing Regulations is as follows: (a) small hotels, a hotel with 100 rooms or less; (b) medium hotels, a hotel with more than 100 rooms and less than 300 rooms; (c) large hotels, a hotel with 300 rooms or more. Hotels pay \$200 (small), \$300 (medium) and \$400 (large) fees for certificate of registration per annum. The fee for a license is \$100 per annum.

The regulation sets very clear and strict standards for hotels. For example, all guest rooms must have a minimum floor area of 11 m^2 and a double room shall have a minimum floor area of 14.5 m^2. Rooms below ground level cannot be use for accommodation of guests. A single room can only accommodate no more than one adult and a double room can accommodate no more than two adults. For light and ventilation, "all rooms used for the accommodation of guests shall be provided with natural light and natural ventilation by means of one or more openings in its external walls having a combined area of not less than 15% of the floor space of such room and at least 50% of the openings shall be capable of allowing free uninterrupted passage of air. For every five rooms there shall be provided for each sex not less than one water-closet and one bathroom" (Hotel Licensing Regulations, 2017).

The Hotel Industry Transformation Map (ITM)

In addition to regulations, Singapore government provides guidance to the hotel industry for future directions. Partnering with the Singapore Hotel Association (SHA), Food Drinks and Workers Allied Union (FDAWU), Workforce Singapore (WSG), and international enterprise (IE), STB developed the Hotel Industry Transformation Map (ITM). The Hotel ITM was announced in November 2016. To transform the hotel industry for sustainable growth, ITM identified four strategies. They are: "building manpower-lean business models; developing new solutions through innovation; growing businesses through internationalization; and building a strong pipeline of quality talent" (Hotel ITM, 2016).

10.2.3 GOVERNMENT INSPIRED INNOVATION: TWO CASE STUDIES OF INNOVATIONS IN SINGAPORE HOTEL INDUSTRY

Singapore government plays important roles in the economic growth of this country. The tourism industry has benefited from the guidance of STB. Similarly, the many innovations in the hotel industry were inspired by STB's innovation projects.

Case 1: Open Hotels Weekend

In 2017, STB, SHA, and Singapore FDAWU initiated a Hotel Careers Campaign that would last for 3 years. The first step of the campaign was to create a website selling the idea that hotel is a "Business of Happiness." In the first round, six hotel employees' stories were posted on this website. The most influential event is the Open Hotels Weekend. The marketing of the event started in July 2017 and the event was held on October 14–15 and 21–22, 2017. Participating hotels arranged hotel visits open to public. The purpose of the event is to introduce the many job opportunities in the hotel industry and change people's prejudices on working in the hotel industry. Students, job hunters, and the public can understand the operation systems of hotels and also the back stage of a hotel. Many can try some of the job duties of hotel employees (STB OHW, 2017). A total of 22 hotels participated in the first round of Open Hotels Weekend (see Table 10.4 for the list).

All participating hotels showed their innovations to the participants. Here are some examples:

Robotics at Park Avenue Rochester

Park Avenue Rochester is a four-star business hotel located at the west of Singapore. As one of the four hotels of the parent company named Park Avenue Hotels & Suites, Park Avenue Rochester has 311 rooms. Similar to other hotels in Singapore, Park Avenue Rochester faces the challenges of manpower shortage, especially in the housekeeping department. The young generation doesn't like the housekeeping jobs. Since the majority of current employees in the house keeping department are in their 50s or 60s, the routine work of pushing a heavy trolley (of 50 kg) becomes very hard for them.

Supported by STB, Park Revenue Rochester started to use two robotics (named Robie1 and Robie2) to collect laundries from guest rooms. The project costs 285,000 SGD with majority been provided by STB. The two robotics can substitute three full time employees. The robotics can take care of monotonous heavy duties, lessen the workload of the department, and increase the morale of the whole department. It also brought high-tech to a traditional department and trained the employees on how to use the robotics. Park Avenue Rochester also developed a guest facing robotic named Cobie. Its main function is to deliver items (such as water and food) to guest rooms. Park Avenue Rochester is the first hotel in Singapore to use robotics. Other Singapore hotels using robotics include Sofitel, Capri, Holiday Inn, Hotel JEN,

TABLE 10.4 Participating Hotels: Open Hotels Weekend October 2017.

Hotels	Number of groups	Duration of the hotel visit (minutes)	Typical activities
Amara Singapore	2	120	Towel animal folding and Ramen noodle demonstration
Fairmont Singapore and Swissotel The Stamford	2	120	Cocktail making
Grand Mercure Singapore Roxy	2	60	Workshop on bartending
Holiday Inn Singapore Atrium	6	45	Auto key tag and RFID tag
JW Marriott Hotel Singapore South Beach	4	60	Tour the ladies-only floor
Lloyd's Inn	4	60	Tatami-style bed making
Marina Bay Sands	8	120	Lunch at staff canteen
Novotel and Mercure Singapore on Stevens	2	60	Guided tours of two hotels
Oasia Hotel Downtown	6	60	Staff sharing
Oasia Hotel Novena	6	60	Shadow housekeeping department duties
Park Avenue Rochester	4	120	Robotics demonstration
Pan Pacific Hotel Singapore	6	60	Bartending and bed making workshop
Regent Singapore, A Four Seasons Hotel	2	90	Housekeeping boot camp
Rendezvous Hotel Singapore	6	60	Shadow housekeeping department duties
Shangri-La Hotel, Singapore	4	90	Cooking workshop, napkin folding, and flower arrangement
Shangri-La's Rasa Sentosa Resort & Spa, Singapore	6	90	Visit of Kids Zone
Swissotel Merchant Court	6	90	Nyonya Kueh workshop
The Fullerton Hotel	1	60	Guided tour
The Fullerton Bay Hotel Singapore	1	60	Uncover events cantered on the pier
The Quincy Hotel	6	60	Bed making
The Warehouse Hotel	4	60	Cock tail making and popiah workshop
Village Hotel Katong	6	60	Shadow housekeeping department duties

Yotel, Frasers, Millennium & Cothorne Hotels, and M hotel. They collaborated with either of the two high-tech companies—Savieke and Techmetics.

Themed Rooms and Cool Zone at Shangri-La's Rasa Sentosa Resort & Spa

Adding kids-friendly facilities has become a trend in hotel industry. Some hotels design themed rooms specifically targeting families. Claimed to be the only beach resort in Singapore, Shangri-La's Rasa Sentosa Resort & Spa opened its Cool Zone to Open Hotels Weekend participants. This is one of the largest Kids club in Singapore. It has Toots Club (for children aged 5–12) and Mini Toots Club (for toddlers aged 4 and below). Cool Zone activities include face painting, water volley ball, balloon twist fun, mermaid lesson, family bingo, archery tag challenge, aqua aerobics, water Polo, and Zig Zag obstacles. Access to Toots Club's programme is only open to guests who have purchased certain resort packages. Rasa Sentosa also has outdoor children's swimming pool and water slides.

RFID at Marina Bay Sands

MBS, the largest hotel in Singapore, has 2,561 rooms, 270 stores and 80 restaurants. The total investment of MBS is 5.7 billion USD. In 2018, the occupancy rate of MBS is 96.7% with an average daily rate (ADR) of 441 USD. The revenue per available room (RevPar) is 426 USD, which is the highest among all Sands properties in Macao, Singapore and USA. MBS has over 9,000 employees and each employee has multiple sets of uniforms, which makes the whole laundry services very tedious. MBS offers 24-hr free laundry services for all its employees. The use of radio frequency identification (RFID) technology automates the identification of uniforms with a chip for each uniform. The location of the uniform can be easily recognized by the information system. To pick up the clean uniforms, employees only need to scan their ID. The uniforms will be automatically distributed to the pick up spot.

Engaging Activities at Different Hotels

All 22 hotels designed very interesting activities to engage visitors (Table 10.4). At Amara Singapore, the cook from China demonstrated the process of making Ramen noodles using local flour mixed with green vegetable juice. Visitors can practice the noodle making process instructed by the cook. Amara also offered a towel animal workshop, teaching participants how to fold a swan. All visitors were divided into groups of twos for towel animal folding competition. The winner can get a buffet voucher as a prize.

At Fairmont and Swissotel, a couple working there used their own story to demonstrate that the hotel is a caring organization and it always creates opportunities for employees. After the lady had demonstrated cocktail making, the gentlemen used liquid nitrogen to make ice creams. Participants can win prizes when they correctly answer questions on the history of Fairmont and Swissotel.

At Warehouse Hotel, the tour included a short cooking workshop on popiah making. Guided by a hotel cook, visitors can make their own popiahs and taste them on site. At Shangri-La hotel, all visitors were divided into three groups, each working on a different theme. Group one is on cooking a cold dish; group two is on napkin folding; and group three is on flower arrangement.

Case 2: Hotel Innovation Challenge Series

In Singapore, many innovations in hotel industry are inspired and directed by STB. STB provides development grants and initiates innovation competitions to engage hotels and high-tech companies to develop innovative solutions to practical problems facing Singapore hotel industry.

Tourism Innovation Challenge for Hotels 2017

In 2017, STB initiated the Tourism Innovation Challenge for Hotels 2017 (Table 10.5). It was in partnership with SHA and the event was supported by Exploit Technologies Pte Ltd—the commercialization arm of the Agency for Science, Technology and Research (A*STAR), Infocomm Media Development Authority (IMDA), NUS Enterprises and Singapore Infocomm Technology Federation (SiTF). More than 35 projects were tested in more than 30 hotels. The main areas of solutions are optical character recognition, data analytics, Bluetooth, IOT, sensors, artificial intelligence, autonomous guided vehicle and equipment (STB, 2017). Successful participants have received financial support from STB (up to 70% of qualifying costs), accessed hotel industry partners facilitated by STB, and created more awareness of the solutions through industry platforms.

Innovation Pitch Challenge 2018

In 2018, as a continuity of the Innovation Challenge 2017, STB called for companies and universities to participate in the Innovation Pitch Challenge 2018.

The objective of the challenge is to uncover innovative upstream ideas for housekeeping and F&B (STB, 2018c). The Innovation Pitch Challenge 2018 listed many practical problems for companies to work on: (a) Food and Beverage: how to quickly set up and tear down a banquet hall; solutions for efficient napkin folding; how to fasten food serving and dish cleaning processes; how to automate the dishwashing process, and how to retrieve items from storage areas. (b) Housekeeping: fast ways for bed making; how to deal with heavy mattresses; how to clean many glass surfaces; better ways for cleaning different types of surfaces and better ways for vacuuming carpets.

TABLE 10.5 Selected Projects of Tourism Innovation Challenge for Hotels 2017.

Company	Solutions	Status
DROP Positioning	Use an automated LIVE location system to simplify the luggage storing and locating process	Completed
RFCOM	Simplifies guests' luggage storing and locating	Completed
Vouch	Intelligent digital concierge	Completed
Evercomm Uni-Tech Singapore	Provides a virtual assistant to effectively maintain critical equipment operation	Completed
Affle	Chat concierge product	In development
Chemistry	Solutions to easier and faster turning over beds in guestrooms	In development
GTRIP Global	Automating the check-in process	In development
Laundry Network	Automated Mobile Robots to pick up/deliver linen	In development
New Doone Technology	Self-service kiosk for automated check-in	In development
Tuple Technologies	A platform for hotels to collect customers' itineraries and preferences	In development

In Round 1, industry practitioners described their problems and shared more details on the challenges. Participants provided their institutions' details to STB, selected problems from the list, introduced the technology to be used to solve the problems and then, submitted their proposals. The majority of the participants are high tech companies. In 2018, 40 hospitality business (HB) students from Singapore Institute of Technology (SIT) formed 8 groups to join the Innovation Pitch Challenge 2018. Their proposals include using robotic carts to serve food, developing automated table and conveyor belt to save labor, setting a bed elevation pole to help with bed making, and splitting current mattresses into layers and reducing the weight of mattresses.

In Round 2, teams passed round 1 screening were paired up with hotels for site visits so that those teams can better understand the hotel challenge areas. Four out of eight SIT student teams reached this round. The student teams were paired up with Hilton, M hotel, Pan Pacific Hotel, and Singapore Hotel Productivity Center for additional information.

After submitting the revised proposal, another round of screening determined the final winning projects. All seven winning projects were featured at the 4th STB-SHA Hotel Industry Conference in November, 2018. The two SIT student projects are "Autonomous Robot to Deliver Collapsed Tables to Designated Spots and Unloads to Floor for Manual Setup" and "Mobile and Space Efficient Napkin Folding Machine to Produce 3 Types of Napkin Folds." The other five projects are from high-tech companies. They are: (1) Glass cleaning solution that delivers anti-fingerprint protection to glass and mirrors surfaces; (2) Portable bed-maker device to help loosen and tighten bedsheets; (3) Autonomous robotic carrier that can be retrofitted over existing carts and trolleys; (4) Autonomous robot to move fully setup banquet tables to designated spots within banquet venue and (5) Stewarding: Space efficient Cartesian robot that can be retrofitted over existing modular dishwashers to provide a fully automated washing process.

Hotel Innovation Challenge 2019

In April 2019, STB launched Hotel Innovation Challenge 2019, calling for innovative proposals for challenges similar to those listed for Innovation Pitch Challenge 2018 (STB, 2019b). Here are a few listed challenges: (1) how to automate the setting up of tableware and the setting up and keeping of chairs after banquets; (2) how to autonomously vacuum hotel room floor; and (3) how to optimize labor allocation and schedule workforce based on dynamic hotel demand.

10.3 DEVELOPMENT AND INNOVATIONS OF AIR TRANSPORT INDUSTRY IN SINGAPORE

10.3.1 HISTORY AND CURRENT DEVELOPMENT OF AIR TRANSPORT INDUSTRY IN SINGAPORE

This section aims to summarize the history of civil aviation industry in Singapore. Table 10.6 shows a timeline of the development of civil aviation industry in Singapore while Figure 10.3 illustrates the passenger movement since 1976 and aircraft movement since 1991 until 2018 in Singapore.

TABLE 10.6 Development of Singapore Aviation Industry.

1920s - 1940s	•Aerodromes in Sembawang and Seletar served civil aircraft and the Royal Air Force (RAF) •Governor Sir Ceil Clementi announced the plan to build a new civil airport on 31 August 1931. •Kallang Airport, the first civil airport was opened on 12 June 1937. •Malayan Airways Limited (MAL) was established on 21 October 1937. It only started its operations 10 years later from Kallang Airport to Kuala Lumpur, Ipoh and Penang in 1947, Bangkok in 1948, North Borneo, Sarawak, Mergui, Rangoon in 1949.
1950s	•Paya Lebar Airport was opened on 20 August 1955 with all commercial flights transferred from Kallang Airport to Paya Lebar Airport on 21 August 1955. •The first plane landed in Paya Lebar Airport was a Garuda flight from Jakarta. •At the end of 50s, passenger movement in Paya Lebar Airport was more than 300,000 while aircraft movement exceeded 30,000 per year.
1960s	•The runway at Paya Lebar Airport was extended from 2.4km to 2.7 km in 1962, which was one of the longest runways in Asia. •Malayan Airways Limited was renamed as Malaysian Airways Limited followed the Federation of Malaysia in 1963. The company became Malaysia-Singapore Airlines (MSA) in May 1966.
1970s	•In 1972, MSA was seperated into Malaysian Airline System and Singapore Airlines. Singapore Airlines set up a subsidiary, Singapore Airport Terminal Services (SATS), to handle the ground services. •Passenger movement in Paya Lebar Airport increased from 1.7 million in 1970 to 4 million in 1975, while the airport was reaching its maximum capacity in 1975. •Singapore Government decided to build Changi Airport. •The construction of Changi Airport started in 1979.
1980s	•Changi Airport started operation on 1 July 1981 with SQ101 as its first arrival flight which carried 140 passengers from Kuala Lumpur. Terminal 1 is designed for handling 22 million passenger a year. •The construction of Terminal 2 Changi Airport began in 1986. •Changi Airport received its first 'Best Airport' award from Business Traveller UK in 1988. •SilkAir, a regional airlines set up by Singapore Airlines, started operations in 1989.
1990s	•Terminal 2 started operations in 1991. •Singapore was the first countries in Asia to sign an Open Skies Agreement with the United States of America in 1997.
2000s	•Changi Airport won Skytrax "World's Best Airport" for the first time in 2000. In the same year, the construction of Terminal 3 began, which was opened on 1 July 2008. •Changi Airport MRT Station was opened in 2002. •Tigerair, the first low cost carrier, established in 2003 and became the first airlines operated in the new Budget Terminal when it was opened on 26 March 2006. •In 2008, the Singapore-Kuala Lumpur route became fully liberalised.
2010s	•Singapore Airlines Group low cost subsidiary, Scoot, was established in 2011. •Budget terminal was closed in 2012. Government planned to convert it into Terminal 4. •Tigerair became a subsidiary of Singapore Airlines Group in 2014 and eventually merged with Scoot in 2017. •Government started the planning of Jewel Changi Airport, Changi East, the three-runway system and Terminal 5 in 2013. •Terminal 4 started operation in 2017 with the installation of the Fast and Seamless Travel (FAST) programme which aim to provide a seamless check-in and immirgration process for travellers.

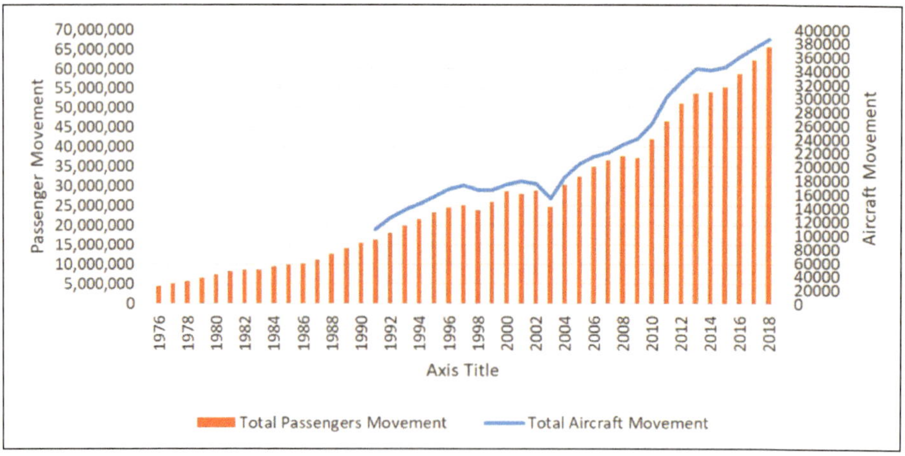

FIGURE 10.3 Volume of air traffic in Singapore.

SINGAPORE CIVIL AVIATION INDUSTRY IN 100 YEARS

1920s–1940s

The history of air transport in Singapore could be traced back to 1920s when aerodromes in Sembawang and Seletar were constructed. Both aerodromes were not meant for civil aviation industry only, instead, they served both civil aircrafts and the Royal Air Force. In early 1930s, the plan of building a civil aviation airport was announced by the Governor of Singapore, Sir Ceil Clementi. On June 12, 1937, the first civil aviation airport in Singapore, Kallang Airport, was opened (Cornelius and Yap, 2013). The opening of Kallang Airport marked the beginning of civil aviation industry in Singapore. In the same year, Malayan Airways Limited (MAL) was established (Chua, 2016). Due to the concern of commercial viability of the air route between Singapore and Malaya, MAL only started its operations 10 years later in 1947 (National Library Board Singapore, 2014). Its initial flight operations were flying from Kallang Airport to Kuala Lumpur, Ipoh, and Penang (Singapore Airlines, 2019).

1950s–1960s

Due to the continuous growth of air traffic and the larger aircraft type, there was a need to build another airport to replace Kallang Airport (Cornelius and Yap, 2013). On August 20, 1955, the new airport, Paya Lebar Airport, was opened (Hussain, 2013) with all the commercial flights transferred from Kallang Airport to the new airport on the next day. A Garuda flight from Jakarta was the first plane landed in Paya Lebar Airport (National Library Board Singapore, 2015). By end of 1950s, the passenger movements of Paya Lebar Airport exceeded 300K while the airport handled more than 30K aircraft movements per year. In order to accommodate bigger aircraft, the runway was extended from 2.4 km to 2.7 km which made it one of the longest runways in Asia in 1960s (Changi Airport, 2019). At the same time, following the Federation of Malaysia in 1963, MAL was renamed as Malaysian Airways Limited. In 1966, it became Malaysia–Singapore Airlines (MSA) (Singapore Airlines, 2019).

1970s

In 1970s, many major events happened in the air transport industry in Singapore which shaped the future development of the industry. MSA was spilt into Malaysian Airlines System and Mercury Singapore Airlines (Singapore Airlines, 2019) in early 1972 (National Library Board Singapore, 2014). Mercury Singapore Airlines was renamed as Singapore Airline (SIA) in the same year (Singapore Airlines, 2019) and has become the national carrier of Singapore since then. In order to concentrate on the airlines business and handle the increasing load of ground handling, SIA set up Singapore Airport Terminal Services (SATS) in 1972 and continues to serve the industry since then (SATS, 2019).

The passenger movement in Paya Lebar Airport was reaching its maximum capacity in 1975. The passenger movement increased by 2.4 times from 1.7 million per year in 1970 to 4 million in 1975 (Changi Airport, 2019). In order to catch up with the expanding volume of air traffic, in late 1970s, Singapore government decided to move the airport away from the city and build a new one in Changi where ample space is available for future expansion (Tan, 2016).

1980s

Paya Lebar Airport had served Singapore for 26 years since 1955 before it ceased operations in 1981. It had a humble start in 1955, the passenger movements increased by 24 times to 7.3 m in 1980 (Department of Statistics

Singapore, 2019), a few months before the airport operations moved to Changi Airport.

Changi Airport started operations on July 1, 1981. The first arrival flight was SQ101 which carried 140 passengers from Kuala Lumpur (Ho, 2019). The passenger capacity of Terminal 1 was 12 million per year (De Cotta, 2018). In 1981, the volume of passenger movement was 8.2 million and grew to 10 million in 1986 (Department of Statistics Singapore, 2019). An average growth rate of 4.4% was observed from 1981 to 1986. In order to cope with the growth of the industry, Singapore government started the construction of Terminal 2 in 1986. 2 years later, Changi Airport received its first "Best Airport" award from Business Traveller UK (Changi Airport, 2019). The civil aviation industry in Singapore continued to expand from 1986 to 1989 when it experienced double-digit annual growth rate of 12% on average. Throughout the whole 1980s, the annual growth rate of passenger movement was 8.2% on average. In 1989, SIA set up a regional airline, SilkAir (Singapore Airlines, 2019), to accommodate the expanding demand.

1990s

Changi Airport Terminal 2 started operations in 1991 (Changi Airport, 2019) with the capacity of 23 million (De Cotta, 2018). According to Department of Statistics Singapore (2019), the passenger movements increased by 10.1% annually on average from 1986, when the construction of Terminal 2 started, to 16 million in 1991. The opening of Terminal 2 released the stress from Terminal 1 which was originally designed for 12 million passengers per year. In 1995, Changi Airport upgraded Terminal 1 and expanded the capacity to 21 million passengers per year (De Cotta, 2018). As a result, the total capacity of Changi Airport reached 44 million passengers per year in 1995 while the annual passenger movement was 23 million. Although the annual growth rate of civil aviation industry slightly slowed down in the era of 1990s when compared to 1980s, it was still a healthy growth. Changi Airport experienced 6.4% annual growth rate in terms of passenger movement from 1990 to 1999 while the aircraft movement increased by 5.6% annually from 1991 to 1999 (Department of Statistics Singapore, 2019). During Asian Financial Crisis in 1997 and 1998, the passenger movements and aircraft movements dropped by 5.4% and 4.3%, respectively, in 1998. Tourist arrivals decreased by 13.28% in the same year (Department of Statistics Singapore, 2018). With the backdrop of unstable economic condition, Singapore became one of the first Asian countries to sign an Open Skies Agreement with the United

States of America in 1997 (Changi Airport, 2019). The volume of passenger movement picked up again in 1999 with an increase rate of 9.5% when compared to 1998 (Department of Statistics Singapore, 2019).

2000s

The statistics from Department of Statistics Singapore (2019) showed that the volume of air traffic in Singapore further slowed down to an annual growth rate of 4% in 2000s. This is mainly due to the outbreak of SARS in 2003 and the global financial crisis from 2007 to 2009. In 2003, the passenger movements and aircraft movements in Changi Airport decreased by 14.9% and 11.7%, respectively, although there was a robust bounce back in 2004. The passenger movement in 2004 increased by 23.1% while aircraft movement increased by 19.8% when compared to 2003. In 2009, Changi Airport experienced a slight dip in passenger movement at 1.3%.

Despite the challenges in the era of 2000s, the air transport industry continues to develop. In 2002, Singapore government enhanced the land and air connectivity by opening the Changi Airport MRT Station (Ministry of Transport, 2002) which provided great convenience to the passengers and visitors. The MRT line connects Changi Airport to the city center and all the way to the West of Singapore. Changi Airport (2019) began the construction of Terminal 3 in 2000, which was opened in 2008 with the capacity of 22 million passengers per year (De Cotta, 2018). In 2006 (Changi Airport, 2019), a new Budget Terminal with the capacity of 7 million passenger per year (Toh, 2012) was opened to compete in the low cost travel market. Tigerair, the first low cost airlines in Singapore which was established in 2003 (CAPA Centre for Aviation, 2015), was the first airline operated in the Budget Terminal. Before the opening of the Budget Terminal, the passenger movement reached 32.4 million passengers per year in 2005 while the capacity utilization rate was 73.6%. In 2007 before Terminal 3 was opened, the passenger movement was 36.7 million with the utilization rate at 71.9%. With the opening of the Budget Terminal, the total capacity in Changi Airport increased to 51 million passengers per year in 2006 and 73 million passengers per year in 2008 when Terminal 3 started operations. The passenger movement reached 35 million and 37.7 million in 2006 and 2008, respectively. With the additional capacity from the Budget Terminal and Terminal 3, the utilization rate dropped to 68.6% and 51.6% in 2006 and 2008, respectively.

In order to facilitate the development of the airline industry, Singapore government concluded an agreement with Malaysia government to fully liberalize the air route between Singapore and Kuala Lumpur in 2008, which allows airlines from Singapore and Malaysia to operate unlimited number of flights between these two cities (CAPA Centre for Aviation, 2018).

2010s

After the global financial crisis, the volume of air traffic increased significantly again in the following 3 years. According to the statistics from Department of Statistics Singapore (2019), the passenger movements in 2010, 2011, and 2012 increased by 13%, 10.7%, and 10%, respectively. In the era of 2010s, the statistics (Department of Statistics Singapore, 2019) showed a healthy growth in terms of the volume of air traffic when compared to the era of 2000s. The passenger movements in Changi Airport increased by 6.6% annually while the aircraft movements increased by 5.5% annually. SIA Group established their very first low-cost subsidiary, Scoot, in 2011 which later on merged with Tigerair in 2017 after SIA acquired it in 2014 (CAPA Centre for Aviation, 2019). Singapore government decided to close Budget Terminal in 2012 and built Terminal 4 on the same piece of land (Changi Airport, 2019). The capacity dropped to 66 million passengers per year in 2012 after the Budget Terminal was closed when the passenger movement was 51 million. The newly built Terminal 4, which can handle 16 million passengers per annum, started operations in 2017. Including Terminal 4, the total capacity of Changi Airport became 82 million passengers per annum. In 2017, the passenger movement was 62 million which was 75.6% of the total capacity. However, before Terminal 4 was opened, the passenger movement reached 58.7 million in 2016 which was 88.9% of the total capacity. It created a stress on Changi Airport to open Terminal 4 on time (De Cotta, 2018).

The Future

When Changi Airport started in 1981, 31 airlines flew there and offered 66 city links to passengers (De Cotta, 2018). After 39 years, there are more than 100 airlines operating at Changi Airport which offer over 400 city links nowadays (Changi Airport 2019). According to the statistics (Department of Statistics Singapore, 2019), from the time Changi Airport started operations in 1981 until 2018, the average annual growth rate of passenger movements is 6.8% while it was 5% for aircraft movements from 1992 to 2018. In 2018, the passenger movements just surpassed 65.6 million while the aircraft movements exceeded 380K. Based on the average annual growth rate of passenger movement from 1981 to 2018, the airport will reach its maximum capacity in a foreseeable future. To cope with the continuously growth of air traffic volume and enhance the status of Singapore as a major air hub, Changi Airport (2019) announced a massive expansion plan. The plan includes the construction of Terminal 5, a three-runway system and other landside and aviation support facilities to increase the capacity of Changi Airport; tunnels and other underground systems to increase the connectivity of the airport to the city; as well as The Changi East Industrial Zone (Airport Technology, n.d.). Meacham (2017) reported that the land preparation for the fifth terminal has already started in 2014 and will continue until 2020. The construction of Terminal 5, which is so far the biggest expansion project in Changi airport, is expected to start in 2020s and complete in 2030s (Airport Technology, n.d.).

10.3.2 CHANGI AIRPORT'S INNOVATIONS AND OTHER INITIATIVES

From 1976, when statistics of passenger movement were recorded for the first time, till 2018, the passenger movements in the airports in Singapore increased by 14.5 times in 42 years (Department of Statistics Singapore, 2019). Singapore government always has a timely plan in order to accommodate the growth of the civil aviation industry. As a result, the airport in Singapore was moved from Paya Lebar to Changi; Changi Airport was expanded from one terminal to two terminals in 1991, three terminals in 2008, four terminals in 2017, and the fifth terminal is expected to operate in 2030s. In addition to cope with the capacity need, Changi Airport also continues to strive for excellence in passenger experience. Right from the beginning, Changi Airport set a very clear goal that the airport should not be a functional facility only, but also provide pleasant and efficient travel experience to passengers (De Cotta, 2018). With this mentality, Changi Airport won the prize of "Best Airport" in the world from Business Travellers UK in 1988 for the first time (Changi Airport, 2019) and also in the following 30 years consecutively (Business Traveller Asia Pacific, 2018). The airport also won Skytrax "World Airport Award" for 10 times, that is, 2000, 2006, 2010, 2013–2019 (Skytrax, 2019). Changi Airport is the first airport to win this award 7 years in a row (Miller, 2019). Until 2019, Changi Airport has already won 595 airport awards (Changi Airport, 2019). This article takes Changi Airport as an example and examines what innovations and other initiatives that the airport has applied in order to catch up with the capacity challenges as well as achieve passenger experience excellence. From the beginning of Changi Airport story in 1980s, the innovations were guided by the principle of pleasing passenger. When Changi Airport was designing Terminal 4, the team had gone even further from passenger experience to guest experience (De Cotta, 2018). Table 10.7 summarized the major innovations and initiatives in Changi Airport and will be discussed in the following sections.

TABLE 10.7 Innovations of Changi Airport.

	Innovations and other initiatives
Delightful and comfortable environment	➢ Pioneered the concept of "airport garden"
	➢ Kinetic Rain
	➢ Taxi queue inside the terminal building
Free leisure facilities and services	➢ Rooftop swimming pool
	➢ Entertainment Deck
	➢ Movie theatres
	➢ Xperience Zone
	➢ TV Lounges
	➢ Free Singapore tour
	➢ Changi Experience Ambassador
	➢ iShopChangi
	➢ Immersive wall
	➢ Heritage zone
Efficiency and convenience	➢ iChangi HD app
	➢ Pioneered automated check-in system
	➢ FAST system
	➢ 3D CT X-ray machine
Stay connected	➢ Free local calls in transit area
	➢ Internet terminals
	➢ WIFI connections

Delightful and Comfortable Environment

Back to 1980s, Changi Airport had already put its effort into designing a delightful physical environment to passengers and visitors. Ho (2019) reported that Changi Airport aimed to create Terminal 1 as a lifestyle attraction. In

1980s, it was the first airport introduced the concept of "airport garden" with an iconic water feature which vertically cut through the departure and arrival level. Following the concept of "airport garden," Changi Airport decorated each terminal with various themed gardens, they include Arrival Garden, Cactus Garden, Discovery Garden, Piazza Garden, and Water Lily Garden in Terminal 1; Orchid Garden, Sunflower Garden, and Enchanted Garden in Terminal 2; Butterfly Garden and Crystal Garden in Terminal 3; Steel in Bloom in Terminal 4; and Canopy Park, Hedge Maze, Rain Vortex, and Petal Garden in Jewel.

In 2012, Changi Airport presented the world's largest moving sculpture, Kinetic Rain, in the departure Hall in Terminal 1 (National Art Council, 2015). This new decoration was part of the upgrade project of Terminal 1. According to Changi Airport (2019), the sculpture consists of 1216 droplets which are made of bronze. The whole sculpture is 9.8 m long by 4 m height. The 1216 bronze droplets can form 16 different shapes. Some of these decorations are located in the public areas while some are inside the transit areas in the airport. All of these decorations created a delightful environment and became very popular photograph taking points among passengers and visitors.

Passenger comfort was one of the key concerns in the design of Changi Airport. In 1980s, Changi Airport (2019) decided to build the taxi stand inside the building of Terminal 1. By doing so, passengers can enjoy the air-conditioning environment from disembark all the way to the taxi stand. The design of Terminal 2, 3, and 4 followed the same principle. This arrangement was particularly essential in Singapore as the weather is all year around hot and humid.

In the design of Terminal 4, Changi Airport made a breakthrough by creating a walk through duty-free zone right after the departure immigration and security areas. The walk through zone design is an open concept where there is no clear barrier of the shop; and no separation between wines and spirits, and perfumes and cosmetics. This innovative design aimed to allow passengers to walk freely between various duty free offerings. Passenger comfort was the major consideration in the design; therefore, Changi Airport ensured a spacious environment for the passengers along the walk through zone (De Cotta, 2018). With this design concept, Changi Airport aimed to minimize passengers' feeling about duty free shops foisted on them, and also allow the passengers to feel the excitement of duty-free shopping immediately after security areas as well as present the commercial offerings to the passengers.

Free Leisure Facilities and Services

In 1995, Changi Airport wanted to offer transit passengers an innovative way to refresh after long haul flights, therefore a rooftop swimming pool was opened in the transit area in Terminal 1. Changi Airport has also designed other free leisure facilities and services to ensure passengers are taken care of and entertained when they are waiting for their flights. These facilities include computer games in Entertainment Deck and latest sports events show in large screens in Xperience Zone in Terminal 2; latest movies for free in Movie Theatres in Terminal 2 and 3; as well as TV lounges in Terminal 1, 3, and 4. The transit passengers who have 5.5 h–24 h between flights might not want to stay within the airport terminal for the whole period of time. Changi Airport offers them free guided heritage tour and city sights tour in Singapore.

Changi Experience Ambassadors were introduced in 2010s. They are a group of customer service officers who serve passengers across all four terminals. They aim to provide assistances to passengers and visitors. The service from ambassadors is particularly important to the passengers who need to transit from one terminal to another.

Shopping is one of the favorite activities among passengers. There are 405 retail outlets throughout four terminals (De Cotta, 2018). In 2013, Changi Airport launched iShopChangi, a new online shopping portal (Changi Airport, 2019), in order to provide convenience to the passengers who might be arriving the airport late and also to catch up with the trend of online shopping. iShopChangi (2019) allows passengers to shop online in advance of their departure and collect their order from the collection centers in the transit lounges of the terminals. For arrival passengers, they can also shop online and the airport provides delivery service to their homes or hotels. Jewel Changi, a massive shopping mall with 280 retails as well as food and beverage outlets, was opened on April 17, 2019 (Kaur, 2019). This new initiative added a wide variety of options to passengers and the public.

Queuing is always an issue in airports. Customers normally feel waiting time is longer when they are unoccupied or if they are anxious (Haksever et al., 2000). Changi Airport installed a massive LED Immersive Wall at the security screening area in Terminal 4. The Immersive Wall is 70 m long by 5 m high which stretches across the whole security screening area (Toh, 2017). The wall displays 16 different digital shows to provide entertainment to passengers when they are queuing for security screening. Changi Airport claimed that some passengers now are not rushing to get the security screening done. Instead, they watch and wait till the digital shows to finish before they continue security screening. Another digital show in Heritage Zone in the transit area of Terminal 4 also keeps the passengers entertained when they are waiting for their flights. Heritage Zone is decorated as a settlement of Peranakan shophouses which shows the evolution of the architecture style from 1880s to 1950s in Singapore (Wong, 2017). In the middle of the shophouse façade, a 6-min Peranakan Love Story (Changi Airport, 2019) is screened to provide passengers a glimpse of life in the old days in Singapore.

Efficiency and Convenience

Passengers want to finish the check-in, security screening, and immigration process as fast as possible. They also want to obtain up-to-date flight information in order to plan their time better. An efficient airport process and real-time flight information allows passengers a better planning to enjoy the entertainment, leisure facilities, shopping, and dining facilities.

Changi Airport launched iChangi HD app in 2012 with an aim to enhance passenger experience (Singapore Business Review, 2012) by providing them real-time information and convenience. The app provides real-time departure and arrival flight information, transportation guide to and from the airport, dining, shopping, and attraction guide in all four terminals and Jewel, airport guide for arrival, departure, and transit passengers, way-finder to check-in row, gate, shop, restaurants, and other facilities in the airport, as well as a Chatbot.

The automated check-in system, Fully Automated Seamless Travel (FAST), that Changi Airport conducted a 6-month trial run in 2014 exemplified its' willingness to provide a more efficiency airport processes to travelers (Ministry of Transport, 2004). Changi Airport was the first airport in the world to provide automated check-in system which integrated biometric recognition technology into the check-in and immigration process for departing passengers (Meacham, 2017). The traditional check-in and immigration process require passengers to approach the airline check-in counter first to obtain boarding pass and check-in luggage. Passengers should then go through police check for cross-checking names on boarding passes and passports before they can proceed to immigration hall. After that, officers at the immigration counters verify passengers' passports. Ministry of Transport (2004) claimed that the traditional process takes 8 to 15 min while the new FAST system can complete the whole process in less than 2 min. Changi Airport cooperated with SIA and invited around 9000 frequent travelers from KrisFlyer programme to conduct the trial of FAST. Each of the invited traveler was given a card with a microchip which contained his facial features and fingerprint details. At a one-stop kiosk, the traveler can authenticate his identify, check-in, select seat, collect boarding pass, and clear immigration. As Mr. Chew Choon Seng, SIA Chief Executive Officer, stated that FAST is a good example of providing benefits to passengers by applying new technology without compromising security and effectiveness (Ministry of Transport, 2004).

With the mind of minimizing queues, offer seamless airport process to passengers without compromising security standard, Changi Airport introduced a new generation of automated check-in system together with the opening of T4 in 2017. This new system is also called FAST but with a different full name, Fast and Seamless Travel. The new FAST system-integrated check-in, bag drop, immigration, and boarding processes (De Cotta, 2018). Facial recognition technology is the key to this new system. At check-in kiosk, passengers scan their passport, and print the boarding pass and luggage tag. The passengers then weight their luggage at the automated bag drop machine which also measures the dimension of the luggage. A camera in the machine take a photo of the passengers and compare with the photo in the passport which the passengers scanned previously at the check-in kiosk, this process is to verify the identity of the passengers. After the verification, the machine sends the luggage into the baggage handling system. At departure immigration, passengers scan their passport and boarding pass, then the automated immigration gates verify both face and fingerprint of the passengers. When the flight is ready for boarding, passengers scan their boarding pass at the automated boarding gate which also take a photo of the passengers to verify against the photos taken earlier at the immigration gate. Changi Airport Terminal 4 adopted self-service approach for the whole departure process which free up passenger service agents to provide assistance to the passengers who have more specific needs. Changi Airport claimed that the FAST system processes 20–30% more passengers per hour while Cathay Pacific also stated that the automated boarding system speeds up the boarding process from 30 min to 15 min (Teo, 2018).

Terminal 4 also adopted new 3D CT X-ray technology in the security screening process. With this new 3D machine, passengers no longer need to take out their electronic devices, such as laptop, to be scanned separately which speed up the whole screening process.

In order to make the new technologies and innovations work properly, robust planning, test and experiments are needed. Along with the opening of Terminal 4, Changi Airport is planning for another project in a much bigger scale, Terminal 5. Changi Airport had generated 70 versions of blueprint for Terminal 4 before the finalized one and made Terminal 4 as a testing bed of the new technologies and innovations for Terminal 5 (De Cotta, 2018). This is to ensure the innovativeness and functionality of Terminal 5 which is planned for the capacity of 50 million passengers per year (Aviation Week Network, 2018).

Stay Connected

Changi Airport realized passengers' demand on staying connected. In 1980s, Changi Airport was the first airport in the world which allowed passengers to make local telephone calls for free in the transit area (Changi Airport 2019). In 2000s, Internet connection became more important than making local telephone calls. Changi Airport consolidated broadband Internet access in 2001. By 2004, Changi Airport provided 200 Internet terminals across all two terminals, which was a record for airports worldwide (Changi Airport 2014). Nowadays, passengers and visitors can enjoy free WIFI connection throughout the whole airport and Jewel.

10.4 CONCLUSION

Singapore ranks high on many global lists. In October 2019, World Economic Forum (WEF) announced that Singapore had overtaken the United States and become the most competitive nation in the world (Taylor, 2019). Singapore also becomes the smartest city in the world and tops IMD smart City Index (Yip, 2019). The smartness and competitiveness of the city have driven the innovations and growth of the tourism industry in Singapore. Most recently, the launching of E-Visitor Authentication (EVA) on Nov 6, 2019 allows Singapore hotels to use facial recognition technologies to automate hotel check-in process (Heng, 2019). We foresee that Singapore's many government-led projects such as Smart Nation initiative will turn it into one of the top cities in terms of improving the livelihoods and experiences of international tourists.

KEYWORDS

- **Singapore**
- **air transport**
- **hotel**
- **development**
- **innovation**
- **case study**

REFERENCES

Airport Technology. Terminal 5, Changi International Airport. [Online] (n.d.). https://www.airport-technology.com/projects/terminal-5-changi-international-airport/ (accessed Jun 28, 2019).

Aviation Week Network. Changi's Terminal 5 for Future Growth. [Online] Feb 6, 2018. https://aviationweek.com/singapore-airshow-2018/changi-s-terminal-5-future-growth (accessed Jul 4, 2019).

Budget Direct. Singapore Tourism Statistics 2019. [Online] Sep 24, 2019. https://www.budget-direct.com.sg/travel-insurance/research/singapore-tourism-statistics.

Business Traveller Asia Pacific. Business Travelller Awards 2018. [Online] Oct 15, 2018. https://www.businesstraveller.com/business-travel/2018/10/15/business-traveller-awards-2018/ (accessed Jul 1, 2019).

CAPA Centre for Aviation. Singapore Airlines Proposed Takeover of Tiger Air a Scene Changing Strategic Move for the SIA Group. [Online] Nov 10, 2015. https://centreforaviation.com/analysis/reports/singapore-airlines-proposed-takeover-of-tigerair-a-scene-changing-strategic-move-for-the-sia-group-252358 (accessed Jul 13, 2019).

CAPA Centre for Aviation. Kuala Lumpur-Singapore air route: HSR no longer a threat. [Online] Jun 26, 2018. https://centreforaviation.com/analysis/reports/kuala-lumpur-singapore-air-route-hsr-no-longer-a-threat-424321 (accessed Jul 13, 2019).

CAPA Centre for Aviation. Scoot. https://centreforaviation.com/data/profiles/airlines/scoot-tr (accessed Jul 13, 2019).

Changi Airport. A History of Firsts. http://www.changiairport.com/corporate/about-us/our-story.html#aem-ajaxpopup (accessed Jun 22, 2019).

Changi Airport. Awards and Accolades. http://www.changiairport.com/content/cag-corp/en/about-us/awards-and-accolades.html (accessed Jul 4, 2019).

Changi Airport. Heritage Zone. http://www.changiairport.com/en/discover/attractions/heritage-zone.html (accessed Jul 2, 2019).

Changi Airport. Partnership. http://www.changiairport.com/corporate/partnerships/airlines.html (accessed Jul 4, 2019).

Chua, A. Singapore Airlines. National Library Board Singapore [Online] 2016. http://eresources.nlb.gov.sg/infopedia/articles/SIP_1705_2010-08-10.html (accessed Jul 12, 2019).

Cornelius, V.; Yap, J. Former Kallang Airport Building. National Library Board Singapore [Online] Jun 11 2013. http://eresources.nlb.gov.sg/infopedia/articles/SIP_2013-06-14_120419.html (accessed Jun 22, 2019).

De Cotta, I. Taking Flight The Changi Airport Terminal 4 Story, Changi Airport Group (Singapore) Pte Ltd: Singapore, 2018.

Department of Singapore Statistics. Tourism. https://www.tablebuilder.singstat.gov.sg/publicfacing/createDataTable.action?refId=1991 (accessed Feb 12, 2018).

Department of Singapore Statistics. Transport. https://www.singstat.gov.sg/find-data/search-by-theme/industry/transport/latest-data (accessed Jun 22, 2019).

Fullerton Hotel. The Fullerton Heritage. https://thefullertonheritage.com/?_ga=2.101681729.1382654558.1572139435-1518212667.1572139435 (accessed Oct 27, 2019).

Goodwood Park Hotel. Our Heritage. https://www.goodwoodparkhotel.com/heritage (accessed Oct 27, 2019).

Haksever, C.; Render, B.; Russell, R. S.; Murdick, R. G. Service Management and Operations, 2nd ed.; Prentice Hall: New Jersey, 2010.

Henderson, J. C. Conserving Colonial Heritage: Raffles Hotel in Singapore. *Int. J. Herit. Stud.* **2001,** *7*(1), 7–24.

Heng, M. Faster Check-In at Singapore Hotels with New Automated Facial Recognition System. Straits Times. [Online] https://www.straitstimes.com/singapore/speedier-check-in-process-for-hotels-possible-with-new-automated-facial-recognition-system (accessed Nov 10, 2019).

Ho, J. Before Jewel: Remembering Changi Airport Terminal 1's debut 38 years ago. Channel News Asia [Online] Apr 13, 2019. https://cnalifestyle.channelnewsasia.com/trending/jewel-changi-airport-terminal-1-1981-history-11425356 (accessed Jul 1, 2019).

Hotel Act, 1999. https://sso.agc.gov.sg/Act/HA1954 (accessed Sep 24, 2019).

Hotel ITM, Transforming the Hotel Industry for Sustainable Growth, 2016. https://www.mti.gov.sg/-/media/MTI/ITM/Lifestyle/Hotels/Hotels—STB-FactSheet.pdf (accessed Sep 24, 2019).

Hotel Licensing Regulations, 2017. https://sso.agc.gov.sg/SL/HA1954-RG1?DocDate= 20170724 (accessed Sep 24, 2019).

Hussain, A. Paya Lebar Air Base to Move to Changi. Today [Online] Aug 19, 2013. https://www.todayonline.com/singapore/paya-lebar-air-base-move-changi (accessed Jul 12, 2019).

iShopChangi. How to Shop. https://www.ishopchangi.com/How-to-Shop (accessed Jul 2, 2019).

Kaur, K. Jewel Changi Airport to Open on April 17 with 280 Shops and F&B Outlets; Public Preview from April 11 to 16. The Strait Times [Online] Mar 7, 2019. https://www.straitstimes.com/singapore/transport/jewel-changi-airport-to-open-on-april-17 (accessed Jul 13, 2019).

Las Vegas Sands (LVS). Las Vegas Sands Annual Report 2018.

Liu, G.; Rai, R.; Lim, A. K. *Raffles Hotel.* Landmark Books; Tien Wah Press: Singapore, 2006.

Mastercard. Global Destination Cities Index 2019. [Online] Sep 24, 2019. https://newsroom.mastercard.com/wp-content/uploads/2019/09/GDCI-Global-Report-FINAL-1.pdf

Meacham, S. World's Best Airport: What it's Like to Spend 24-Hours Straight Inside Singapore's Changi Airport. Traveller. [Online] Jul 7, 2017. http://www.traveller.com.au/a-day-in-the-life-of-changi-airport-gx49xf (accessed Jul 2, 2019).

Miller, P. The World's Best Airport in 2019 are Announced. Skytrax [Online] Mar 27, 2019. https://www.worldairportawards.com/worlds-best-airports-announced-2019/ (accessed Jul 1, 2019).

Ministry of Transport. Take The Fast Lane At Changi Airport World's First Fully Automated Check-In And Immigration Clearance System That Uses Biometric Recognition

Technology, Goes On Trial. [Online] Nov 25, 2004. https://www.mot.gov.sg/news-centre/news/Detail/Take%20The%20Fast%20Lane%20At%20Changi%20Airport%20World's%20First%20Fully%20Automated%20Check-In%20And%20Immigration%20Clearance%20System%20That%20Uses%20Biometric%20Recognition%20Technology,%20Goes%20On%20Trial/ (accessed Jul 2, 2019).

Ministry of Transport. Completion of the Changi Airport MRT Extension. [Online] Feb 6, 2002. https://www.mot.gov.sg/news-centre/news/Detail/Completion%20Of%20The%20Changi%20Airport%20MRT%20Extension/ (accessed Jul 13, 2019).

National Art Council. Public Art Trust Kinetic Rain. https://www.publicarttrust.sg/Public-Art/Kinetic-Rain (accessed Jul 13, 2019).

National Library Board Singapore. Singapore Airlines is Incorporated. [Online] 2014. http://eresources.nlb.gov.sg/history/events/64aa59aa-7a46-4bd5-a543-e259ffb97666 (accessed Jun 26, 2019).

National Library Board Singapore. Paya Lebar Airport Officially Opened. [Online] Oct, 2015. http://eresources.nlb.gov.sg/history/events/eb6a4bf3-d75c-49de-a010-9a4c716a9329 (accessed Jun 22, 2019).

SATS. Our History. https://www.sats.com.sg/AboutUs/OurHistory/Pages/OurHistory.aspx (accessed Jul 12, 2019).

SHA. *The Heritage of Hospitality: A History of the Singapore Hotel Industry;* KWF Printing: Singapore, 2011.

Singapore Airlines. Our Heritage. https://www.singaporeair.com/en_UK/sg/flying-withus/our-story/our-heritage/ (accessed Jun 22, 2019).

Singapore Business Review. Changi Airport launches iChangi HD app for iPad. [Online] Oct 15, 2012. https://sbr.com.sg/aviation/more-news/changi-airport-launches-ichangi-hd-app-ipad (accessed Jul 2, 2019).

Skytrax. World Airport Awards. https://www.worldairportawards.com/airport-of-the-year-winners/ (accessed Jul 1, 2019)

STB, Tourism Innovation Challenge for Hotels, 2017.

STB OHW, Open Hotels Weekend 2017. https://workforahotel.sg/events/ohw (accessed Oct 4, 2019).

STB, Tourism Sector Performance Q4 2018 Report, 2018a.

STB, Singapore Hotel Guide, 2018b.

STB, Innovation Pitch Challenge 2018, 2018c.

STB, 2018 Year-in Review, 2019a. https://www.stb.gov.sg/content/dam/stb/documents/mediareleases/annex-a-yir-presentation-deck-final-003.pdf (accessed Oct 17, 2019)

STB, Hotel Innovation Challenge 2019, 2019b.

STB Hotel, Singapore Tourism Board Hotel Statistics 2011.

STB Hotel, Singapore Tourism Board Hotel Statistics 2012.

STB Hotel, Singapore Tourism Board Hotel Statistics 2013.

STB Hotel, Singapore Tourism Board Hotel Statistics 2014.

STB Hotel, Singapore Tourism Board Hotel Statistics 2015.

STB Hotel, Singapore Tourism Board Hotel Statistics 2016.

STB Hotel, Singapore Tourism Board Hotel Statistics 2017.

STB Hotel, Singapore Tourism Board Hotel Statistics 2018.

STB Hotel, Singapore Tourism Board Hotel Statistics 2019.

Sun, L. The Core Competences and Strategic Management of Raffles—A Case Study of Singapore Hotel Industry. *Int. J. Innov. Manag. Technol.* **2011,** *2*(4), 270–273.

Taylor, C. Singapore Overtakes the US to Become World's Most Competitive Country, WEF Says. MSN. [online] Oct 8, 2019. https://www.msn.com/en-sg/money/topstories/singapore-overtakes-the-us-to-become-worlds-most-competitive-country-wef-says/ar-AAIu0eV?li=BBr8Cnr (accessed Oct 14, 2019).

Tan, B. M. Changi Airport. National Library Board Singapore [Online] 2016. http://eresources.nlb.gov.sg/infopedia/articles/SIP_574_2004-12-23.html (accessed Jul 12, 2019).

Teo, G. Automated services at Changi Airport's T4 improve efficiency, but not all airlines on board. Channel News Asia [Online] Nov 1, 2018. https://www.channelnewsasia.com/news/singapore/automated-services-at-changi-airport-s-t4-improve-efficiency-but-10885590 (accessed Jul 13, 2019).

Toh, M. Singapore's Changi Airport to Build New LCC Terminal. Flight Global [Online] Mar 1, 2012. https://www.flightglobal.com/news/articles/singapores-changi-airport-to-build-new- lcc-terminal-368982/ (accessed Jul 4, 2019).

Toh, W. L. T4 Marries Art and Technology to Reduce Travel Stress for Passengers. Straits Times [Online] Jul 25, 2017. https://www.straitstimes.com/singapore/transport/t4-marries-art-and-technology-to-reduce-travel-stress-for-passengers (accessed Jul 2, 2019).

Wong, W. Changi Airport's T4 to Showcase Singapore Culture and Heritage. Channel News Asia [Online] Jul 25, 2017. https://www.channelnewsasia.com/news/singapore/changi-airport-s-t4-to-showcase-singapore-culture-and-heritage-9061838 (accessed Jul 2, 2019).

Yip, W. Y. Singapore Tops World Smart City Index. Straits Times [online] Oct 4, 2019. https://www.straitstimes.com/tech/singapore-tops-world-smart-city-index (accessed Oct 8, 2019).

CHAPTER 11

WILDLIFE TOURISM IN URBAN DESTINATIONS: SINGAPORE'S URBAN OTTERS AND "BIODIVERCITY" STORY

EUNICE TAN

College of Arts, Business, Law and Social Sciences, Murdoch University, Australia

E-mail: E.Tan@murdoch.edu.au

ABSTRACT

This chapter explores the role of charismatic native wildlife as attractions in urban tourism. Specifically, it examines the dimensions of wildlife tourism within the context of Singapore's sustainability and urban development agendas. It shifts the discourse on wildlife tourism from prevailing ecologically sensitive, rural/regional tourism destination toward a sustainable urban tourism destination context. The chapter also reviews key narratives and semantic representations associated with wildlife tourism, natural heritage conservation and crowdsourced advocacy for nature in Singapore, with its urban otters playing the leading roles. A conceptual framework for sustainable urban wildlife tourism is also proposed.

11.1 INTRODUCTION

While Singapore is a highly urbanized metropolis, it is also rich in biodiversity and natural heritage. In a span of just over five decades, the island city-state has transformed itself from a young independent republic with limited natural resources, high unemployment and squalid urban conditions to being one of the world's most livable and sustainable cities. Its accolades include being ranked as the most livable city in Asia, second among global cities for green buildings, and the most sustainable Asian city (Toh, 2019). A small, densely populated land-scarce country, Singapore's population of over 5 million live, work and play in a space of just over 700 km^2 (Beatley, 2012; Tham, 2019). From its independence, its leaders opted for a model of sustainable development (Balakrishnan, 2016), in which environmental conservation and greening agendas are integral to the country's socioeconomic and national infrastructure development (Er, 2018). The delicate balance of fulfilling economic development agendas while ensuring natural habitat conservation is challenging. For Singapore, being "clean and green" is not just about enhancing residents' quality of life; it is also about presenting a positive image to impress visitors and potential foreign investors (Geh, 2016; Toh, 2018).

Founding Prime Minister Lee Kuan Yew was passionate about greening the city, and efforts to achieve green urbanism have evolved over the years: from the 1960s "Singapore, Garden City" motto to its current "City in a Garden" vision and "BiodiverCity" initiatives (Beatley, 2012; NParks, 2016). Consequently, there is increased attention given to Singapore's sustainability and biodiversity agendas, particularly the conservation of endemic, native, or extant flora and fauna. Today, Singapore is a haven of rich biodiversity, supporting thousands of flora and fauna species in its four nature reserves, 350 parks, almost 400 km of nature ways and park connectors, and 2 million urban trees (Er, 2018; Low, 2012; NParks, 2011). Correspondingly, there are greater opportunities for ecofriendly and nature-based pursuits in urban Singapore. One such pursuit gaining popularity is "otter-watching"—where passionate otter-fans, local and international, trail wild urban otter families across the island's parks, shores, and waterways. The smooth-coated otters are listed as critically endangered in Singapore and were thought to be extinct following a 30-year absence (Theng and Sivasothi, 2016). The Singapore urban otters (hereafter, SGOtters) have gained notoriety as one of its iconic wildlife attractions and mascots of the city-state's sustainability and conservation movement.

As competition for tourism dollars intensify in Southeast Asia, Singapore is accelerating its nature-based and sustainable urban tourism development agenda in an effort to diversify and enhance its tourism portfolio. With increased urbanization and urban populations, cities today are not only important gateways and destinations, but have also become significant environments for tourism and leisure (Ashworth and Page, 2011). However, with greater tourism concentration and consumption, there are pressing concerns to address sustainability agendas generally and the sustainability of urban tourism specifically (Yuan et al., 2018). As Lim and Bouchon (2017) suggest, tourism and sustainable development are key measures legitimizing global cities within ranking attributes evaluating quality of life to economic development. Thus, with the right management policies and strategies in place, the urban and natural environments may not always be incompatible (Henderson et al., 2001). Sustainable urban tourism aims to minimize tourism's environmental footprints and impacts, while ensuring a harmonious and productive relationship among visitors, industry, and host communities (Scott and Cooper, 2010). For a highly compact and urbanized city-state like Singapore, sustainable urban tourism offers a prudent policy direction since its urban areas perform multiple roles and have competing uses of space catering to diverse needs of different stakeholders (Savage et al., 2004).

This chapter investigates the role of charismatic native wildlife as a destination attraction and tourism icon. Specifically, it examines the dimensions of wildlife tourism within the context of Singapore's urban tourism environment. This context is unique in that it moves the discourse on wildlife tourism from prevailing ecologically sensitive rural/regional tourism destinations to that of a sustainable urban tourism destination. Similarly, while there has been significant research on wildlife tourism over the years, there is limited discourse on wildlife tourism in urban cities. Using folksonomy as an approach, the chapter reviews key narratives and semantic representations associated with wildlife tourism, natural heritage conservation, and crowdsourced advocacy for nature in Singapore, with its urban otters playing the leading roles. A framework for the sustainable development of wildlife tourism in urban destinations is also presented.

11.2 WILDLIFE TOURISM IN URBAN DESTINATIONS

11.2.1 WILDLIFE AS TOURISM ATTRACTIONS

With expanding urbanization, people are increasingly living in isolation from nature and wildlife. For some, being far removed from nature triggers a greater prolific human desire to reconnect with the natural world (Ballantyne et al., 2011; Curtin, 2009). Consequently, there is keen demand for people seeking interactions with nature and wildlife, particularly noncaptive wildlife considered to be attractive, endangered, threatened, or rare. In turn, this has led to the emergence of a diverse range of wildlife tourism activities and attractions worldwide (Ballantyne et al., 2009; Cong et al., 2014; Higginbottom, 2004).

In particular, charismatic and iconic species are considered highly appealing by many tourists, especially if there is an opportunity to see them in the natural habitat (Hausmann et al., 2017; Knight, 2010).

Wildlife tourism refers to "tourism based on encounters with non-domesticated (non-human) animals…in either the animals' natural environment or in captivity" (Higginbottom, 2004, p. 2). Thus, one of the core experiential elements of wildlife tourism is the interaction between humans and wildlife (Cong et al., 2014). Essentially, wildlife tourism aims to manage wildlife interactions that increase the probability of positive encounters with wildlife for visitors while protecting the wildlife resources. In this regard, both human and ecological needs must be adequately understood and considered (Reynolds and Braithwaite, 2001). From a conservation and sustainability standpoint, the nonconsumptive utilization of wildlife (e.g., feeding, viewing, and photography) is considered as significantly less impactful than other forms of consumptive activities involving wildlife (e.g., hunting and fishing) (Higginbottom, 2004; Ballantyne et al., 2011).

As opportunities for wildlife tourism activities expand, there is a corresponding demand for more intimate interactions with wildlife in their natural habitat. The convergence between tourists (social) and wildlife (environmental) within the wildlife tourism experience is a vital consideration in its sustainability. The main challenge in wildlife tourism is often human management (tourists) rather than wildlife management (Cong et al., 2014; Reynolds and Braithwaite, 2001). Within this context, the applicability of environmental interpretation and visitor education comes into focus. Tourism experiences that offer opportunities for direct contact with nature and/or wildlife are well placed to deliver positive and powerful educational messages. Beyond the educational and visitor learning aspects, effective interpretation during wildlife tourism experiences can also influence future behavior and contribute to overall satisfaction with the experience (Ballantyne et al., 2011). Ultimately, interpretation is a process of communicating, knowledge acquisition, and visitor education that encourages proconservation attitudes and sustainable behaviors (Tan and Law, 2016). Effective interpretation is paramount in the crafting of meaningful and engaging human-wildlife encounters, while at the same time protecting animals and their habitats. Consequently, many wildlife tourism experiences today are supported by conservation-themed interpretation that aims to increase visitors' appreciation and awareness of conservation practices (Ballantyne et al., 2009).

Reynolds and Braithwaite (2001) posit that six factors determine the quality of wildlife tourism experiences: (1) authenticity, (2) intensity, (3) uniqueness, (4) duration, (5) species popularity, and (6) species status. These six quality factors are in turn influenced by modifiers such as context variables (e.g., time and space factors) and service variables (e.g., management and customer service factors). From the perspective of wildlife watching, Valentine and Birtles (2004) further suggest that desirable destinations for wildlife watching require: (1) species richness, density, and concentration; (2) endemism; and (3) designated protected areas. From the perspective of wildlife interactions in urban areas, these issues intensify as human–wildlife intersections become inevitable due to urban development intensity and human population density (Soulsbury and White, 2016). Hence, understanding visitors' (and the community's) attitudes and responses toward wildlife and its encounter becomes a crucial component in managing the wildlife–human interaction nexus. As Curtin (2009) suggests, one essential distinction is how people perceive themselves in relation to the natural environment. This combined with the heightened perceived intrinsic value of animals, results in a greater sentimental response and emotional attitude toward wildlife and nature.

To enhance the memorability of wildlife tourism experiences and positive outcomes on future intentions and conservation behavior, there is a need to consider the visitor's response to the encounter in terms of what they (1) see and hear, i.e., sensory perceptions; (2) think, i.e., reflective response; (3) feel, i.e., emotional connection; and (4) do, i.e., behavioral response and intentions (Cong et al., 2014). Similarly, Tan and Law (2016) recommend the need to holistically manage the visitors' cognitive–affective–conative trilogy to encourage visitor mindfulness and conservation advocacy. If judiciously designed and implemented, wildlife tourism can influence the pro-environmental attitudes, conservation knowledge, and behavior of visitors (Ballantyne et al., 2009).

11.2.2 SUSTAINABLE URBAN TOURISM

While it is commonly acknowledged that wildlife tourism occurs predominantly in rural/regional areas—since wildlife is usually most abundant away from major urban zones (Edgell, 2016; Higginbottom, 2004), this is not the case for highly urbanized cities like Singapore. As mentioned, this chapter focuses on the notions of wildlife tourism experiences in an urban setting. Hence, this section of the chapter examines the concepts of sustainable urban tourism. Although cities are not conventionally considered to be fragile and endangered, urban environments are certainly not immune to the impacts of development and tourism activity (Savage et al., 2004; Yuan et al., 2018). For

Singapore, its compact size and lack of hinterland means that its residents are forced to live, work, and play in one small island city-state, compelling the adoption of sustainable development from its inception (Balakrishnan, 2016). Sustainable development is defined as "development that meets the needs of the present without compromising the ability of future generations to meet their own needs" (WCED, 1987, p. 8). From an urban tourism context, sustainable development is a multidimensional concept that advocates solutions supporting the effective management of human activities, environmental resources, and impacts of production/consumption (Zamfir and Corbos, 2015). As Singapore is a microcosm of global trends concerning the environment, economy, and society, it is particularly important to be cognizant of the triple bottom line impacts of people, planet, and profit (Balakrishnan, 2016).

The sustainable development of tourism is a much-debated topic in contemporary tourism (Ballantyne et al., 2011). Edgell (2016) suggests that sustainable tourism can be "a major vehicle for the realization of man-kind's highest aspirations in the quest to achieve economic prosperity while maintaining social, cultural and environmental integrity" (p. 12). Thus, sustainable tourism development in urban destinations requires consideration of the triple bottom lines of economic, social-cultural, and environmental impacts (Higginbottom, 2004; Savage et al., 2004). There is also a need to acknowledge the complexities of sustainable urban tourism that differentiates it from other forms of contemporary urban tourism. An integrated city-wide application of sustainability within urban tourism destinations is a complex policy domain, requiring the consideration of both supply- and demand-side factors, as well as stakeholder networks involved in its policy development, management, and governance (Lim and Bouchon, 2017; Miller et al. 2015; Scott and Cooper, 2010; Zamfir and Corbos, 2015). As Ashworth and Page (2011) highlight, the contemporary city is a multifaceted entity—urban tourism is but one of its many economic and social constitutions. It is an entity that offers "a variety of products and experiences to people who have a wide range of motivations, preferences, and cultural perspectives," while simultaneously being "involved in a dialectic engagement with the host community" (Ashworth and Page, 2011, p. 3). Therefore, from a destination management and policy perspective, there is a need for synergy and integration between key stakeholders (e.g., government/policy makers, industry operators, and local community, etc.) to ensure the development and maintenance of effective sustainable urban tourism strategies. Such a bottom-up, consultative approach serves to inspire deeper stakeholder convictions and achievement of mutually agreed outcomes (Tham, 2019). The following section of this chapter explores Singapore's urban "BiodiverCity" journey and its efforts in developing biophilic urbanism.

11.3 SINGAPORE'S "BIODIVERCITY" STORY

11.3.1 BUILDING A CITY IN A GARDEN AND NURTURING BIOPHILIC URBANISM

From its early days as a nation, Singapore's pioneering leaders believed that the economy and environment were inextricably linked; wherein a clean and green city was considered as beneficial for its citizens' morale, visitors, and investors (Balakrishnan, 2016; Geh, 2016). Henderson et al. (2001) observes a positive tourist reception toward Singapore's green agenda; wherein three out of four visitors articulated the clean and green environment as the strongest appeal. Hence, the protection of natural heritage and provision of green spaces are seen as blueprints for enhancing quality of life in the city. With escalating urban populations, cities today play an increasing vital role in conserving biodiversity (Low and Lim, 2012; Ngiam et al., 2017). In this regard, the Singapore Index for Cities' Biodiversity (Chan et al., 2014) recommends three components to help cities achieve their biodiversity goals: (1) native biodiversity in the city, (2) ecosystem services provided by biodiversity, and (3) governance and management of biodiversity. The ecological urbanization of cities can be pursued as solutions to addressing sustainable development (Zamfir and Corbos, 2015). This section of the chapter briefly reviews some of Singapore's green initiatives and biodiversity conservation efforts undertaken in its journey to becoming a biophilic city in a garden.

The greenery of Singapore and its evolution to becoming a "BiodiverCity" is a result of five decades of deliberate planning, commitment, and stakeholder partnerships. Beginning in 1963 with the inaugural tree planting campaign by Lee Kuan Yew and establishment of the Garden City Action Committee in 1970, Singapore has since capitalized on every opportunity to green the city-state (Er, 2018; NParks, 2011). As a result, its green cover has increased from 36% in 1986 to over 56% in 2010 (Geh, 2016). However, just greening up the island alone is not sufficient. Increasingly, there is advocacy for a more holistic and biodiverse approach to conservation. The aim is to not just be a garden city, but a city in a garden and haven for biodiversity. Consequently, NParks (2009) maps its national biodiversity strategy to that of the convention on biological diversity by: (1) safeguarding and conserving its biodiversity; (2) including biodiversity concerns in policy and decision making; (3) improving knowledge of biodiversity and the natural environment; (4) enhancing public awareness and education; and (5) strengthening stakeholder partnerships and promoting international collaborations. In an effort to infuse biodiversity into spaces where its people live, work, and play, Singapore's leaders have introduced a number of key national programs, which include, but are not limited to:

- Singapore National Biodiversity Strategy and Action Plan (2009)
- Singapore Green Plan (2012)
- Singapore 2030 Agenda for Sustainable Development (2013)
- Singapore Sustainable Blueprint (2015)
- NParks SGFutures—"Our BiodiverCity, Our Future" initiative (2016)

Over the years, these programs have encompassed a multitude of sustainability initiatives developed to: (1) forward its urban sustainable development goals; (2) expand the greening of urban landscapes; (3) conserve its native endemic or extant flora and fauna (especially threatened and endangered species); and (4) safeguard its rich, diverse natural heritage and biodiversity. As a result, there have been increased opportunities for nature-appreciation and ecofriendly pursuits in urban Singapore, including wildlife watching and nature walks. Today, Singapore has journeyed past its garden city depiction and aims to nurture a biophilic city in a garden. Biophilia (i.e., a love for living systems) considers the inherent affinity and necessity people have for the natural system, and its value in ensuring physical and mental well-being (Kellert, 2016). Biophilic urbanism is achieved in Singapore by infusing greenery, landscaping, and natural systems into every aspect of the city, its buildings, roads, waterways, and other built environment (Newman, 2014). As Er (2018) shares:

> "After 50 years of greening, Singapore is today not only just a Garden City, but a Biophilic City in a Garden–a green oasis comprising an interconnected network of verdant streetscapes, gardens, parks, nature reserves and vertical greenery. Singapore has transitioned from greenery provision to the development of a sustainable urban ecosystem and a closer affinity with nature in the city" (p. 100)

11.3.2 WILDLIFE AS NATURE'S AMBASSADORS: THE SINGAPORE URBAN OTTERS

This section of the chapter introduces the SGOtters and their return to Singapore following a long period of absence. Specifically, it explores the role of charismatic wildlife like the SGOtters as mascots for wildlife tourism in urban destinations. Within wildlife tourism encounters, noncaptive wildlife deemed attractive, rare, or vulnerable is considered highly appealing to many tourists (Cong et al., 2014). Native wildlife can represent exceptional peak tourism experiences at a destination (Curtin, 2009), particularly in urban environments where the mingling of human-wildlife habitats presents a unique habitat mosaic (Soulsbury and White, 2016). This can in turn present a distinctive sense of place for experiential consumption of nature systems, its flora and fauna. Beyond their value as wildlife attractions, otters are also considered as sound indicators for healthy aquatic habitats and wetlands conservation (Theng and Sivasothi, 2016).

Rapid urbanization and development have led to extensive alteration of Singapore's natural landscape and biodiversity—both terrestrial and aquatic. In tandem with its City in a Garden initiatives, Singapore has also initiated efforts to preserve the biodiversity of its waterways and aquatic habitats. One such initiative is its ABC (Active, Beautiful, Clean) Waters Program launched in 2006 as a strategic plan to improve the health, quality, and potential of Singapore's waterbodies (Khoo, 2016; PUB, 2019). Through the process of naturalization, integration and landscaping, Singapore's vast network of storm drains, canals, and reservoirs were reimagined as appealing community spaces and nature connectors (Beatley, 2012). These transformed spaces improve flood management facilities, water quality, and biodiversity. It also brings people closer to nature. Today, the ABC Water Program has enhanced Singapore's green ecosystem with a blue map of some 17 reservoirs, 32 rivers and 8000 km of waterways, boosting its livability factor (Khoo, 2016; Theng and Sivasothi, 2016). These nature sites displaying effective integration of the green–blue matrix also offer a rich biodiversity of flora and fauna, attracting native wildlife like the otters.

The smooth-coated otters are listed as critically endangered in Singapore and thought to be extinct following a 30-year absence. There are two species currently existing in Singapore: the smooth-coated otter (*Lutrogale perspicillata*) and the smaller small-clawed otter (*Aonyx cinereus*) (Theng and Sivasothi, 2016). It is commonly thought that the smooth-coated otters are more prevalent and commonly spotted in Singapore's wetlands and waterways. However, a recent study found that Singapore's otters are actually a hybrid of the two species—which is a first record of hybridization worldwide (Seow and Teng, 2017). Following a first sighting in the late 1990s, the population has since grown to include about 70 otters from approximately 10 family groups, who live in various territories along Singapore's shores and waterways

(Kotwani, 2017). Unlike their counterparts in other parts of the world, the SGOtters have adapted and thrived in a challenging urban environment that is contrary to their usual habitats. Sharing the urban city-scape with their human counterparts, the SGOtters travel along Singapore's waterways, nature corridors, parks, and surrounding urban spaces as they forage for food, play, and raise families. As a result of these close encounters and regular sightings, the SGOtters have amassed thousands of otter-fans locally and internationally. Beyond its appeal as an urban wildlife attraction, celebrity SGOtter families across Singapore also draw dedicated fans, international wildlife photographers, filmmakers, researchers, and international events to its shores. For instance, one of the most popular (and largest) celebrity SGOtter families, the Bishan Otters, have found fame in numerous media platforms and coverage, including Sir David Attenborough's documentary "Wild City", welcomed visiting foreign dignitaries and even photobombed the marriage proposal of an otter-loving couple from the UK! (Ishak, 2019; The Economist, 2016). The SGOtters were also chosen as a symbol to represent Singapore's 51st birthday (Kok, 2016).

With their popularity and unique juxtaposition as free-range wildlife in one of the world's densest cities, the SGOtters can offer exceptional branding opportunities as the city-state's mascot and tourism attraction (Ishak, 2019), as well as representation for the nation's biodiversity and environmental conservation efforts. Moreover, charismatic native wildlife like the SGOtters can be a positive tool for public outreach, conservation advocacy, visitor education, and interpretation for wildlife tourism and local natural heritage. The following section of the chapter explores some of the narratives and semantic representations associated with the SGOtters and habitats they call home.

11.4 WILDLIFE GOING VIRAL: CAPTIVATING URBAN CHARISMATIC WILDLIFE

11.4.1 WILD TALES, DRAMAS, AND FANFICTION: FOLKSONOMY OF THE SINGAPORE URBAN OTTERS

As with any celebrity personality or popular characters in today's highly public, user-centric digital world, the SGOtters too have their share of dedicated fans, followers, and stories going viral. The prevalence and ubiquity of mobile-driven devices and social media platforms (Budge, 2017; Tan and Law, 2016) offer opportune access and a rich source of narratives and information for investigating tourist experiences. Cong et al. (2014) observe that with the pervasiveness of websites and social network systems (SNSs), tourists today are more willing and able to share their travel experiences with the digital world and communities of interest. These SNSs are changing how people find, share and index information, and communicate with each other. Consequently, this social behavior has transformed consumers into content producers (Kim et al., 2012). Teng and Sivasothi (2016) observe an upward trend for naturalist blogging and online communications regarding the SGOtters and related topics of interest through SNSs. The proliferation of social media services and volume of user-generated content available today pose challenges for researchers in finding and interpreting the appropriate media content (Budge, 2017; Kim et al., 2012).

Since the appreciation, receptiveness, and meaning of nature (and its flora and fauna) vary between segments of visitors, the public, and societies (Ngiam et al., 2017), it is useful to examine the narratives surrounding the subjects and phenomenon under investigation from their (user) perspectives. This study uses a folksonomy-based approach to examine collaborative social tagging and narratives associated with the SGOtters. Folksonomy relates to the collaborative (folk) and informal creation, categorization, and organization of content (taxonomy) (Uddin et al., 2013). A folksonomy-based approach facilitates the retrieval and clustering of extensive data resources, descriptive tags, anecdotal evidence, and user-generated content on a particular issue and/or associated social behaviors. Within the context of research framing this chapter, folksonomy involves the coding and analysis of: (1) collaborative social tags—wherein users annotate and share resources and (2) hashtags – illustrating users' collaborative cognition of information (Budge, 2017; Kim et al., 2010; Uddin et al., 2012).

In this study, data captures (using NVivo 11) for the analysis were collected between September and October 2018, via the following popular SNSs relevant to the SGOtters: (1) Instagram: #Otterwatch (n = 790 posts); #Singaporeotters (n = 585 posts); #Sgotters (n = 447 posts); #ottercity (n = 103 posts) and (2) Facebook: Otterwatch (n = 1000 [most recent] posts); Ottercity (n = 366 posts). Facebook and Instagram were selected as the SNS platforms for data capture as these are two SNSs dominating social media landscape today (Budge, 2017). Moreover, Otterwatch and Otercity were selected as data capture sources, since these are the two most prominent SNS platforms dedicated to the SGOtters—Otterwatch has over 40,000 Facebook followers from around the world, while Ottercity has over 10,000 and growing by 1000 new followers every month (Ishak, 2019). Hashtags, photos, and associated content posted were coded and analyzed following the data capture. At this preliminary stage of the research, only primary posts were analyzed, excluding replies/comments within the post threads. The following section of the chapter presents a brief discussion of the (1) tag semantics and relationships, and (2) synonymy and polysemy in social tagging (i.e., semantic similarities between tags/word sets) (Kim et al., 2010; Uddin et al., 2012) relating to the SGOtters.

Figure 11.1 illustrates sample word clouds from the data captures, while Figure 11.2 below presents the prime thematic clusters illustrating the stories and narratives being shared on the otters. Subsequently, these thematic categories are further grouped into observed popular pairwise clusters (Table 11.1) within posts and/or images.

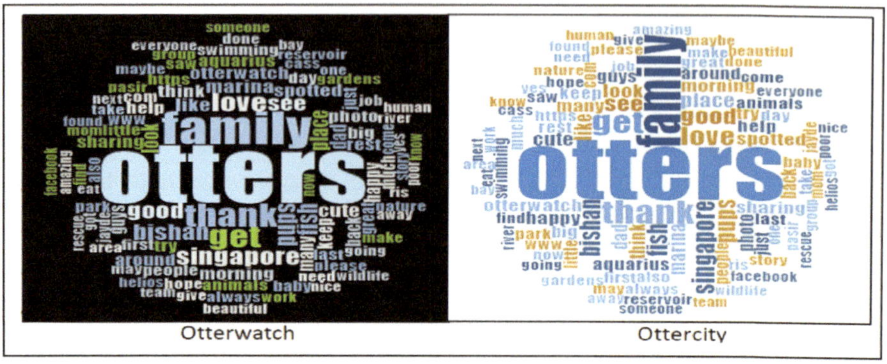

FIGURE 11.1 Sample word clouds from SNS posts.

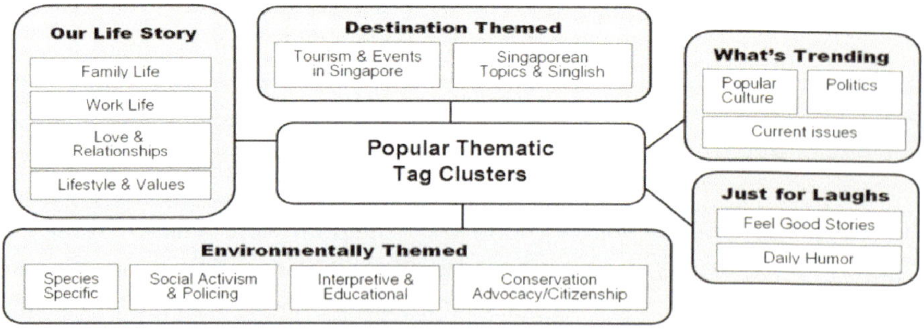

FIGURE 11.2 Prime thematic clusters of narratives.

Within the realm of digital storytelling and information sharing, the exploration of tag semantics and associated visuals within key thematic clusters provide an understanding of primary motives and narratives that impact, appeal, or relate to people most. It provides an influential platform through which people account for and document all aspects of their life-world and experiences (Budge, 2017). For the SGOtters, their stakeholder networks of dedicated otter-fans, otter-watchers, and communities of interest keep the stories, fandom, and obsession alive within the digitized social media space. These stories and information shared in turn provide useful narratives for environmental interpretation and education. As discussed, environmental interpretation for sustainability communicates messages and content that serve to enhance visitor experience, learning, and appreciation, while protecting and sustaining the environment through conservation advocacy (Ballantyne et al., 2009; Tan and Law, 2016). Within the context of sustainable wildlife tourism, it educates visitors about the wildlife, the threats it faces and initiatives for protecting their habitat and biodiversity, and encouraging responsible behaviors during wildlife encounters (Cong et al., 2014; Teng and Sivasothi, 2016).

TABLE 11.1 Popular Pairwise Topics.

Most Popular Pairwise Topics	
Pairwise Theme 1	Family & Love
Pairwise Theme 2	Singapore & Humor
Pairwise Theme 3	Popular Culture & Tourism/Events
Pairwise Theme 4	Politics & Current Issues
Pairwise Theme 5	Environmental Advocacy & Social Activism
Pairwise Theme 6	Interpretive Eduction & Species Specific

It was observed that stories and narratives relating to: (1) family, love, and life; (2) excitement, delight, and/or gratefulness for rare otter-encounters; and (3) stories about Singapore culture and humor are the most prevalent. There are also an increasing number of posts related to (1) environmental advocacy and social activism, and (2) information on the species, their habitats and the environment. The appeal of the SGOtters and passionate narratives shared about this endearing wildlife illustrate their charm and relatability, mirroring aspects of the human lifeworld, and as inspirations for togetherness and tight-knit family bonds (Kok, 2016). For example, within the most prevalent family-love pairwise, stories relating to the real (and humanistic) struggles of daily lives, trails, and tribulations (e.g., loss and death, celebrations of birth and humorous tactics of teenagers in their growing up years) paralleling that of people seems particularly relatable. Additionally, stories on Singapore's culture and way of life (including Singlish!) gives audiences an insider glimpse into life on this little island city-state.

11.4.2 THE CITY IN NATURE, AND NATURE IN THE CITY: EMBRACING WILDLIFE IN THE CITY

Mounting concerns for the environment and conservation in recent decades have fortified the relationship between people, tourism, and the environment (Reynolds and Braithwaite, 2001). Consequently, there is a need for a more holistic consideration of the biodiversity of nature, and not merely the development of green spaces. Earlier sections of the chapter had articulated the need to get away from the mere aesthetics and sterility of clean and green gardens. Thus, Singapore should move beyond just being a city in a garden to being a city in nature; wherein ecological considerations and advancing the biodiversity of nature becomes the goal. In line with its quest to build a biophilic city, Singapore strives to sustain a model for nature conservancy that champions environmental sustainability in its compact urban setting (Low and Lim, 2012), in which biophilic designs in buildings and landscapes are adopted to ensure the connection and integration between the built and natural environments (Kellert, 2016). Singapore's model for urban biodiversity reveals how dense urbanity can be expressed more naturally (Newman, 2014). As Kellert (2016) explains, "the goal of biophilic urbanism is to ameliorate the contemporary urban disconnect from nature, making the experience of the natural world an integral part of ordinary city life" (p. 6). Thus, nurturing a biophilic city is not just about establishing nature within a city's built environment; it is also about how its communities and stakeholders become part of the ecosystem and stewards of nature (Er, 2018).

Consequently, there are increasing efforts in Singapore to foster a collective culture that connects with and care about nature. In this regard, the notion of citizenry participation becomes valuable, wherein enthusiastic members of the public and fervent eco-warriors communicate and share a common passion to build a sustainable future for Singapore (Beatley, 2012; Teng and Sivasothi, 2016; Tham, 2019). As suggested by Geh (2016), people do not just want and appreciate more public parks, space and amenities, but also more say in it too. Beyond the role of stakeholders in promoting community stewardship and crowdsourced advocacy, effective environmental education is also important in championing conservation and encouraging mindfulness about potential human–wildlife conflicts. This bottom up approach (Tham, 2019) to managing environmental causes should thus be nurtured and encouraged. With reference to the SGOtters, these individuals perform vital roles not only as mediators and intermediaries, but also as wildlife warriors and storytellers who document, produce, and share information about the SGOtters and Singapore's "BiodiverCity" story to their fans and the wider community.

Based on the literature reviewed and observations discussed, a framework for wildlife tourism in urban destinations (Fig. 11.3) is proposed. The framework illustrates that when considering the nonconsumptive leisure and tourism use of wildlife, there is a need to manage the convergence of: (1) the city, (2) the wildlife, and (3) the visitor. As Soulsbury and White (2016) emphasize, it is vital to prioritize sustainable management of human–wildlife encounters in urban spaces, and acknowledge that urban wildlife are part of the urban ecosystem. Within this interaction trilogy, destination managers need to be acutely aware of the unique dimensions that characterize each component. These interacting elements subsequently influence the overall quality of the experience, wherein the visitor–wildlife interaction constitutes the central component of the wildlife tourism product (Higginbottom, 2004). For many, the human–wildlife–nature encounter

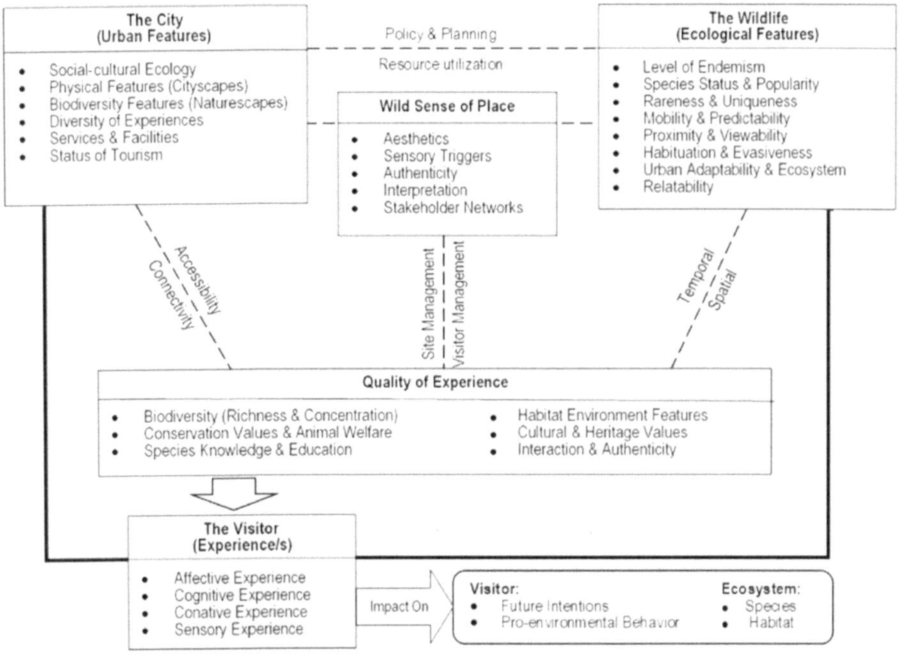

FIGURE 11.3 Conceptual framework: wildlife tourism in urban destinations.

is considered to be peak aesthetic experiences and individuals may feel a heightened sense of happiness, fulfillment, and even cathartic inspirations (Curtin, 2009). Accordingly, understanding visitor responses and preferences for nature-based experiences are crucial for developing effective marketing communications strategies (Hausmann et al., 2017). Therefore, the final step in the framework considers the visitor experience outcomes and impact on future intentions, as well as consequences on wildlife and their habitats. Finally, to ensure the development of sustainable wildlife tourism, there is a need to implement policies, planning, and management strategies that help mitigate impacts (Ballantyne et al., 2009). The framework highlights that multiple interrelated elements make the urban wildlife tourism experience unique – and this is certainly the case for wildlife tourism in urban Singapore.

11.5 CONCLUDING REMARKS

In an effort to diversify and enhance its tourism portfolio, Singapore has enhanced its sustainable urban tourism development agenda and offerings of nature-based experiences. Notwithstanding its intense urbanization, Singapore endures as a haven for biodiversity surrounding its inhabitants, in nature reserves, parks, gardens, and local neighborhoods. Consequently, native wildlife can become mascots for the city-state and its evolution within the environmental conservation and sustainability movement. Singapore can capitalize on the prevailing otter-fascination and international curiosity to communicate a uniquely Singaporean brandmark. Through investigating the role of charismatic native wildlife like the SGOtters, this chapter discusses the underexplored concept of wildlife tourism in urban destinations. It rethinks the predominant discourse on nature-based/wildlife tourism, moving it away from activities occurring in predominantly rural, peripheral, or ecologically sensitive areas to focus on the urban environmental context. Within the context of wildlife tourism, the chapter recognizes that in areas where suitable wildlife exists, the development of nonconsumptive wildlife tourism experiences can provide positive benefits and outcomes for economy, society, and ecology. However, in cities, these wildlife tourism encounters would entail significantly different experiential and social-ecological conditions, requiring a distinctive destination management strategy for optimal outcomes. Within the notion of wildlife tourism in sustainable urban tourism, the observations in this chapter highlight the need to focus on: (1) positive attributes of urban tourism—opportunities for nature-based/wildlife tourism that traditional nonurban areas do not (e.g., budget and accessibility); (2) the importance of crafting the story–interpretation, digital storytelling, and animating charismatic wildlife characters; and (3) adapting the principles of sustainable development to the characteristics of urban tourism. Thus, whether the SGOtters are deemed to be *"otterly-cute"*, or *"otterly-annoying"*, their fans or critics from both sides of the wildlife–human–city debate will continue to be fought within the digital realm and Singapore's uniquely *"BiodiverCity"* story.

KEYWORDS

- **sustainable development**
- **urban wildlife tourism**
- **digital storytelling and environmental interpretation**
- **native wildlife as tourism attractions**

REFERENCES

Ashworth, G.; Page, S. J. Urban Tourism Research: Recent Progress and Current Paradoxes. *Tourism Manag.* **2011,** *32*(1), 1–15.

Balakrishnan, V. Sustainable Singapore: Taking the Long Term View. In *Singapore 2065: Leading Insights on Economy and Environment from 50 Singapore Icons and Beyond*; Quah, E., Ed.; World Scientific: Singapore, 2016; pp 23–28.

Ballantyne, R.; Packer, J.; Hughes, K. Tourists' Support for Conservation Messages and Sustainable Management Practices in Wildlife Tourism Experiences. *Tourism Manag.* **2009,** *30*(5), 658–664.

Ballantyne, R.; Packer, J.; Sutherland, L.A. Visitors' Memories of Wildlife Tourism: Implications for the Design of Powerful Interpretive Experiences.*Tourism Manag.* **2011,** *32*(4), 770–779.

Beatley, T. Singapore: How to Grow a High-Rise City in a Garden.*SiteLINES: J. Place* **2012,** *8*(1), 14–17.

Budge, K. Objects in Focus: Museum Visitors and Instagram.*Curator: The Museum J.* **2017,** *60*(1), 67–85.

Chan, L.; Hillel, O.; Elmqvist, T.; Werner, P.; Holman, N.; Mader, A.; Calcaterra, E. *User's Manual on the Singapore Index on Cities' Biodiversity;* National Parks Board: Singapore, 2014.

Cong, L.; Wu, B.; Morrison, A. M.; Shu, H.; Wang, M. Analysis of Wildlife Tourism Experiences with Endangered Species: An Exploratory Study of Encounters with Giant Pandas in Chengdu, China.*Tourism Manag.* **2014,** *40,* 300–310.

Curtin, S. Wildlife Tourism: The Intangible, Psychological Benefits of Human–Wildlife Encounters.*Curr. Issues Tourism* **2009,** *12*(5-6), 451–474.

Edgell Sr, D. L. *Managing Sustainable Tourism: A legacy for the Future*; Routledge: Abingdon, OX, 2016.

Er, K. Growing a Biophilic City in a Garden. *Ethos* **2018,** *19,* 98–113, https://www.csc.gov.sg/ethos.

Geh, M. Ecolink: Future Challenges to the Linking of Singapore's Economy and Environment. In *Singapore 2065: Leading Insights on Economy and Environment from 50 Singapore Icons and Beyond*; Quah, E., Ed.; Singapore: World Scientific, 2016; pp 23–28.

Hausmann, A.; Slotow, R.; Fraser, I.; Di Minin, E. Ecotourism Marketing Alternative to Charismatic Megafauna can also Support Biodiversity Conservation.*Anim. Conserv.* **2017,** *20*(1), 91–100.

Henderson, J. C.; Koh, A.; Soh, S.; Sallim, M. Y. Urban Environments and Nature-Based Attractions: Green Tourism in Singapore.*Tourism Recreat. Res.* **2001,** *26*(3), 71–78.

Higginbottom, K. Wildlife tourism: An Introduction.In *Wildlife Tourism: Impacts, Management and Planning*; Higginbottom, K., Ed.; CRC Centre for Sustainable Tourism: Alton, VIC, 2004; pp 1–14.

Ishak, N. S. 2019. Singapore Otters a Hit Overseas. *The New Paper*, https://www.tnp.sg/news/singapore/singapore-otters-hit-overseas

Kellert, S. Biophilic Urbanism: The Potential to Transform.*Smart Sustain. Built Environ.* **2016,** *5*(1), 4–8.

Khoo, T. C. Stormwater Management: City of Gardens and Water. *Urban Sol.* **2016,** 54–63, https://www.clc.gov.sg/docs/default-source/urban-solutions/urb-sol-iss-8-pdfs/essay-city-of-gardens-and-water.pdf

Kim, H. N.; Rawashdeh, M.; Alghamdi, A.; El Saddik, A. Folksonomy-Based Personalized Search and Ranking in Social Media Services.*Inform. Syst.* **2012,** *37*(1), 61–76.

Knight, J. The Ready to View Wild Monkey: The Convenience Principle in Japanese Wildlife Tourism. *Ann. Tourism Res.* **2010,** *37*(3), 714–762.

Kok, L. M. ST Readers Choose Bishan Otters to Represent Singapore in her 51st year. *The Straits Times*, https://www.straitstimes.com/singapore/st-readers-choose-bishan-otters-to-represent-singapore-in-her-51st-year (accessed 7 August, 2013).

Kotwani, M. Can Singapore Growing Otter Population Continue to Thrive in an Urban Landscape? Channel News Asia. https://www.channelnewsasia.com/news/singapore/can-singapore-s-growing-otter-population-continue-to-thrive-in-8981816.

Lim, S. E. Y.; Bouchon, F. Blending in for a Life Less Ordinary? Off the Beaten Track Tourism Experiences in the Global City.*Geoforum* **2017,** *86,* 13–15.

Low, J.; Lim, L. J. In *Conservation of Our Natural Heritage: The Singapore Experience*, Proceedings of the 12th International Coral Reef Symposium, James Cook University, Cairns, 2012; pp 10–13.

Miller, D.; Merrilees, B.; Coghlan, A. Sustainable Urban Tourism: Understanding and Developing Visitor Pro-Environmental Behaviors.*J. Sustain. Tourism* **2015,** *23*(1), 26–46.

National Parks Board (NParks). 2009. *Conserving Our Biodiversity: Singapore's National Biodiversity and Action Plan*, https://www.nparks.gov.sg/biodiversity/our-national-plan-for-conservation//-/media/nparks-real-content/biodiversity/national-plan/nbsap_2009.pdf

National Parks Board (NParks). 2011. *NParks Annual Report 2010/11*, https://www.nparks.gov.sg/about-us/annual-reports/nparks-annual-report-archives

National Parks Board (NParks). 2016. *Our Biodiversity, Our Future*, https://www.sg/en/sgfuture/Past-Engagements/our-biodivercity-our-future.aspx

Newman, P. Biophilic Urbanism: A Case Study on Singapore.*Aust. Plann.* **2014,** *51*(1), 47–65.

Ngiam, R. W. J.; Lim, W.L.; Collins, C.M. A Balancing Act in Urban Social-Ecology: Human Appreciation, Ponds and Dragonflies.*Urban Ecosyst.* **2017,** *20*(4), 743–758.

Public Utilities Board *Active, Beautiful, Clean Waters Program*, https://www.pub.gov.sg/abcwaters/about

Savage, V. R.; Huang, S.; Chang, T. C. The Singapore River Thematic Zone: Sustainable Tourism in an Urban Context.*Geograph. J.* **2004,** *170*(3), 212–225.

Scott, N.; Cooper, C. Innovation for Sustainable Urban Tourism: Some Thoughts on Best Practice.*Rev. Adm. Pública* **2010,** *44*(5), 1171–1190.

Seow, G.; Teng, M. Singapore Smooth-Coated Otters are Hybrids. *Nature News,* **2017,** 9, https://nss.org.sg/newsletter/902d1c3a-2Nature%20News%20Mar&%20Apri%202017C.pdf

Soulsbury, C. D.; White, P. C. Human–Wildlife Interactions in Urban Areas: A Review of Conflicts, Benefits and Opportunities.*Wildl. Res.* **2016,** *42*(7), 541–553.

Tan, E.; Law, R. mLearning as a Softer Visitor Management Approach for Sustainable Tourism.*J. Sustain. Tourism* **2016,** *24*(1), 132–152.

Tham, A. Envisioning Eden: The Manufactured Ecotourism Environment of Singapore.*J. Ecotourism* **2019,** *18*(1), 1–24.

The Economist. 2016. Move over, Merlion, https://www.economist.com/asia/2016/08/25/move-over-merlion (accessed 25 August 2016).

Theng, M.; Sivasothi, N. The Smooth-Coated Otter Lutrogale Perspicillata (Mammalia: Mustelidae) in Singapore: Establishment and Expansion in Natural and Semi-Urban Environments.*IUCN Otter Special. Group Bull.* **2016,** *33*(1), 37–49.

Toh, B. K. Developing a Livable and Sustainable Singapore. *Ethos* **2018,** *19,* 4–15, https://www.csc.gov.sg/ethos.

Uddin, M. N.; Duong, T. H.; Nguyen, N. T.; Qi, X. M.; Jo, G. S. Semantic Similarity Measures for Enhancing Information Retrieval in Folksonomies.*Exp. System. Appl.* **2013,** *40*(5), 1645–1653.

Valentine, P.; Birtles, A. Wildlife Watching.In *Wildlife Tourism: Impacts, Management and Planning*; Higginbottom, K. Ed.; CRC Centre for Sustainable Tourism: Alton, VIC, 2004; pp 15–34.

World Commission on Environmental and Development (WCED) *Our Common Future: From One Earth to One World*; Oxford University Press: NY, 1987.

Yuan, J.; Deng, J.; Pierskalla, C.; King, B. (2018). Urban Tourism Attributes and Overall Satisfaction: An Asymmetric Impact-Performance Analysis.*Urban Forest. Urban Green.* **2018,** *30,* 169–181.

Zamfir, A.; Corbos, R. A. Towards Sustainable Tourism Development in Urban Areas: Case Study on Bucharest as Tourist Destination.*Sustainability* **2015,** *7*(9), 12709–12722.

CHAPTER 12

PHILIPPINE TOURISM VIS-À-VIS ASEAN: A REVIEW OF KEY ISSUES

LEONARDO (DON) A.N. DIOKO

Macao Institute for Tourism Studies, Macao, China

E-mail: don@iftm.edu.mo

ABSTRACT

This chapter reviews contemporary issues in Philippine tourism development and seeks to identify some key underlying factors driving its long-term trajectory and the challenges the country faces in attracting international visitors. The review broadly examines scholarly studies on Philippine tourism, the country's major tourism policies and planning instruments, and evaluates recent high-profile events indicative of the country's tourism performance relative to its ASEAN counterparts. The review then distills its findings into four key fundamental factors posited to account for the long-term outcome of Philippine tourism and how these may drive future challenges ahead. The influence of these factors in accounting for international arrivals is then empirically tested using a time series cross-sectional analysis the results of which lend some support to the review's arguments.

12.1 INTRODUCTION

One of the most prominent recent developments in Philippine tourism was the abrupt announcement of the closure of the country's premier tourist destination of Boracay Island in 2018. Most famous for its white powdery beach and coastlines and often popularly ranked as among the most beautiful islands in the world by travel magazines (CNT Editors, 2018), its closure, on orders by the country's President (Villamor, 2018), was in response to growing environmental degradation wrought by years of over tourism and unchecked growth in visitor numbers. The drastic measure raised many questions not only about the state of tourism in the Philippines but also in Southeast Asia in general. Among these, the case highlighted the severity with which rapid and sustained increase in regional tourism growth has affected the fragile ecosystem of Boracay Island and other islands like it in the region. It also revealed the mismanagement of tourism growth and weaknesses in planning, governance, and oversight of coordinated development. Although a single but high-profile incident, the event may be taken as symptomatic of the broader as well as profound issues surrounding tourism development in the Philippines. Interestingly, the Philippines is not the only destination in the world that has taken severe measures to control tourism. Among other destinations recently plagued by over tourism are Thailand (Cripps, 2018), the Cinque Terre islands in Italy (Korey, 2017), and Barcelona (Plush, 2017).

Is this episode merely symptomatic of underlying forces affecting Philippine tourism development? Or are there more profound antecedents steering various factors simultaneously toward a critical convergence manifesting its effects? To help answer this, this chapter aims, therefore, to outline the recent trajectory of Philippine tourism development, evaluate the country's tourism performance relative to its Southeast Asian neighbors, delve into major policies guiding Philippine tourism, and posit some key issues regarding where Philippine tourism stands and the factors that influence it. The chapter then examines in greater depth four hypotheses regarding fundamental factors that may have strategically impacted Philippine tourism and which will determine its future potential and directions.

12.1.1 PHILIPPINE TOURISM IN SCHOLARLY LITERATURE

It is challenging to identify relevant factors and issues influencing Philippine tourism development—or any country's tourism development, for that matter—if not much validated research has been published on it. In a Boolean search conducted for the terms "Philippine" and "tourism" in the Science Direct database, restricting the search to titles, abstracts, and subject keywords, only 14 peer-reviewed scientific articles emerged spanning a period from 1990 to 2019. That is but a fraction of the total 601 research articles on tourism published for all 10 ASEAN members combined over the same 30-year period. The same search terms yielded 165 scientific articles for Malaysia, 136 for Indonesia, 135 for Thailand, and 77 for Singapore (Table 12.1). This strongly suggests that tourism has not figured very prominently nor is it considered important enough in the overall development of the Philippines for it to warrant careful attention via the funding and conduct of scientific and systematic internationally peer-reviewed studies. That neglect may be on the mend, however, as 11 of the total 14 articles on Philippine tourism were published in the current decade (2010–2019), a five-fold jump from the two articles published in the decade prior. Still, the 11 articles form just a fraction of the 425 total number of articles on all ASEAN countries published in between 2010 and 2019. The same pattern emerges for the Philippines in a similar search undertaken in other scientific databases (Taylor and Francis and ProQuest) wherein tourism as a subject of scholarly study remains generally overlooked when compared to its Southeast Asian neighbors of comparable size, population, and touristic endowment.

TABLE 12.1 Number of Peer-Reviewed Scholarly Articles About Southeast Asian Countries in Major Database Providers Generated by Boolean Search Term "(country)" and "tourism," After Deleting Duplicate Search Results.

	Science Direct					Routledge All years	ProQuest All years
	Up to 1989	1990–1999	2000–2009	2010–2019	Total		
Malaysia	5	6	13	141	165	30	260
Indonesia	0	24	22	90	136	13	220
Thailand	0	14	18	103	135	37	285
Singapore	0	13	28	36	77	13	156
Vietnam	0	6	10	24	40	19	119
Philippines	*0*	*1*	*2*	*11*	*14*	*21*	*83*
Cambodia	0	1	5	5	11	10	53
Myanmar	0	1	1	7	9	1	24
Laos	0	1	3	5	9	1	37
Brunei	0	1	1	3	5	0	5
Total	5	68	103	425	601	145	1242

Several broad themes can be drawn from the relatively limited scientific literature surrounding Philippine tourism. The impacts of tourism appear a clear issue, with several studies tackling conservation and protection (Ong et al., 2011; Oracion et al., 2005), ecotourism, and sustainability issues (Ocampo et al., 2018; Tamayo et al., 2018). Demand side issues on visitor segments and the type of attractions that draw tourists to the Philippines have also been the focus of some studies, albeit sporadically (Alejandria-Gonzalez, 2016; Saverimuttu and Varua, 2014; Smith et al., 2011; Smith, 1990). Several articles shed some light on the country's tourism planning, policy development, and formulation (Chon and Oppermann, 1996; Dela Santa, 2013, 2015; Dela et al., 2016), and advocate the importance of governance issues and shortcomings thereof (Carter, 2004; Manuela and de Vera, 2015; Mirabueno and Yujuico, 2014; Trousdale, 1999). Of particular interest is a study by Maguigad (2013) that highlights an oft-ignored factor specific to the geography of archipelagic nations such as the Philippines (and Indonesia) and the lack of a strong policy guiding national tourism planning and development that such a challenging geography poses.

A notable feature in several studies on Philippine tourism is the narrow but recurring focus on coastal and island tourism, especially of Boracay Island, which is often regarded as a microcosm of the state of Philippine tourism as well as a test case for factors affecting the entire country's tourism development (Carter, 2004; Maguigad, 2013; Ong et al., 2011; Smith et al., 2011; Smith, 1990; Trousdale, 1999). One of the two earliest scholarly references to Philippine tourism (Smith, 1990) was about the famed tourist island, which continues to feature in recent studies (Maguigad, 2013). Indeed, few other destinations within the Philippines receive the same scholarly attention as Boracay has over decades. It should come as no surprise, therefore, that Boracay was back in the news spotlight yet again in the recent heavy-handed directive by the Philippine president to close the island temporarily for 6 months to help its ecosystem recover from over tourism. It is apparent that Philippine tourism development narrowly revolves around or mainly anchored on Boracay. This is contrary to reality, of course, but it would not be incorrect for anyone reviewing the scientific literature that Boracay Island is often the object of case studies and perhaps the most celebrated example of Philippine tourism development.

Many of the above studies (and others) are characterized heavily by their advocacy objectives especially in tourism policy, planning, and governance matters, and involve systematic reviews, grounded interviews with stake-holders, and critical analyses of domestic policies. Few studies are empirical in nature, with a few exceptions such as those examining crime levels and tourism (Palanca-Tan, Garces et al., 2015) and the influence on weather on visitation levels (Saverimuttu and Varua, 2014).

Overall, the sparse scientific literature on Philippine tourism suggests that Philippine tourism development has hardly undergone a serious, systematic, and sustained investigation over the last 30 years. Furthermore, the few identifiable issues suggest that tourism development in the Philippines has not progressed far from addressing common initial developmental questions faced by developing destinations. This is paradoxical when juxtaposed against the growth in visitor numbers the country has witnessed recently and which its policy makers aim to increase further. In contrast, other ASEAN countries have seen more scholarly studies address a host of other factors beyond sustainability, sociocultural, and environmental as well as governance or policy issues.

Perhaps tourism doesn't feature significantly enough relative to the entire range of socioeconomic issues concerning the Philippines. There are of course other priority areas which the developing country has sought to intractably address such as poverty alleviation, disaster mitigation, and crime. This is not reasonable justification, nevertheless, nor should it diminish what arguably is an important economic sector for the country. According to the World Travel and Tourism Council (WTTC, 2018), travel and tourism comprised 8.7% of direct contributions to the Philippines' GDP, ranking 27th in the world and a rate significantly higher compared to the average of 4.9% for Southeast Asian countries as a group. Travel and tourism total contribution to the Philippines GDP was 21.1%, ranking 35th in the world and much higher than the average 12% for Southeast Asian nations. In terms of employment, the estimated direct contribution of travel and tourism in the Philippines is 5.8%, while the average for Southeast Asia is 4.7%. The scarcity of research on Philippine tourism thus stands in stark contrast to how important the industry is to the Philippine economy.

12.1.2 PHILIPPINE TOURISM PERFORMANCE VIS-A-VIS ASEAN

Between 2004 and 2017, international visitor arrivals to the Philippines trebled from 2.3 million in 2004 to 6.6 million in 2017, a growth rate of 289% over the period, or an annualized rate of 9% (UN World Tourism Organization). Though impressive, this rate of growth falls somewhat below the ASEAN average growth rate of 328% over the same period, though it matches the group's annualized growth rate of 9%. The highest growth in international visitors over the period, in both simple and annualized rates, was recorded by the Mekong subgrouping of countries (Cambodia, Myanmar, Vietnam, Laos, and Thailand) which are geographically contiguous countries. Archipelagic countries such as Indonesia and the Philippines performed in the middle of the whole Southeast Asian grouping. Smaller states such as Brunei and Singapore logged below average growth in international arrivals while Malaysia, which is half peninsular (Western Malaysia) and half insular (Eastern Malaysia) recording the lowest growth rate over the 13-year period. Still, as a group, ASEAN countries achieved an impressive rate of growth over the period, with countries having the biggest share of international visitors being Thailand (35.5 million visitors), Malaysia (25.9 million), Indonesia (14 million), followed closely by Singapore (13.9 million) (UN World Tourism Organization).

TABLE 12.2 Visitor Arrivals, ASEAN Countries.

	2004	2017	2004–2017 growth rate (%)	2004–2017 annualized growth rate (%)
Thailand	11,737,000	35,592,000	303	9
Malaysia	15,703,000	25,948,000	165	4
Indonesia	5,321,000	14,040,000	264	8
Singapore	6,553,000	13,903,000	212	6
Vietnam	2,928,000	12,922,000	441	12
Philippines	2,291,000	6,621,000	289	9
Cambodia	1,055,000	5,602,000	531	14
Myanmar	657,000	3,443,000	524	14
Laos	(No data)	3,257,000	*n.a.*	*n.a.*
Brunei	119,000	259,000	218	6
ASEAN	46,364,000	121,587,000	328	9

Sources: UN World Tourism Organization, Yearbook of Tourism Statistics, Compendium of Tourism Statistics (UN World Tourism Organization) and data files.

With 6.6 million international visitors, the Philippines' performance in attracting international visitors appears very unremarkable and may even be considered lackluster relative to the performance of its neighbors. Figure 12.1 plots the growth rates of and relative share of international arrivals data among ASEAN countries along a growth-share matrix, with a third dimension (international arrivals) represented by the size of the circles in the matrix, each of which represents an ASEAN country. From the size of the circles and the height representing regional share of all international arrivals to Southeast Asia, Thailand, Malaysia, Singapore, Indonesia, and Vietnam appear to be most dominant very competitive. In terms of growth rate from 2004 to 2017, which is plotted along the horizontal (x-) axis, Myanmar, Cambodia, and Vietnam are the fastest growing destinations in ASEAN for international tourists. At current rates, these countries will likely join the currently dominant four countries in the medium term. The Philippines' performance, as depicted in its plot location, suggests that it is neither growing competitively nor achieving sustainable and sizeable share of international arrivals compared to its peers. Cambodia almost has the same percentage share (4.6%) of international arrivals as the Philippines

(5.4%) yet it is growing at about twice the rate the Philippines is. The most puzzling aspect regarding the Philippines' tourism performance in Figure 12.1 is that it is in the same quadrant as Brunei, a country only 2% of the Philippines' geographic size. Clearly, in terms of attracting international visitors, Philippine tourism is suboptimal if not inferior relative to the performance of its neighbors but also relative to its own potential.

The noteworthy and recent closure of Boracay Island is, therefore, perplexing as it suggests that over tourism is plaguing the Philippine tourism industry. Boracay is of course only one of the thousands of islands in the country open to tourism, though certainly the most well-known and visited. Such an assessment may be misplaced. When viewed in the broader context of regional tourism and the country's performance relative to its Southeast Asian neighbors, it is evident that over tourism is hardly the central issue for Philippine tourism development. On the contrary, Figure 12.1 suggests it is the relative *underperformance* of tourism development in the Philippines that should be the focal issue of investigation.

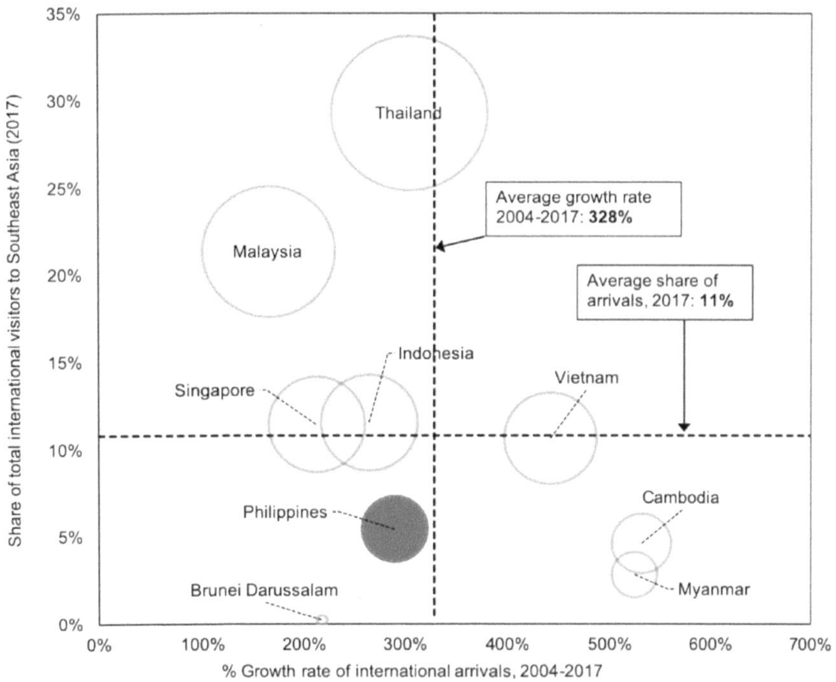

FIGURE 12.1 Growth-share matrix, international arrivals, ASEAN countries, 2004–2017.

Source: Based on Table 12.2 data from UN World Tourism Organization, Yearbook of Tourism Statistics, Compendium of Tourism Statistics and data files.

Although there appears to be a gap between the research literature on the Philippines, which tends to highlight tourism sustainability, governance, and management issues on one hand, and the underwhelming competitive and market performance of its tourism industry on the other, both matters are not mutually exclusive. Problems concerning ecological sustainability do not arise exclusively from demand and market-driven trends. It can just as well originate from inadequate absorptive capacity of destinations to receive growing visitor numbers and the attendant planning, governance, and management it requires on the supply side. For example, the singular and continuous prominence of Boracay in Philippine tourism and, similarly, Bali in Indonesia imply a severe lack of long-term tourism product development and diversification of offerings for both countries. Shortcomings in tourism product development and diversification, in turn, may be symptomatic of deeper rooted inadequacies such as inabilities to attract and sustain investments in infrastructure (which promotes accessibility, reach, and convenience for visitors as well as investments in new attractions), weak governance and planning (which requires policy making, management, and constant monitoring of regulatory issues), and general safety and risk issues (which are constant concerns for travelers).

12.1.3 TOURISM POLICY SHAPING AND PLANNING

There have been three major policy or planning milestones guiding tourism development in the Philippines since 2009. The first of these is the enactment of the Tourism Act of 2009 or Republic Act 9593, which set forth the path for modernization and restructuring of Philippine tourism development (Dela Santa and Saporsantos, 2016), and which mandated the formulation of tourism development plans. From this arose the Philippine National Tourism Development Plan (NTDP) 2011–2016 and continued in the plan currently in force, the NTDP 2017–2022. NTDP 2011–2016 envisioned the Philippines as the “must-see” destination in Asia with the goal of attracting 6.6 million international visitor arrivals (actual arrivals, 2016: 5.9 million (UNWTO)) and enhancing income and employment opportunities. The plan included an objective diagnosis of several fundamental problems hindering the country’s tourism progress (Philippine Department of Tourism, 2011). These include severe infrastructure inadequacies in its airports and destination sites, inadequacies in its air, sea, and land transport connectivity thereby inhibiting accessibility and the potential for market growth and expansion, inefficient operating environment for travel businesses and cumbersome governance frameworks, and the general perception of the country as an unsafe destination with a reputation for poor compliance with safety standards and substandard security.

NTDP 2017–2022 maintains the previous plan's goals to "develop a globally competitive, environmentally sustainable and socially responsible tourism industry that promotes inclusive growth through employment generation and equitable distribution of income…" (Philippine Department of Tourism, n.d., p. 9). The current plan aims to attract 22 million international visitors by 2022, primarily by intensifying development of transport infrastructure, improving access and facilitation of travel, and more crucially, to attract investments in tourism through improved business environment and governance—an aspect neglected in the preceding plan. The current strategy continues to address and seek to overcome the recurrent fundamental weaknesses earlier identified in NTDP 2011–2016. These include underdeveloped transport infrastructure and tourism facilities, inconsistent standards and quality of facilities and services, pervasive concerns surrounding personal safety of travelers and political and economic stability. One striking detail in the plan underscores how severely deficient airport capacity is for the entire country. In 2015, total passenger capacity for airport terminals was estimated at 42.8 million but actual passenger traffic for all gateways was 56.5 million, a severe operational overcapacity of 32%.

From earlier plans drafted prior to 2009 up to the Tourism Act of 2009 and the subsequent NTDPs emerging from it, successive tourism policy and planning in the Philippines has constantly emphasized issues of environmental sustainability, ecological conservation, social inclusiveness, and equitable growth. A substantial portion of the corpus of scholarly research studies (earlier reviewed) on Philippine tourism have revolved mainly around these themes. Alampay's (2005) dedicated volume on managing and fostering sustainable tourism in the Philippines, and the various articles therein, count prominently among these tomes. Intriguingly, factors underlying the underperformance of Philippine tourism growth (relative to its neighbors) remain seriously under researched. This despite several NTDPs identifying and underscoring several critical measures the country needs to undertake to improve tourism performance, the most notable being substantial improvement and investments in infrastructure to promote transport accessibility and tourist facilities. It is tempting to argue that the Philippines has prioritized sustainability and balance over growth as the essence of its tourism development strategy. In comparing the tourism planning and development of Thailand and the Philippines over several decades, Rodolfo (2005) admonished that destinations like the Philippines can "...learn from the negative effects of tourism development in Thailand, as in the case of Pattaya, Koh Samui, and even Bangkok." Another possible explanation regards the Philippines as having focused more on the quality of visitors it attracts, rather than quantity. But using tourism receipts per visitor as a metric, this possibility is negated in the Philippines' case by Figure 12.2, which shows that from 2004 to 2017, tourism receipts per international visitor has been flat and low for the Philippines relative to its ASEAN neighbors. In contrast, Thailand not only increased the quantity of its visitors significantly over the same period, but also its per visitor receipts as well.

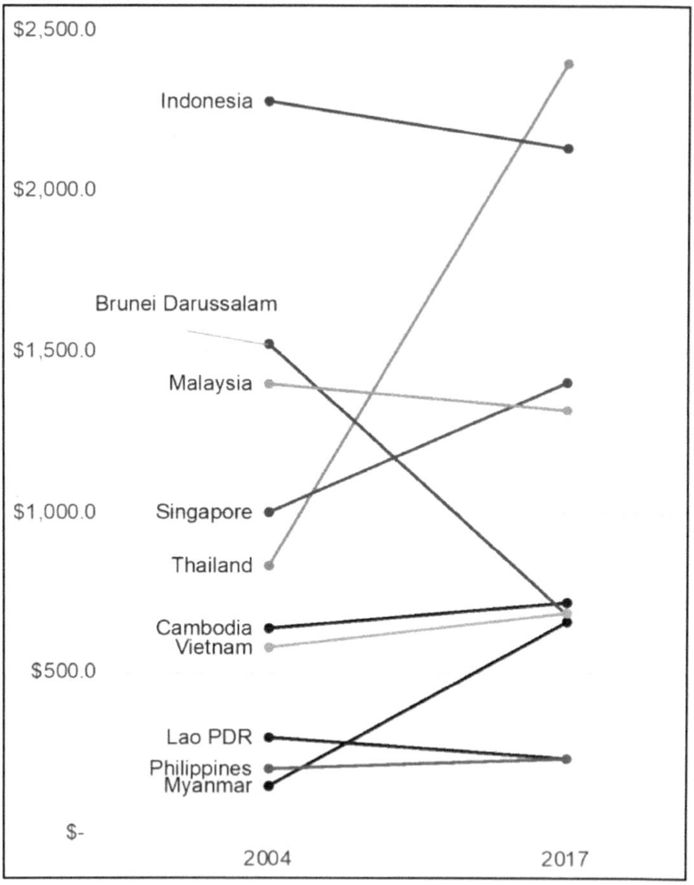

FIGURE 12.2 International tourism receipts, in current US$, 2004 vs 2017.

Source: UN World Tourism Organization, Yearbook of Tourism Statistics, Compendium of Tourism Statistics and data files.

12.1.4 SUMMARY

Overall, tourism development in the Philippines presents an intriguing case. Despite its lackluster performance over the years in growing and attracting volumes of international visitor arrivals (relative to its ASEAN neighbors), but consistent with what scholarly research on the Philippines have emphasized, the country confronts long-standing sustainability issues brought on by tourism development, amplified even by moderate rates of historical visitor growth. All these present an unfavorable combination of circumstances culminating in the high profile and heavy-handed intervention of its President in closing Boracay Island in 2018. Ironically perhaps, the Philippines has become an inadvertent embodiment of the principle that even relatively low numbers and moderate rates of growth in visitor arrivals can asymmetrically trigger unsustainable tourism development issues. The Philippine example suggests that visitor numbers and sustainable tourism management can progress independently of each other at great peril. Although environmental sustainability and socially responsible development have featured prominently and centrally in the country's overall tourism planning and policies, these are not impediments to what should be higher-than-actual average rates of growth and actual number of visitor arrivals to the Philippines, given the comparative performance realized by its neighbors with similar conditions. It is likely, therefore, that other relevant but influential factors give rise to the peculiar situation of Philippine tourism that feature acute sustainability problems despite lower rates of growth and actual volume of visitors.

12.2 GEOGRAPHY AND GOVERNANCE

An obvious explanatory factor in the case of the Philippines is the uneven development of tourism, leading to tourist overconcentration is some areas, in the most popular tourist islands of Boracay, Palawan, and Cebu. The same circumstances are somewhat comparable to the development of Indonesian tourism which tends to see visitor growth concentrate in or mostly clustered in the islands of Bali, Lombok, and other peripheral island destinations (Hampton and Hampton, 2009).

There is certainly some degree of complicity in this regard, attributable to the country's popularity as a tourist destination, particularly in social media as well as in traditional travel media outlets which often publish ratings of the best islands in the world every year, in which several Philippine islands consistently rank high (CNT Editors, 2018). But market popularity alone is insufficient a factor to account for the relative position, performance, and issues that altogether characterize the current state of Philippine tourism. Taking into consideration, therefore, the above review of tourism development in the Philippines and its performance, policies and plans, as well as what scholars have published in their studies, several factors are hereby posited to be influential in steering the long-term trajectory of tourism in the country and its ASEAN neighbors in general.

12.2.1 AN ARCHIPELAGIC GEOGRAPHY

Island destinations that are heavily reliant on tourism (or "insular tourism") often see uneven socioeconomic development, inequality, and social polarization (Apostolopoulos et al., 2002). As Connell (2007) points out, because of their geography, island destinations are often isolated or difficult to access, small, and suffer from poor connectivity. These deficiencies are often hard to overcome, even if island destinations possess natural and unique features such as pristine beaches, natural ecological shoreline features and topologies, as well as enviable surf conditions—all of which enable islands to become famous tourist destinations. Though island destinations enjoy strong comparative resource advantages for developing and benefitting from tourism (Scheyvens and Momsen, 2008), their insularity leaves them particularly vulnerable to unsustainable development as well as prone to resource management conflicts (Buckley, 2002).

Archipelagic nations such as the Philippines thus countenance a unique set of circumstances and challenges in developing tourism. Despite having well-laid out policies and tourism plans likely and conceivably drawn from developmental templates of successful but non-archipelagic, geographically contiguous, tourist destinations, the Philippines faces demanding conditions specific to its geography not unlike those of Caribbean and Pacific Island States, as well as countries like Greece and its Aegean islands (Coccossis, 2001; Hampton and Hampton, 2009) or Indonesia, the only other archipelagic nation in Southeast Asia (Ku, 1991). Mirabueno and Yujuico (2014) highlight how the strategic and organizational burdens of highly insulated and segregated islands that make up archipelagic nations such as the Philippines altogether work to hamper the coordinated development of important road and transport networks, even if a sizable user demand such as those stimulated by tourism encourages it.

12.2.2 GOVERNANCE AND POLICY

Maguigad (2013) underscored the difficulty of tourism planning and policy implementation in archipelagic tourism destinations such as the Philippines, where the onus of development rests mostly and devolved in favor of local governments. Though welcomed by local communities, such a governance arrangement can lead to discordant and under coordinated development of tourism at the strategic and national levels. The interplay of many complex factors affecting a country's tourism development—and how to balance national versus local priorities and issues—is far too numerous and beyond the scope of this article to enumerate. However, some evidence in the Philippines' case point toward tourism-related interests being favored more, inadvertently or otherwise, compared to other local community livelihood interests. As Oracion et al. (2005) observed, the establishment of marine protected areas in certain Philippine communities benefitted tourism constituencies much more than the residents reliant on fishing livelihood, even if the establishment of such conservation areas was intended to benefit all.

It is important to distinguish planning processes and policy making from actual policy implementation and the governance mechanisms required for the sustained management and monitoring necessary for tourism policies to yield their intended benefits. An island community rich in tourism potential may be geographically disadvantaged because of its inherent insular features (e.g., the problem of "peripherality"; Buhalis, 1999) yet still overcome such disadvantage via effective and prudent supervision, efficient governance structures, and competent policy administration. An early study by Trousdale (1999) of Boracay Island which observed the effect of its then governance structures and administration on its tourism development stressed the importance of good governance in the then nascent island destination.

12.2.3 INFRASTRUCTURE PLANNING, DEVELOPMENT, AND SECURITY

Among the immediate and proximate consequences of good governance is the emergence and maintenance of well-developed public infrastructure systems and security, which serve not only to firmly underpin stability and foster sustained economic and livelihood of local communities, but also facilitate the actualization of tourism potential, furthering the robust growth of visitors and investment in tourism resources and facilities, including the structures needed to protect tourism resources and its long-term sustainability (e.g., development of adequate waste disposal systems, roads, and transport amenities such as airports). Controlling and preempting infrastructure development in favor of ecological or sociocultural conservation were earlier touted as ideal policy directions but now considered impractical and less achievable in the face of unrelenting and inevitable tourism growth, as was seen in the Galapagos Islands (Kenchington, 1989). But some evidence of effective tourism development in island destinations such as Mauritius (Khadaroo and Seetanah, 2007) and the Maldives (Domroes, 2001) serve as counterfactual cases of adequate and proper infrastructure planning and development leading to the competitive development of tourism in those island destinations.

In the case of the Philippines, the burden of an expansive archipelagic geography combined with governance inadequacies may also have given rise to security and safety shortcomings, which may altogether have handicapped international visitor arrival growth. Security and safety conditions of a destination are primal concerns of international tourists and, therefore, influence major trends in visitor arrivals (Pizam, 1999). Recognizing this as a threat, the NTDP for 2011–2016 acknowledged explicitly the need to address safety and security of visitors within the context of more effective tourism governance (Philippine Department of Tourism, 2011); the 2016–2022 plan did so as well (Philippine Department of Tourism, n.d.). However, it appears that the Philippines suffers from higher rates of endemic crime and violence relative to its ASEAN counterparts. Figure 12.3 plots the rate of intentional homicide per 100,000 population for ASEAN countries (except Laos, for which no data was available), from 2000 to 2016 (UNODC).

12.2.4 SUMMARY

Utilizing Heath and Wall's (1991) broad strategic planning framework to evaluate the state of Philippine tourism vis-à-vis its ASEAN neighbors, the foregoing review identified four fundamental factors that likely account for the current state and performance of visitor arrivals to the Philippines. These include (a) an expansive insular geography that makes investments in infrastructure development paramount yet challenging to implement and build; (b) a highly devolved political system—concomitant with the country's archipelagic nature—that cripples effective governance and the planning and management systems needed to sustain and improve (c) infrastructure for tourism development; and (d) deficiencies in safety and security, also engendered by weak governance, that has dampened growth in visitor arrivals compared to ASEAN neighbors with similar tourism potential and developmental opportunities.

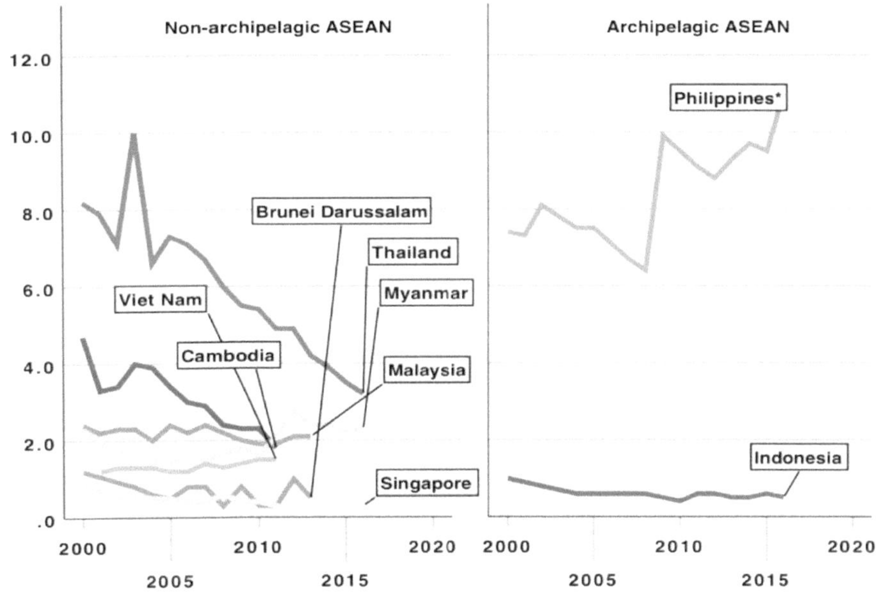

FIGURE 12.3 Intentional homicide rate per 100,000 population, 2000–2016.

Source: United Nations Office on Drugs and Crime (UNODC). Intentional homicide victims by sex, counts, and rates per 100,000 population. https://dataunodc.un.org/crime/intentional-homicide-victims-by-sex.

Considering that tourism growth has been historically robust but mainly concentrated on the more famous island destinations such Boracay, Palawan, Cebu, and Bohol, these four factors and their complex interplay likely precipitated the paradoxical high-profile closure of Boracay island even if, overall, the Philippines has generally underperformed compared to its ASEAN counterparts.

A review of its tourism policies and plans reveals that Philippine tourism planning and policy has been comprehensive, objective, and circumspect in the weaknesses it must overcome. On the other hand, a review of published scholarly research on Philippine tourism suggests an extensive emphasis on sustainability and community involvement issues, providing adequate counter balance to government marketing efforts and plans toward increasing visitor arrivals.

Although much of the tourism performance of any country such as the Philippines rests on factors and circumstances specific to its internal conditions and unique setting, destinations such as those comprising ASEAN nevertheless compete for a substantial share of the international visitors and revenues with its peers, who are equally subject to the same fundamental factors of geography, governance, infrastructure, and safety. It is incumbent, therefore, for such factors to be tested comparatively across countries comprising ASEAN.

12.3 STUDY

12.3.1 METHOD

Using publicly available macro-level data, the following analysis was conducted to examine the effects of the above four factors in accounting for the Philippines' tourism performance over time and relative to its regional ASEAN neighbors. The objective is to examine the effects on long-term inbound international visitor arrivals of (a) an archipelagic geography, (b) government effectiveness, (c) efficiency of infrastructure, and (d) prevalence of crime. Based on the review and evaluation of its tourism performance and situation over recent years, the four factors are posited to explain the performance of the Philippines in terms of attracting international visitors relative to its ASEAN neighbors. Hence, the analysis proceeded by adopting a pooled cross-sectional time-series model, whereby international visitor arrivals is treated as the outcome variable, an archipelagic geography is treated as a time-invariant factor, while government effectiveness, infrastructure efficiency, and prevalence of crime (a proxy for the country's level of security and safety) are treated as random explanatory factors. All variables in the analysis comprised 16 years of annual observations from 2000 to 2016.

Data for international visitor arrivals come from the UN World Tourism Organization, World Tourism Organization, Yearbook of Tourism Statistics, Compendium of Tourism Statistics and related data files (UN World Tourism Organization). Data used as proxy for government effectiveness is based on an indicator called "Government Effectiveness" which "captures perceptions of the quality of public services, the quality of the civil service and the degree of its independence from political pressures, the quality of policy formulation and implementation, and the credibility of the government's commitment to such policies" (Kaufmann et al., 2010). The indicator is an estimate which gives a country's aggregate score in units of a standard normal distribution, that is, ranging from approximately −2.5 to +2.5 standard deviation, and normalized to assume positive values for modeling purpose. Data approximating infrastructure efficiency is based on the World Economic Forum's Executive Opinion Survey, which asks business executives to rate their perception of a country's port facilities coded as "Quality of port infrastructure," with scores ranging on a scale from 1 (port infrastructure considered extremely underdeveloped) to 7 (port infrastructure considered efficient by international standards) (World Economic Forum).

To measure and compare between-country levels of safety, security, and the prevalence of crime, data on homicide rates per 100,000 population collected by the UNODC was used. This variable defines "intentional homicide" as unlawful death inflicted upon a person with the intent to cause death or serious injury (UNODC). Diprose (2007) advocates that rates of homicide are commonly and comparatively indicative of the frequency of violence and threats to security and, therefore, can be a useful proxy for the relative safety and security of tourist destinations. Finally, the analysis included GDP per capita as a control variable and sourced from the World Bank and OECD national accounts data, compiled by the World Development Indicators of the World Bank (The World Bank and OECD national accounts).

Separate models were run in the analysis, one for archipelagic ASEAN countries, comprising of the Philippines and Indonesia, and the other for the remaining non-archipelagic countries (except Laos, for which data on the above variables was sparse). In addition to specifying the main effects of government effectiveness, quality of port infrastructure, and homicide rate per 100,000 population on visitor arrivals, interaction effects of (a) government effectiveness and quality of port infrastructure, as well as (b) government effectiveness and homicide rate, were included. The rationale for including interaction terms is that infrastructure development and planning, as well as general levels of safety and security, are posited to be codependent on the efficiency and effectiveness of government structures. Government effectiveness, in turn, is posited (and previously argued) to be handicapped to a greater degree by an archipelagic geography.

12.3.2 FINDINGS

Table 12.3 reports the descriptive statistics for the variables used in the analysis. Results of the pooled cross-sectional time series analysis for nine ASEAN countries are shown in Table 12.4, comparing non-archipelagic from archipelagic countries (namely, the Philippines and Indonesia). Between 2000 and 2016 and after controlling for per capita GDP (in current US$), international visitor arrivals for non-archipelagic ASEAN countries was significantly and positively influenced by the quality of port infrastructure, 4.83 mn visitors (SE = 2.38 mn, $p = .042$) and a positive crossover interaction between government effectiveness and homicide rate (per 100,000), 5.95 mn visitors (SE = 2.87 mn, $p = .038$). Neither government effectiveness nor the rate of homicides alone was significantly associated long-term with international visitor arrivals for non-archipelagic ASEAN countries.

TABLE 12.3 Mean and Standard Deviation for Variables used in the Pooled Cross-Sectional Time Series Analysis, 2000–2016.

	International visitor arrivals	**Quality of port infrastructure, Scale of 1 to 7**	**Government effectiveness, score in sd units from −2.5 to 2.5**	**Homicide rate per 100,000**	**Per capita GDP (in current US$)**
Brunei	191,769	4.65	0.86	0.66	30,671.7
Darussalam	(40,040)	(0.22)	(0.21)	(0.29)	(10,502.4)
Cambodia	2,919,462	3.74	−0.86	3.12	707.3
	(1,348,944)	(0.28)	(0.10)	(0.84)	(317.3)
Indonesia	7,393,692	3.54	−0.28	0.59	2247.7
	(2,147,138)	(0.41)	(0.14)	(0.15)	(1159.5)
Malaysia	22,800,000	5.59	1.06	2.17	7527.4
	(3,970,039)	(0.10)	(0.11)	(0.18)	(2661.4)
Myanmar	1,502,846	2.61	−1.43	1.93	668.9
	(1,290,340)	(0.01)	(0.18)	(0.41)	(454.3)
Philippines	3,812,077	3.14	0.00	8.39	1858.0
	(1,133,553)	(0.24)	(0.12)	(1.31)	(754.2)
Singapore	9,524,615	6.76	2.15	0.45	39,846.5
	(2,246,396)	(0.06)	(0.16)	(0.18)	(13,571.0)
Thailand	19,400,000	4.61	0.31	6.02	4130.3
	(7,132,797)	(0.18)	(0.08)	(1.84)	(1595.5)
Vietnam	5,655,000	3.45	−0.24	1.33	1154.6
	(2,215,656)	(0.40)	(0.15)	(0.11)	(638.8)

For the two archipelagic ASEAN countries of Indonesia and the Philippines, neither government effectiveness nor quality of port infrastructure proved influential in explaining the long-term trend (i.e., from 2000 to 2016) in international visitor arrivals for both countries but homicide rate was and negatively, −0.284 mn visitors (SE = 0.13mn, p = .026), after controlling for per capita GDP. Interestingly, evidence of a weak but significant negative crossover interaction between government effectiveness and homicide rates on visitor arrivals was detected, −1.12mn (SE = 0.63mn, p = .079.), after controlling for per capita GDP.

TABLE 12.4 Regression Estimates from the Pooled Cross-Sectional Time Series Analysis of International Arrivals, 2000–2016.

	Non-archipelagic ASEAN countries		**Archipelagic ASEAN countries**	
Arrivals	**Coef. (in mn)**	**SE (in mn)**	**Coef. (in mn)**	**SE (in mn)**
Constant	−9.46	(9.31)	−1.91	(4.73)
Homicide rate per 100,000	0.31	(0.94)	**−0.284**	**(0.13)***
Quality of port infrastructure, Scale of 1–7	**4.83**	**(2.38)***	1.34	(1.69)
Government effectiveness, in *sd* units from −2.5 to 2.5	−16.0	(12.60)	−13.1	(23.20)
Government effectiveness × homicide rate	**5.95**	**(2.87)***	**−1.12**	**(0.63)**^
Government effectiveness × quality of port infrastructure	1.64	(1.38)	5.31	(6.59)
Per capita GDP (current US$) *control variable*	−0.00014	(0.00015)	0.00205	(0.00077)**
Wald chi-square:	91.0, $p < .001$		106.25, $p < .001$	
Overall R-sq:	0.6946		0.8985	

*$p < .05$, **$p < .01$, ^$p = 0.079$

12.3.3 DISCUSSION

It appears that for ASEAN countries, government effectiveness has not been significant in accounting for the substantial increase in international visitor arrivals over time. As earlier posited, because of the geographic contiguity of insular ASEAN countries (i.e., non-archipelagic countries), the quality of infrastructure sustained by continuous investment and improvements is a strategic factor in stimulating visitor arrivals. Sadly, because of their insular and isolated geographies, archipelagic ASEAN countries such as the Philippines and Indonesia are handicapped when it comes to infrastructure improvements and investments and thus cannot capitalize on gainable efficiencies that can stimulate tourism arrivals.

Of more concern for insular countries like the Philippines is the well-established deleterious effect of shortcomings in general security and safety, which affects negatively long-term visitor arrivals. Why this is not the case for non-archipelagic ASEAN countries as the foregoing analysis shows, is not clear, except for the possibility that a good infrastructure quality (development and efficiencies for which may be facilitated by a contiguous geography) likely works toward sufficiently mitigating any safety and security concerns they may harbor for geographically contiguous countries.

The inconsequential role of government effectiveness in explaining the remarkable growth of visitor arrivals for ASEAN is intriguing because it implies that international visitor arrivals to the region has flourished despite the level of government effectiveness, mode of governance (e.g., central- vs devolved styles), or degree of intervention. It also suggests how tourism growth progresses independent of political systems, structures, and varying levels of stability, as well as geographical circumstances. When considered interactively with a country's quality of port infrastructure (specified as an interaction term in the model), government effectiveness appears to dampen the main effect of port infrastructure in attracting visitor arrivals in the case of non-archipelagic ASEAN countries or does not significantly affect it all in the case of archipelagic destinations such as the Philippines.

When considered interactively with rates of homicide, however, the role of government effectiveness exhibits a more significant influence in visitor arrivals. For non-archipelagic ASEAN nations, it appears that whenever homicide rates are high, having higher rates of government effectiveness helps significantly in boosting visitor arrivals. For archipelagic nations, however, the reverse seems to be the case, as the model results shows a hint that the interactive effect of government effectiveness with homicide rates appears to negatively affect visitor arrivals (i.e., a crossover interaction). This suggests that for insular nations like the Philippines, high government effectiveness may help diminish homicide rates but at the same time also negatively affects international arrivals, perhaps due to tighter and more cumbersome security measures in place and the associated negative signaling deters visitors from coming.

12.4 CONCLUSION

This review of contemporary tourism developments in the Philippines opened with the contradictions brought to light and triggered by the recent high-profile closure of its popular beach island of Boracay, on orders of the highest authorities, which cited long-standing mismanagement and neglect of environmental issues threatening the island's ecology and community. On the one hand, the incident suggested that tourism in the Philippines has already attained visitor volume and vigorous growth rates far more than sustainable levels allow. But on the other, it conflicts with the rather middle-of-the-road growth rate the country's tourism performance has achieved relative to its ASEAN neighbors from 2000 to 2016. Even if similar drastic closures have also been declared elsewhere in ASEAN such as in Maya beach, Thailand, which too was prompted in reaction to over tourism concerns, the vast difference in international visitor volume (as well as other factors) between the Philippines and Thailand (which receives far more visitors at high annual growth rates) fails to adequately explain why similar outcomes are occurring across neighboring ASEAN countries and particularly the Philippines. Though they belong to the same geographic region and offer somewhat similar attractions in culture, nature, exotic food, etc., for visitors, ASEAN countries also have significant dissimilarities in factor conditions affecting long-term tourism performance.

The factor conditions posited in this study to explain the between and within general trends and trajectories of ASEAN countries' international visitor arrivals include geography (archipelagic versus non-archipelagic), ratings of government effectiveness, ratings on the quality of port infrastructure, and the level of general security and safety (as approximated by homicide rates), with the latter two factors additionally posited to interact with government effectiveness. Cross-sectional regression analysis over a long time-period reveals that the archipelagic geography of the Philippines (along with Indonesia) differentiates the degree with which international visitor arrivals can be attracted, because of geography's intermediate and differential effects the effectiveness of policies to enhance port infrastructure (which can support tourism growth and development) and on general levels of security and safety (as approximated by homicide rates). Perhaps because of its highly insular features, where island destinations can be isolated, cut-off, and inaccessible both in terms of infrastructure development and effective central governance, managing and improving security and safety has proven to be challenging, thus affecting long-term visitor arrivals to the Philippines negatively. In contrast, among non-archipelagic ASEAN countries, analysis reveals that security and safety concerns does not significantly influence visitor arrivals. Indeed, the analysis shows that their sustained high rates of visitor arrivals over the years (from 2000 to 2016) was likely boosted by high ratings of the quality of port infrastructure.

Why should the quality of port infrastructure have a differential impact on international visitor arrivals between archipelagic and non-archipelagic

ASEAN countries? The argument posed in this study is that a highly insular geography, versus one that is more contiguous, preempts greater efficiencies in investments, development, coordination, and security. It is an argument evidently supported by the foregoing analysis. The belief that more effective governance can overcome such difficulties for archipelagic nations and yet be a stimulus for non-archipelagic nations seems negated by the analysis, which shows that ratings of government effectiveness, whether alone or interacting with the quality of port infrastructure, do not have any bearing whatsoever on trends in international arrivals. It seems, therefore, that among ASEAN countries, archipelagic or otherwise, it is best for government agencies to simply play an indirect or less intrusive role.

In only one respect does rating for government effectiveness apparently impact international arrivals over time for ASEAN countries as a whole: When it interacts with safety and security as measured by homicide rates per 100,000 population. However, this influence varies between archipelagic and non-archipelagic countries. For non-archipelagic ASEAN countries, the evidence suggests that better government effectiveness may dampen increases in homicide rates (when both are considered to interact), thus helping to increase visitor numbers. For archipelagic countries, however, evidence of a negative crossover interaction between government effectiveness ratings with homicide rates suggests an opposite outcome whereby higher ratings of government effectiveness may still dampen homicide rates but nevertheless results in lower visitor arrivals, perhaps due to signaling cues being misinterpreted by potential tourists or overly strict measures to curb crime and violence affecting visitor intentions or experience negatively.

This review finds that the relatively few scholarly studies published on Philippine tourism with their predominant themes surrounding sustainability, livelihood, and equity, tie in very much with the foregoing analysis and the influence of the hypothesized fundamental factors in accounting for the Philippines' tourism performance in attracting international visitor arrivals over time. Alampay (2005, p 18) may be spot on in concluding that tourism in the Philippines revolves around economic sustainability, ecological sustainability, and social equity. But so is the tourism of many countries, including those among its close neighbors in the ASEAN. What previous studies have not accounted for adequately is the profound influence of an archipelagic endowment in geography on promoting and sustaining a vibrant and growing tourism industry.

Maguigad's (2013) concerns on the ability of local governments and the need to build linkages, working relationships, as well as smoothing power relations among stakeholders resonate more in the context of this study's empirical analysis, as these derive from the archipelagic burden with which Philippine tourism need to strive under. Previously published studies on Philippine tourism highlighting governance failures and shortcomings on road and transport development (Manuela and de Vera, 2015) and advocating the need for better governance collaboration and coordination (Mirabueno and Yujuico, 2014) can now be regarded as being overshadowed by the influence of a difficult geography.

This review finds tourism policy and planning in the Philippines, evidenced in its two NTDPs, to be very forthwith and very much aware of the strategic challenges and issues. The most recent Philippine tourism development plans are explicit and cognizant of the gaps in infrastructure development and much needed improvements in safety and security. Policy implementation is of course vastly different from policy making. Indeed, the empirical study reported above shows that government effectiveness is generally ineffectual in improving visitor arrival growth, except when it comes to promoting general safety and security. What is imperative for tourism authorities in the Philippines to recognize, based on this study's results, is that it is probably better to leave tourism development planning to local authorities and strategically empower them. If the analysis of this study is valid, archipelagic nations such as the Philippines need to overcome the strategic handicap and burden of their geography, even if gaps in governance and infrastructure aren't necessarily a disadvantage as much as gaps in security and safety are.

Overall, there appears to be four interrelated factors shaping the long-term trajectory of Philippine tourism development: Underinvestment, weaknesses in policy implementation and continuity in governance (planning is not in itself weak—the country knows its weaknesses), unintended marketing, and an archipelagic geography from which it draws its strengths and weaknesses.

KEYWORDS

- **archipelagic geography**
- **insular tourism**
- **tourism governance and planning**
- **tourism infrastructure**
- **crime and safety**

REFERENCES

Alampay, R. B. The Challenge of Sustainable Tourism Development in the Philippines. In *Sustainable Tourism Challenges for the Philippines*; Alampay, R. B., Ed.; Philippine APEC Study Center Network (PASCN) and the Philippine Institute for Development Studies (PIDS): Makati City, Philippines, 2005; pp 1–22.

Alejandria-Gonzalez, M. C. P. Cultural Tourism Development in the Philippines: An Analysis of Challenges and Orientations. *J. Qual. Assur. Hosp. Tour.* *17*(4), 496–515, 2016.

Apostolopoulos, Y.; Apostolopoulos, Y.; Gayle, D. J. Island Tourism and Sustainable Development: Caribbean, Pacific, and Mediterranean Experiences; Greenwood Publishing Group, 2002.

Buckley, R. Surf Tourism and Sustainable Development in Indo-Pacific Islands. II. Recreational Capacity Management and case Study. *J. Sustain. Tour.* **2002,** *10*(5), 425–442.

Buhalis, D. Tourism on the Greek Islands: Issues of Peripherality, Competitiveness and Development. *Int. J. Tour. Res.* **1999,** *1*(5), 341–358.

Carter, R. W. Implications of Sporadic Tourism Growth: Extrapolation from the Case of Boracay Island, The Philippines. *Asia Pac. J. Tour. Res.* **2004,** *9*(4), 383–404.

Chon, K. S.; Oppermann, M. Tourism Development and Planning in Philippines. *Tour. Recreat. Res.* **1996,** *21*(1), 35–43.

CNT Editors. The Best Islands in the World: 2018 Readers' Choice Awards. Conde Nast Travelers. https://www.cntraveler.com/galleries/2014-10-20/top-30-islands-in-the-world-readers-choice-awards-2014 (accessed April 18, 2019).

Coccossis, H. Sustainable Development and Tourism in Small Islands: Some Lessons from Greece. *Anatolia* **2001,** *12*(1), 53–58.

Connell, J. 'The Best Island on the Globe': Constantly Constructing Tourism on Niue. *Aust. Geogr.* **2007,** *38*(1), 1–13.

Cripps, K. Thailand Bay Made Popular by 'The Beach' Closes Indefinitely. https://edition.cnn.com/travel/article/maya-bay-closure-thailand/index.html (accessed April 19, 2019).

Dela Santa, E. The Politics of Implementing Philippine Tourism Policy: A Policy Network and Advocacy Coalition Framework Approach. *Asia Pac. J. Tour. Res.* **2013,** *18*(8), 913–933.

Dela Santa, E. The Evolution of Philippine Tourism Policy Implementation from 1973 to 2009. *Tour. Plan. Develop.* **2015,** *12*(2), 155–175.

Dela Santa, E.; Saporsantos, J. Philippine Tourism Act of 2009: Tourism Policy Formulation Analysis from Multiple Streams. *J. Policy Res. Tour. Leis. Events* **2016,** *8*(1), 53–70.

Diprose, R. Physical Safety and Security: A Proposal for Internationally Comparable Indicators of Violence. *Oxford Develop. Stud.* **2007,** *35*(4), 431–458.

Domroes, M. Conceptualising State-Controlled Resort Islands for an Environment-Friendly Development of Tourism: The Maldivian Experience. *Singap. J. Trop. Geogr.* **2001,** *22*(2), 122–137.

Hampton, M.; Hampton, J. Is the Beach Party Over? Tourism and the Environment in Small Islands: A Case Study of Gili Trawangan, Lombok, Indonesia. Tourism in Southeast Asia: Challenges and New Directions, 2009; p 286.

Heath, E.; Wall, G. *Marketing Tourism Destinations: A Strategic Planning Approach*; John Wiley & Sons, Inc: New York; 1991.

Kaufmann, D.; Kraay, A.; Mastruzzi, M. Government Effectiveness (Estimate). In *The Worldwide Governance Indicators: Methodology and Analytical Issues: The World Bank*; W. G. I., Ed.; 2010.

Kenchington, R. A. Tourism in the Galapagos Islands: The Dilemma of Conservation. *Environ. Conserv.* **1989,** *16*(3), 227–232.

Khadaroo, J.; Seetanah, B. Transport Infrastructure and Tourism Development. *Ann. Tour. Res.* **2007,** *34*(4), 1021–1032.

Korey, A. How Cinque Terre Is Handling the Pressure of Being on Every Traveler's Bucket List. https://www.travelandleisure.com/trip-ideas/cinque-terre-tourist-overflow (accessed April 19, 2019).

Ku, C. The Archipelagic States Concept and Regional Stability in Southeast Asia. *Case W. Res. J. Int. L.* **1991,** *23*, 463.

Maguigad, V. M. Tourism Planning in Archipelagic Philippines: A Case Review. *Tour. Manag. Perspect.* **2013,** *7*, 25–33.

Manuela, W. S.; de Vera, M. J. The Impact of Government Failure on Tourism in the Philippines. *Transp. Policy* **2015,** *43*, 11–22.

Mirabueno, J.; Yujuico, E. Paving the Way for Philippine Tourism via Interagency Collaboration on Road Networks. *Transp. Policy* **2014,** 36, 306–315.

Ocampo, L.; Ebisa, J. A.; Ombe, J.; Geen Escoto, M. Sustainable Ecotourism Indicators with Fuzzy Delphi Method – A Philippine Perspective. *Ecol. Indic.* **2018,** *93*, 874–888.

Ong, L. T. J.; Storey, D.; Minnery, J. Beyond the Beach: Balancing Environmental and Socio-Cultural Sustainability in Boracay, the Philippines. *Tour. Geogr.* **2011,** *13*(4), 549–569.

Oracion, E. G.; Miller, M. L.; Christie, P. Marine Protected Areas for Whom? Fisheries, Tourism, and Solidarity in a Philippine Community. *Ocean Coast. Manag.* **2005,** *48*(3), 393–410.

Palanca-Tan, R.; Garces, L. P. D. M.; Purisima, A. N. C.; Zaratan, A. C. L. Tourism and crime: Evidence from the Philippines. Southeast Asian Studies **2015,** *4*(3), 565–580.

Philippine Department of Tourism (Producer). National Tourism Development Plan 2011–2016 Draft Action Plan. Powerpoint Slides. https://studylib.net/doc/5499623/the-philippine-national-tourism-development-plan-2011-2016

Philippine Department of Tourism (Producer). (n.d.) National Tourism Development Plan, 2016–2022.

Pizam, A. A Comprehensive Approach to Classifying Acts of Crime and Violence at Tourism Destinations. *J. Travel Res.* **1999,** *38*(1), 5–12.

Plush, H. Barcelona Unveils New Law to Keep Tourists Away. The Telegraph. https://www.telegraph.co.uk/travel/destinations/europe/spain/catalonia/barcelona/articles/barcelona-unveils-new-law-to-keep-tourists-away/ (accessed April 16, 2019).

Rodolfo, M. C. L. S. A Comparison of Tourism Policy Frameworks: The Philippines and Thailand. In *Sustainable Tourism Challenges for the Philippines. Makati City, Philippines: Philippine APEC Study Center Network (PASCN) and the Philippine Institute for Development Studies (PIDS)*; Alampay, R. B., Ed.; 2005.

Saverimuttu, V.; Varua, M. E. Climate Variability in the Origin Countries as a "Push" Factor on Tourist Arrivals in the Philippines. *Asia Pac. J. Tour. Res.* **2014,** *19*(7), 846–857.

Scheyvens, R.; Momsen, J. Tourism in Small Island States: From Vulnerability to Strengths. *J. Sustain. Tour.* **2008,** *16*(5), 491–510.

Smith, R. A.; Henderson, J. C.; Chong, V.; Tay, C.; Jingwen, Y. The Development and Management of Beach Resorts: Boracay Island, The Philippines. *Asia Pac. J. Tour. Res.* **2011,** *16*(2), 229–245.

Smith, V. L. Geographical Implications of "Drifter" Tourism Boracay, Philippines. *Tour. Recreat. Res.* **1990,** *15*(1), 34–42.

Tamayo, N. C. A.; Anticamara, J. A.; Acosta-Michlik, L. National Estimates of Values of Philippine Reefs' Ecosystem Services. *Ecol. Econ.* **2018,** *146*, 633–644.

The World Bank and OECD National Accounts. World Development Indicators: The World Bank.

Trousdale, W. J. Governance in Context: Boracay Island, Philippines. *Ann. Tour. Res.* **1999,** *26*(4), 840–867.

UN World Tourism Organization. International Tourism, Number of Arrivals, Yearbook of and Compendium of Statistics The World Bank.

United Nations Office on Drugs and Crime (UNODC). Intentional Homicide Victims by Sex, Counts and Rates per 100,000 Population. https://dataunodc.un.org/crime/intentional-homicide-victims-by-sex

Villamor, F. Idyllic Philippine Resort Island of Boracay Is Closed to Tourists. https://www.nytimes.com/2018/04/04/world/asia/boracay-philippines-tourists-closed.html (accessed March 25, 2019).

World Economic Forum. Quality of Port Infrastructure, Global Competitiveness Report: The World Bank.

World Travel & Tourism Council. Travel & Tourism Economic Impact 2018 Philippines, 2018.

CHAPTER 13

A PORTRAIT OF INDONESIAN TOURISM: PRESENT AND FUTURE

HERA OKTADIANA

James Cook University, Australia

Trisakti School of Tourism, Indonesia

E-mail: hera.oktadiana@jcu.edu.au

ABSTRACT

This chapter reviews the development of the tourism industry in Indonesia. It focuses on four aspects. The first section describes the current state of Indonesia's tourism industry. This includes documenting the growth of tourist arrivals, the roles of tourism as the country's economic generator, and tourism competitiveness. The second part focuses on the government initiatives to boost tourism, including the proposal for the 10 New "Bali" destinations. The third component discusses tourism sustainability programs which comprise three strategic initiatives: sustainability tourism destinations (STD), sustainable tourism observatory (STO), and sustainable tourism certification (STC). The fourth facet portrays Indonesia as a halal destination. In a final section, the future direction of Indonesia tourism is considered. Data were gathered through secondary information obtained from the government reports, documents produced by national and international organizations, media/news, and other relevant papers.

13.1 INTRODUCTION

Who does not know Bali—the Island of Gods? A (famous) destination that is named first among the top 25 destinations in Asia and the fifth in the world according to 2019 TripAdvisor Travelers' Choice. But do most people know that Bali is an island in Indonesia? Probably not. On multiple occasions the author has met people who claimed that they have been to Bali but not to Indonesia. An amusing question asked by people to the author was, "So, you are from Indonesia. Is it close to Bali?." Similarly, when the author wanted to buy a travel adaptor in Australia for a friend who wanted to go to Indonesia, she saw different types of travel adaptors for various countries and regions. On one box of adaptors it was written: suitable for Europe and Bali. Bali is not a country. It is part of Indonesia, the world's largest archipelago with more than 17,000 islands, the fourth most populous country in the world with a population of more than 250 million, and the world's biggest Muslim nation (Bowden, 2015; Ministry of Tourism Republic of Indonesia, 2016).

Indonesia, the string of emerald islands around the equator, has a beautiful natural landscape ranging from the tropical rainforest, beaches and reefs, rice terraces, lakes, rivers, waterfalls, valleys, and volcanoes. Indonesia lies in the Pacific Rim of Fire with around 150 volcanic sites and is the center of the Coral Triangle which is known as a global center for biodiversity. The country is also a home of various species of flora and fauna where some are rare and fascinating such as the giant flower of Rafflesia Arnoldi, the black orchid, and the prehistoric giant lizard Komodo. As a multicultural nation with 350 ethnic groups and 483 local languages, Indonesia is very rich and diverse in culture and traditions (Bowden, 2015; Ministry of Tourism Republic of Indonesia, 2016). Bowden (2015, p 5) stated that "for those interested in culture there is probably no other country in the world that offers such diversity from remote jungle communities in Papua to urban sophisticates in Jakarta." It can be argued that there are many (under the radar) destinations in Indonesia that are worth exploring beyond Bali (Ariffin, 2018).

Lonely Planet named Indonesian its list of top 10 countries for 2019. Nevertheless, there are probably few contemporary studies that portray Indonesia tourism holistically by covering the current state of the sector, the government initiatives to boost tourism activity, the national tourism sustainability program, and a consideration of Indonesia as a halal destination. It can be argued that such information is mainly available from the news and reports. Moreover, studies on Indonesia tourism are predominantly conducted (around 79%) by foreign scholars, emphasizing topics specifically on tourism planning and development, tourism impacts and sustainability, culture and change, tourism marketing, and tourism employment (Oktadiana and Pearce, 2017).

This chapter aims to review the development of Indonesia tourism. Secondary data from the last 5 years were obtained to inform the work. They derive from the government reports, national and international organizations documents, media/news, academic journal articles, and other relevant papers. Indeed it is challenging to find updated academic papers that depict the current state and future directions of Indonesia tourism. The chapter is comprised of four parts. The first part describes the profile of Indonesia tourism. The second segment concerns the approaches taken by the Indonesian government to boost Indonesia tourism. The third section discusses Indonesia tourism sustainability programs. The fourth part depicts Indonesia as halal tourism destination.

13.2 INDONESIA: A TOURISM PROFILE

According to the Travel & Tourism Benchmarking Reports 2017 issued by World Travel & Tourism Council (2017) and based on the year 2016, the contribution of travel and tourism to Indonesia's GDP was USD 58 billion or 6.2%. The sector generated (directly or indirectly) 6.7 million jobs or 5.6% of the total employment. It was expected that Indonesia's travel and tourism GDP would grow 5.6% on average per annum over the 10-year period of 2017–2027. This number is higher than the growth of the mining and agricultural sectors, automotive and chemical manufacturing industries, and the expected rate of the total economy of the country. The travel and tourism sector also plays a vital role in the country's service exports with visitor exports of USD 13 billion or 55.6% of all service exports and 8% of all exports (goods and services). This sector is interconnected with the others and provides spillover benefits across the entire spectrum of economy. For example, every USD 1 million in travel and tourism sales generated USD 133,0000 GDP in the agricultural sector, while wholesale and retail were boosted by USD 84,000. For employment, 198 jobs were created for every USD 1 million in travel and tourism spending. This is much larger than the overall generative power of the economy (12 jobs) and the other sectors such as auto manufacturing, communications, the petrochemical industry (93 jobs each), and financial services (88 jobs).

Based on the Travel & Tourism Competitiveness Report 2017 (World Economic Forum, 2017), the ranking of Indonesia moved from 50 to 42. Since 2019, Indonesia's ranking has been improving from rank 81 in 2009 to 74 in 2011, then 70 in 2013, and further to 50 in 2015. Indonesia is excelling in its price competitiveness—taxes and airport charges, hotel price, fuel price, purchasing power parity (5th), prioritization of travel and tourism—such as government support/prioritization, expenditure for travel and tourism, marketing and branding strategy (12th It is also ranked well for natural resources—number of World Heritage natural sites, species, protected area, (14th), international openness—visa requirements, air service agreements, regional trade agreements (17th), and cultural resources and business environment—such as number of World Heritage cultural sites, cultural heritage, sports stadiums, and international association meetings (23rd). In several respects, however, Indonesia needs to improve its services and infrastructure. They include environmental sustainability (131st), health and hygiene (108th), tourist service infrastructure (96th), safety and security (91st), ICT readiness (91st), ground and port infrastructure (69th), human resources and labor market (64th), and business environment (60th).

The earliest record of foreign tourist arrivals to Indonesia as indicated in Indonesia Statistics or *Badan Pusat Statistik* was dated in 1974 with 313,452 visitors. The number has been increasing, particularly in the last 10 years. The number jumped from 6.2 million foreign tourists visiting Indonesia in 2008 to 15.8 million in 2018. After four decades, Indonesia eventually received more than 10 million foreign tourists to the country in 2015. It can be noted that in the first decade (1974–1983), there were 4.9 million foreign tourists to Indonesia. From 1984–1993 there were 17.5 million foreign tourists who visited the country. The number increased to 47.6 million within 1994–2003 and 64.8 million within 2004–2013. Compared to the 10-year period of 2004–2013, the number of foreign tourist arrivals to Indonesia in the last 5 years (2014–2018) were almost similar, amounting to 61 million visitors (Badan Pusat Statistik, 2017, 2019). In some years, Indonesia tourism has suffered from declines in foreign tourist arrivals. These downturns can be traced to the Asian financial crisis 1997–1998 and the Bali bombings in 2002 and 2005. There was only a miniscule increase in the number of foreign tourist arrivals in 2009 which can be attributed to the effect of the Ritz Carlton/Marriott bombing (Indonesia Investments, 2016).

Most of the visitors coming to Indonesia within the period of 2000–2017 were from the Asia Pacific region (around 78%) which was further dominated by the ASEAN countries. Other sources included Europe, America, and Middle East and Africa with the number of arrivals around 15%, 4%, and 2%, respectively. The top 10 inbound to Indonesia from 2000–2017 were Singapore, Malaysia, Australia, Japan, South Korea, Taiwan, British, the United States, India, and Germany. While those 10 countries still hold the largest market for Indonesia tourism, there has been an increase in the number of visitors from China and Russia within the 5-year period of 2013–2017 (Badan Pusat Statistik, 2018).

These statistical highlights about Indonesian tourism are presented in Table 13.1.

TABLE 13.1 Profile of Indonesia Tourism.

Contribution to GDP in 2016[a]	USD 58 billion (6.2%)
Contribution to employment in 2016[a]	6.7 million jobs (5.6% of the total employment)
Travel and tourism GDP average growth per annum (2017–2027)[a]	5.6%
Foreign tourist arrivals 2014–2018[b]	9.4 million (2014), 10.2 million (2015), 11.5 million (2016), 14 million (2017), 15.8 million (2018)
Top 10 foreign tourist arrivals in global (2000–2017)[b]	Singapore, Malaysia, Australia, Japan, South Korea, Taiwan, the United Kingdom, the United States, India, and Germany.
Top five foreign tourist arrivals (2013–2017)[b]	
Asia Pacific	Malaysia, Singapore, China, Australia, Japan,
Europe	the United Kingdom, France, Germany, the Netherlands, Russia
WTTC ranking[c]	50 (2015), 42 (2017)
Competitive advantages (based on WTTC 2017 ranking)	Price competitiveness, national prioritization of travel and tourism industry, rich natural resources, international openness, and cultural resources and business environment
Improvement required (based on WTTC 2017 ranking)	Environmental sustainability, health and hygiene, tourist service infrastructure, safety and security, ICT readiness, ground and port infrastructure, human resources and labor market, and business environment

Sources: [a]World Travel & Tourism Council (2017), [b]Badan Pusat Statistik (Statistic Indonesia) (2017, 2018, 2019), [c]World Economic Forum (2017).

13.3 ADVANCING THE INDONESIAN TOURISM SECTOR

The significant contributions of travel and tourism have led the Indonesian government to develop the sector as a key pillar of Indonesia's economic growth and the main source of foreign exchange revenues. Under the presidency of Joko Widodo, tourism is viewed as one of country's priority sectors. After Joko Widodo was elected President in 2014, a stand-alone Ministry of Tourism was formed for the first time. The Ministry has set a target of 20 million foreign tourist arrivals in 2020 as a part of the overall plan for the country to become a tourist powerhouse. This number is double that of the 2015 tourist arrivals (Lee, 2018; Tarahita and Rakhmat, 2017). Tourism Minister Arief Yahya has endorsed Indonesia's three tourism missions: mobile positioning data (MPD), sustainable tourism programs (sustainable tourism destination (STD)—sustainable tourism observatory (STO)—sustainable tourism certification (STC), and homestay rural tourism (Sofia, 2019). Specific strategies to support and sustain the expansion of Indonesia tourism have been noted as vital to achieve these ambitious goals (Tarahita and Rakhmat, 2017).

To enhance the contribution of tourism to the economy to 7.5%, the government initiated the 10 new destinations or 10 New Bali. They include Lake Toba (North Sumatra), Tanjung Kelayang Beach (Bangka Belitung), Tanjung Lesung (Banten, Java), Kepulauan Seribu (Thousand Islands) (Jakarta), Borobudur Temple (Central Java), Mount Bromo (East Java), Wakatobi (Southeast Sulawesi), Mandalika (West Nusa Tenggara), Labuan Bajo (East Nusa Tenggara), and Morotai Islands (North Maluku) (Ariffin, 2018) (Fig. 13.1). The selection of the new destinations was based on three A's: attraction (which the destinations already have), access and, amenities (requirements and benefits related to infrastructure development). The 10 destinations have distinctive features, ranging from cultural experiences at the world-renowned heritage Borobudur temple (Fig. 13.2), observing Komodo dragons in Labuan Bajo, to diving in Wakatobi (Chan, 2017).

FIGURE 13.1 Map of Indonesia and the 10 new destinations.

Source: Author's illustration.

These 10 new destinations are designated to replicate the tourism success in Bali through improved infrastructure. It is expected that these new destinations will escalate the number of jobs from 11.8 million to 13 million. The infrastructure development plan includes building new hotels, resorts, airports, and docks. For example, Jakarta's Kepulauan Seribu which consists of a cluster of 150 islands will need 10 new docks by 2019 (Ariffin, 2018). Such development undoubtedly requires large investment. The Ministry of Tourism calls for 10 million tourist arrivals and USD 20 billion in investment in the 10 new destinations by 2019 (Bain, 2016).

FIGURE 13.2 Borobudur Temple—UNESCO World Heritage Site.
Source: Author's photo.

To promote the 10 new destinations, the Indonesian government has encouraged travel agents and airline companies to offer tour packages to those destinations. The government also has been improving a number of airports such as Adi Sumarmo in Surakarta, Adisutjipto in Yogyakarta, Lombok International Airport. Kualanamu in Medan, and Silangit in North Tapanuli to serve as international hubs (Chan, 2017; The Jakarta Post, 2017). The air connectivity expansion provides easier access for foreign tourists to visit these other parts of Indonesia apart from Java and Bali. Another approach to attract foreign tourists has been the creation of a visa-free policy to 169 countries as regulated in the Presidential Regulation number 21/2016 (Tarahita and Rakhmat, 2017).

To support the development of tourism-related basic infrastructure and services and to reinforce private investment, the World Bank provided a loan of USD 300 million. This will be used initially to finance the projects in three key destinations: Lake Toba in North Sumatra, Borobudur-Yogyakarta-Prambanan triangle in Java, and Lombok island in West Nusa Tenggara. It is expected that the development in those three main areas will increase the tourist arrivals to 27.3 million and enhance the tourist spending to USD 3.3 billion by 2041. It is anticipated that private investment will rise 13-fold to USD 421 million. The project also focuses on human capital development and natural and cultural asset management for the long-term benefits of the local communities (The World Bank, 2018).

In the wake of the Chinese tourist boom, Indonesia is targeting more Chinese tourists to visit the country. As Indonesia is a part of China's One Belt and One Road Initiative, the Minister of Tourism Arief Yahya has made an open invitation to Chinese investors to fund the 10 new destinations (Ariffin, 2018). Data from Statistic Indonesia specified that there has been an increase in the number of Chinese tourists although the majority visit Bali. Since 2014, more than 1 million Chinese tourists have come to Indonesia. In 2017 alone, the number reached more than 2 million, a 34% increase from 2016. This is probably due to the visa waiver applied for the Chinese tourists (Badan Pusat Statistik, 2018; Oxford Business Group, n.d.). Apart from the Chinese tourists, the government also aims to attract more Australian and Indian tourists. It is expected that the operations of low-cost carriers to and from those countries will boost their tourist numbers (Chan, 2017).

Another step taken by the Ministry of Tourism to enhance inbound tourism from ASEAN countries from 40% to 50% is through partnerships with the O2O mobile platform, Grab. This form of partnership focuses on the digital experience, particularly for the millennials. The programs and services include information about the new travel destinations to promote the 10 New Bali, calendars of events, online bookings for trips and packages via the Grab App, GrabCar rental service, GrabFood's food delivery, and GrabChat auto-translate messages. The Ministry also invested substantially (more than 70% of the marketing budget) in other digital media platforms such as Google, TripAdvisor, Expedia, and Baidu (Lee, 2018; Trew, 2018).

To provide more accommodation for the tourists in the relatively short term, the Ministry of Tourism does not just rely on the hotel development as a hotel may require 4–5 years to build. The Minister initiated a homestay program across the archipelago which has 75,000 villages; specifically by concentrating on establishing 10,000 environment-friendly homestay accommodation units in the 10 priority destinations in 2019. This type of accommodation does not only generate economic returns for the local community, but also provides opportunities for the tourists to experience the local tradition and culture (Mufti, 2019; Prambudi, 2019; Sofia, 2019).

To achieve the 2019 tourist arrivals target, the government has set several strategies: to develop border tourism and hub tourism, and to renovate several terminals at Soekarno-Hatta International Airport for low-cost carriers. The aim of border tourism is to offer easy, convenient and low cost access for tourists to visit Indonesia from its neighboring countries such as Malaysia and Singapore. The hub tourism is aimed to attract foreign tourists that travel to the capital cities of Singapore, Kuala Lumpur, and Bangkok as many foreign travelers transit in those cities before continuing their journey. The purpose of renovating terminals for low-cost carriers is to address the increased number of passengers coming to Indonesia using these types of fares (about 55% growth annually) (Anisa, 2019).

The industry generally has responded positively to the government's plan and ambitions. However, major issues faced by many destinations are poor facilities, services, and connectivity. Other issues concern safety and security (including terrorism incidents), health care facilities, infrastructure, and natural disasters (volcanic eruptions, tsunami, earthquakes), and airline incidents (e.g., the crash of Lion Air flight JT610) (Anisa, 2019; Chan, 2017). Despite the problems, the move and initiatives toward the advancement of Indonesia tourism industry is bright news for the country. The improvement will generate growth in employment, the economy, and build business opportunities, as well as a better, easier, and safer environment for both domestic and foreign tourists to enjoy the beautiful nature of Indonesia and its cultural and heritage sites (Tarahita and Rakhmat, 2017).

13.4 TOURISM SUSTAINABILITY INITIATIVES

As reported in the World Economic Forum's Travel and Tourism Competitiveness report (2017), Indonesia tourism sector needs to address the better protection of the environment to sustain its development and future success, as well as advance its citizens' well-being. This includes attention to deforestation, insufficient treatment of wastewaters, and attending to the plight of species listed as threatened. Additionally, the enhancement of clean facilities and provision of road signs to tourist sites should be revamped through better urban and environmental planning (Tarahita and Rakhmat, 2017).

The Indonesian government has a strong commitment toward sustainable development (Sofia, 2019). Sustainability tourism programs and strategies comprise STD, STO, and STC. The scope of STD includes national guidelines on sustainability tourism, a strategic platform, a master plan, MOU and destination contract, a national and international assessor training program, facilitator and local partner training, conferences, and prototyping 10 priority destinations. There are three key measurement indicators of STD: economy (capacity building of the local community and training for trainers to provide skills and knowledge to use tourism resources wisely), community (empowerment of facilitators and local partners), and environment (response to climate change). The STD partnerships include Global Sustainable Tourism Council (GSTC), United Nations Sustainable Solution Network (UNSDSN), national working groups and 20 regions (Teguh, 2017).

The sphere of STO consisted of STO national guidelines and prototyping of five STOs, that is, the destinations being observed for their participation in applying sustainable tourism practices. The first measurement component of this facet is the field and action research. This is aimed to collect information concerning the implementation of tourism sustainable practices and to address issues rising during the implementation. The second indicator is reporting the results of research in STO national conference. The third measurement is providing recommendations at the national conference to Wonderful Indonesia Sustainable National Tourism Observatory (WINSTO) and the International Network of Sustainable Tourism Observatories (INSTO) (Teguh, 2017).

After China, Indonesia is the second country in Asia that has developed a STO through cooperation with universities. A critical point in STO is waste management as well as cleanliness (Prambudi, 2019, Sofia, 2019). Five STO models have been built in five areas: Pangandaran in West Java with Bandung Institute of Technology, Sleman-Yogyakarta with University of Gadjah Mada, Samosir with the University of North Sumatra, Sanur Bali with University of Udayana, and Sasot Lombok with the University of Mataram. Another seven STO models are being developed within the 10 priority destinations: Tanjung Lesung-Banten, Tanjung Kelayang-Belitung, Thousand Islands-Jakarta, Bromo-East Java, Labuan Bajo-East Nusa Tenggara, Wakatobi-Southeast Sulawesi, and Morotai-North Maluku. The models will be supported with University of Indonesia, Bogor University of Agriculture, University of Pancasila, University of Airlangga, University of Flores, University of Hasanuddin, and University of Khairun respectively (Sofia, 2019).

The span of STC comprises STC national guidelines, certification in sustainable destinations, training of assessors, cooperation with certification body/organization, and prototyping of three destinations. The guidelines on STC refer to the Minister of Tourism's regulation number 14/2016. Three key measurements are used for STC. They are the assessment of several destinations to identify the feasibility of the certification process scheme, the readiness of accreditation body/national accreditation agency to conduct the assessment, and recommendations provided based on the assessment. This can be done through several activities such as Indonesia Sustainable Tourism Award (ISTA) which evaluates the participants based on a set of certification criteria (Teguh, 2017). Moreover, the Ministry of Tourism will launch the Indonesian Sustainable Tourism Council to set guidelines for STC program (Sofia, 2019).

Another government initiative concerning sustainable tourism is though ecotourism in small villages. Two examples of the villages that offer unique ecotourism attractions and activities are Pemuteran village in Bali and Tangkahan village in North Sumatra. In Tangkahan, tourists can enjoy riding elephants and jungle trekking, while in Pemuteran (fishing) village they can snorkel, dive, participate in coral conservation, and feed turtles (Mufti, 2019). The Ministry of Tourism has targeted 2000 tourism villages by end of 2019, an increase from 1734 villages in 2018. The tourism village development will be concentrated around the 10 priority destinations. The development of tourism villages is in line with the Minister of Tourism's regulation no 14/2016 that stipulates the empowerment of local communities, environmental conservation, and cultural preservation for tourism sustainability (Mufti, 2019; Prambudi, 2019, Sofia, 2019). Figure 13.3 shows an example of a tourism village in Bali, Penglipuran. This traditional Balinese village offers unique experiences for the tourists. Tourists are welcomed to enter a house or a traditional kitchen (mostly made from bamboo) to observe the interior. They can enjoy traditional culture and interact with the local community by staying at a homestay or a guesthouse. Penglipuran tourism village has received an award as the world's cleanest village (https://www.penglipuranvillage.com/penglipuran-traditional-balinese-village/).

13.5 BECOMING A HALAL DESTINATION

Global Muslim Travel Index (GMTI) 2018 projected that there will be 2.8 billion or around one in three people practicing Islam worldwide by 2050. This makes Muslims the world's fastest-growing religious group. It is also predicted that globally, Muslim visitor arrivals will grow to 156 million in 2020 from 131 million in 2017 and 121 million in 2016. By 2026, Muslim visitors will reach 230 million and spend USD 180 billion on online travel (Mastercard-Crescentrating, 2018, 2019a).

As the world's most populous Muslim country where 87% or 225 million people are Muslims, Indonesia is aiming for halal tourism. The country seeks to lead global halal tourism and be seen as a tourist-friendly Muslim destination. The Minister of Tourism expected the growth in halal tourism of 42% in 2019 or around 5 million Muslim tourists from 3.5 million in

2018 and 1.95 million in 2017. The 5 million Muslim tourists will make up 25% of the target of total arrival of 20 million in 2019. Indonesia Muslim Travel Index (IMTI) was launched on 7th December 2018. This travel index designated 10 major halal tourist destinations: Aceh, Riau and Riau Islands, West Sumatra, Jakarta, West Java, Central Java, Yogyakarta, East Java, Lombok, and South Sulawesi. Among the 10 destinations, Lombok, a home of more than 1000 mosques, is appointed as the main halal tourism destination, followed by Aceh and Jakarta. IMTI is viewed as essential to achieve the halal tourism targets in 2019 to become number one halal destination and to attract 5 million Muslim tourists (Antara and Bhwana, 2019; Mastercard-Crescentrating, 2019b; The Jakarta Post, 2018, 2019).

FIGURE 13.3 Penglipuran tourism village with a homestay in Bali.

Source: Author's photos.

The halal tourism development in Indonesia is adapting the ACES model of GMTI 2019. There are four key components of ACES model: access, communications, environment, and services. The access considers air, rail, and sea access and transport infrastructure. Communication components concerns Muslim visitor guides, outreach initiatives—stakeholder education and market outreach, tour guides, and digital marketing. Environment considers domestic and international tourist arrivals, Wi-Fi connection at airports, and commitment to halal tourism. Services considers halal restaurants, mosques, airports, hotels, and attractions (Antara and Bhwana, 2019; Mastercard-Crescentrating, 2019a, 2019b).

According to GMTI 2019, Indonesia and Malaysia both ranked equal first with the score of 78. Indonesia's position has moved from the sixth rank in 2015, to fourth in 2016, then third in 2017, and on to the second rank in 2018. Indonesia excels in overall communications (in communicating to Muslim travelers) especially in the outreach aspect. Indonesia is also in the top list for the service criteria with high scores in halal dining, prayer places, and airport. However, the country is still lacking in offering unique experiences and hotels (Mastercard-Crescentrating, 2019a, 2019b).

Azhar (2018) observed that based on the State of the Global Islamic Economy 2017–2018, Indonesia ranked fourth on Global Islamic Economy Indicator (GIEI). The three top countries providing well-developed halal tourism are Malaysia, United Arab Emirates, and Turkey. Indonesia is still lagging behind its neighboring country Malaysia.

The development of halal travel ecosystem in Indonesia began in 2012, marked by the soft launch of Shariah tourism and a Shariah tourism guidelines book. This was followed by the grand launching of Shariah tourism in 2013. In the same year, Global Halal Forum was held, theming Wonderful Indonesia as a Muslim-friendly destination. In 2014, the Ministry of Tourism specified Shariah tourism destinations. The law on halal product assurance was passed. Indonesia received three awards in the World Halal Travel Awards in 2015 in Dubai. In 2015, Indonesia also offered Muslim visitor guides in English and Arabic as well as embarking on a destination marketing project with Halal Trip through influencers and blogs and videos. In 2016, the Ministry of Tourism formed Halal Tourism Acceleration and Development Team. That year Indonesia won 12 out of 16 awards at the World Halal Tourism Awards. In 2017, New Halal Products Certification Agency was established, followed by the launch of IMTI in 2018 and 2019 (Mastercard-Crescentrating, 2019b).

The current issues with the selected destinations are mainly concerning communication and services. As noted in IMTI 2019, the destinations should improve air connectivity and road infrastructure for good access. For a welcoming environment, the destinations should enhance Wi-Fi coverage at public areas such as hotels, restaurants, shopping malls, and airports. The destinations also need to offer Muslim-friendly services such as halal certified restaurants and Shariah-compliant accommodation, religious experiences, prayer services and facilities, and gender-segregated facilities (Mastercard-Crescentrating, 2019b). Rahmafitria (2019) contends that currently hotels in Indonesia that brand themselves as a Shariah hotel or a Muslim-friendly hotel are still scanty. She further argues that the implementation of the halal concept in the tourism industry, particularly in the accommodation sector, should apply universal/general Islamic principles which aim to provide comfort and ease for Muslim tourists to perform daily religious practice during travel.

Another point to note is the shift of halal travel 1.0 into halal travel 2.0. Halal travel 1.0 that was marked by services provided by hospitality and travel sectors in regards of halal food and prayer facilities is transforming into halal travel 2.0, the use of technology (such as augmented reality and big data analytic) to shape the Muslim tourists' experiences (Thomas, 2019).

13.6 CONCLUSION

It can be argued that Indonesia tourism is advancing in various aspects, particularly in regards to the price competitiveness, the increase growth of foreign tourists, and its global ranking. This is reinforced with strong government support. Yet Indonesia tourism has not reached its full potential considering how rich Indonesia is in its natural and cultural resources. There are so many destinations and attractions to explore in the country. Indonesia is certainly a contender to be one of the world's leading tourist destinations as well as the top halal tourism destination. Although tourist arrivals to the country are continuously growing, Indonesia is still lagging behind Malaysia, Singapore, and Thailand (Ariffin, 2018; Oxford Business Group, n.d.). To increase its global competitiveness, Indonesia tourism authorities have to pay serious attention to several pivotal aspects that still require improvement such as caretaking the environment and attending to sustainability, health and hygiene, tourist service infrastructure, safety and security, ICT readiness, and building human resources (as identified in Travel & Tourism Competitiveness Report 2017).

Various strategic initiatives have been set by the Indonesia government to boost the country's tourism, such as the development of 10 New Bali and 10 halal destinations, establishment of sustainable tourism development, expansion of tourism villages and homestays, and advancement in digital promotion. These initiatives not only need to consider the country's tourism strengths, but also the issues and challenges facing the industry. It is hoped that Indonesia can increase its foreign tourist arrivals given the rise of the Chinese, millennials, and Muslim tourists. The success of Indonesia tourism industry will certainly contribute to the total GDP, employment, service export, and foreign exchange earnings, as well as benefit other sectors. Figure 13.4 summarizes the current state of Indonesia tourism.

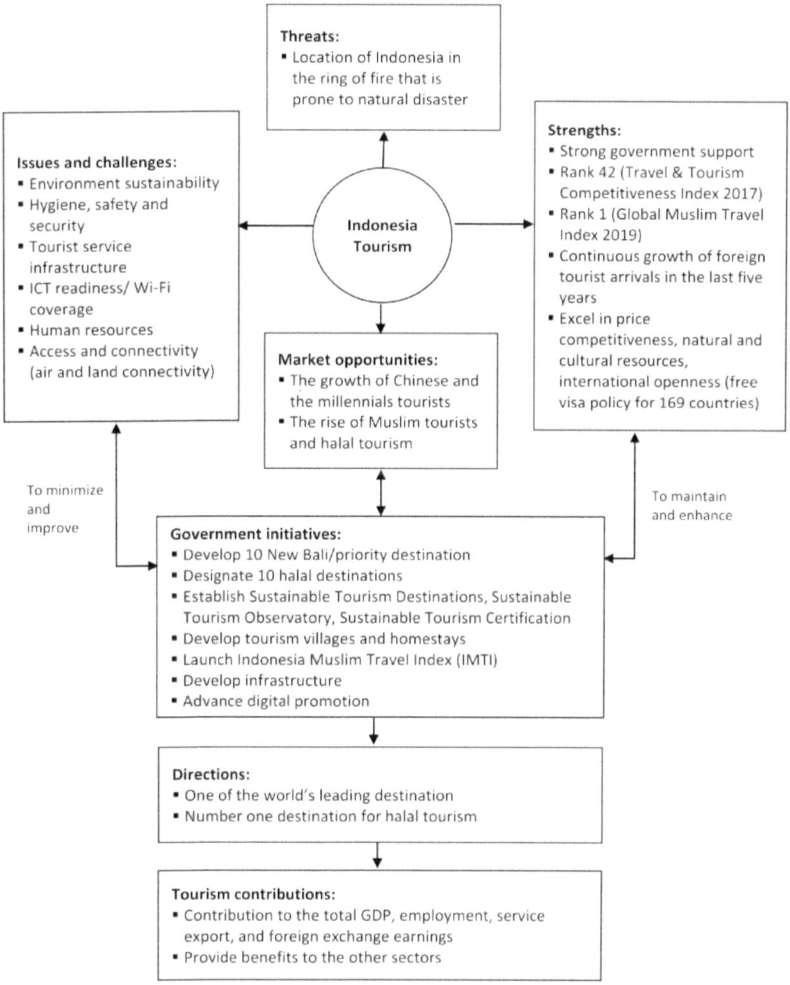

FIGURE 13.4 The state of Indonesia tourism.

Undoubtedly, the Indonesian government has done remarkable work to boost the country's tourism sector. Nevertheless, further attention and priority should be given to the Indonesia tourism and hospitality education as the producer of tourism labor/manpower. Indonesia does not only need people who are competent in operational and managerial skills to provide the service, but also people who are able to think critically about planning and strategies in the broader term and who have the capacity to do quality research. As tourism and hospitality education in Indonesia is comprised of two facets, a vocational approach and the academic mode, there should be a clear distinction in the purpose of education and curriculum design between these two systems (Oktadiana and Chon, 2019). Another important point of consideration is the partnership and link between five key stakeholders—academia, business/industry, community, government, and media (known as pentahelix)—to boost Indonesia tourism (Teguh, 2017). To build and sustain the integration between those five major players, a good communication system and a well-developed work plan should be established. Moreover, it is necessary for the stakeholders to create initiatives and strategies for the natural disaster prevention and mitigation. Their responsiveness and effort in attending to natural disasters will help to revive Indonesia's tourism industry when such shocks occur (Anisa, 2019).

The contribution of this chapter lies in providing contemporary insights into Indonesia tourism through a compilation of updated reports and news. The review can be valuable for those who are interested in understanding and learning about tourism development in the world's biggest archipelago, Indonesia.

KEYWORDS

- **Indonesia tourism**
- **tourism development**
- **Ten New Bali**
- **tourism sustainability**
- **halal destination**

REFERENCES

Anisa, D. F. Indonesia Prepares New Strategy to Meet 2019 Tourist Arrival Target. *Jakarta Globe*. https://jakartaglobe.id/context/indonesia-misses-tourists-arrival-targets-set-up-new-strategy-to-capture-higher-targets-this-year (accessed April 15, 2019).

Antara; Bhwana, P. G. Indonesia Aims to Rank First in Global Halal Tourism. https://en.tempo.co/read/1176596/indonesia-aims-to-rank-first-in-global-halal-tourism (accessed April 15, 2019).

Ariffin, E. Tourism Could Boost Indonesia's Economy. *The Asean Post*. https://theaseanpost.com/article/tourism-could-boost-indonesias-economy-0 (accessed April 1, 2019).

Azhar, R. Indonesia Ranks 4th in Halal Tourism Indicator. *Jakarta Now*. https://nowjakarta.co.id/updates/news/indonesia-ranks-4th-in-halal-tourism-indicator (accessed April 16, 2019).

Badan Pusat Statistik (Indonesia Statistic). Jumlah Kunjungan Wisman Menurut Kebangsaan dan Bulan Kedatangan Tahun 2017–2019 (Foreign Visitors' Arrival by Country and Month 2017–2019). https://www.bps.go.id/dynamictable/2018/07/30/1548/jumlah-kunjungan-wisman-menurut-kebangsaan-dan-bulan-kedatangan-tahun-2017—2019.html (accessed April 2, 2019).

Badan Pusat Statistik (Indonesia Statistic). Wisatawan mancanegara yang datang ke Indonesia menurut kebangsaan, 2000–2017 (Foreign Visitors' Arrival to Indonesia Based on the Nationalities, 2000–2017). https://www.bps.go.id/statictable/2014/09/08/1394/wisatawan-mancanegara-yang-datang-ke-indonesia-menurut-kebangsaan-2000-2017.htmlon (accessed April 2, 2019).

Badan Pusat Statistik (Indonesia Statistic). Jumlah wisatawan asing tahun 1974–2015, devisa wisman 1998–2015 dantingkat penghunian kamar (TPK) hotel bintang dan non bintang tahun 1985–2015 (Number of Foreign Tourists 1974–2015, Revenue Generated 1998–2015, and Room Occupancies of the Star-Rated and Non-Star Rated Hotels 1985–2015). https://www.bps.go.id/statictable/2014/01/30/1393/jumlah-wisatawan-asing-tahun-1974—2015-devisa-wisman-1998-2015-dan-tingkat-penghunian-kamar-tpk-hotel-bintang-dan-non-bintang-tahun-1985–2015.html (accessed April 2, 2019).

Bain, K. Boosting Indonesia's Tourism Sector. *Ancham Indonesia*. http://www.amcham.or.id/en/news/detail/boosting-indonesia-s-tourism-sector (accessed April 3, 2019).

Bowden, D. *Enchanting Indonesia*; John Beaufoy Publishing Limited: Oxford, 2015.

Chan, F. Jokowi Plans to Replicate Bali's Success in 10 Other Indonesian Spots. *The Straits Times*. https://www.straitstimes.com/asia/jokowi-plans-to-replicate-balis-success-in-10-other-indonesian-spots (accessed April 3, 2019).

Dikanaya, T.; Rakhmat, M. Z. Indonesia's Move Towards Tourism. *International Policy Digest*. https://intpolicydigest.org/2017/11/02/indonesia-s-move-towards-tourism/ (accessed April 5, 2019).

Indonesia Investment. Tourism Industry Indonesia. https://www.indonesia-investments.com/business/industries-sectors/tourism/item6051 (accessed April 2, 2019).

Lee, X. E. Indonesia is Looking to Become a Tourist Powerhouse. *CNBC*. https://www.cnbc.com/2018/12/20/indonesia-tourism-minister-arief-yahya-talks-goals.html (accessed April 3, 2019).

Mastercard-Crescentrating. *Global Muslim Travel Index 2018*; Mastercard & Crescentrating: Singapore, 2018.

Mastercard-Crescentrating. *Global Muslim Travel Index 2019*; Mastercard & Crescentrating: Singapore, 2019a.

Mastercard-Crescentrating. *Indonesia Muslim Travel Index (IMTI) 2019*; Mastercard & Crescentrating and the Ministry of Tourism Indonesia: Singapore, 2019b.

Ministry of Tourism Republic of Indonesia. *Journey to Indonesia*; Ministry of Tourism Republic of Indonesia: Jakarta, 2016.

Mufti, R. R. Government to Boost Sustainable Tourism in Small Villages. *The Jakarta Post*. https://www.thejakartapost.com/news/2019/02/01/government-to-boost-sustainable-tourism-in-small-villages.html (accessed April 5, 2019).

Oktadiana, H.; Chon, K. Differentiating Academic and Vocational Hospitality and Tourism Education: The Milieu of Indonesia. In *Tourism Education and Asia*; Liu, C., Schänzel, H., Eds.; Springer Nature Singapore, Pte. Ltd: Singapore, 2019; pp 15–34.

Oktadiana, H.; Pearce, P. L., The "bule" Paradox in Indonesian Tourism Research: Issues and Prospects. *Asia Pac. J. Tour. Res.* **2017,** *22*(11), 1099–1109.

Oxford Business Group, (n.d.). Indonesia Taking a Varied Approach to Bolster Tourism Numbers. https://oxfordbusinessgroup.com/overview/natural-attraction-government-taking-multi-pronged-approach-order-bolster-visitor-numbers (accessed April 22, 2019).

Prambudi, D. UNWTO dukung Indonesia kembangkan Sustainable Tourism Development dan homestay desa wisata (UNWTO Supports Indonesia to Develop Sustainable Tourism and Tourism Village Homestays). *Eljohn News.* https://eljohnnews.com/unwto-dukung-indonesia-kembangkan-sustainable-tourism-development-dan-homestay-desa-wisata/on (accessed April 5, 2019).

Rahmafitria, F. Pariwisata Halal dan Industry (Halal Tourism and Industry). *Pikiran Rakyat*; 2019; p 14.

Sofia, H. Minister Yahya Promotes Three Tourism Missions at FITUR in Spain. *Antara News.* https://en.antaranews.com/news/122003/minister-yahya-promotes-three-tourism-missions-at-fitur-in-spain (accessed April 5, 2019).

Teguh, F. *Resources Efficiency, Safeguarding Natural and Cultural Resources: Sustainable Tourism Policy and Practices.* http://cf.cdn.unwto.org/sites/all/files/pdf/4_mr_frans_teguh_-_png_21_march_2017_0.pdf (accessed April 15, 2019).

The Jakarta Post. Travel Agents, Airlines Told to Offer Government's 10 Tourist Destinations. https://www.thejakartapost.com/news/2017/04/03/travel-agents-airlines-told-to-offer-governments-10-tourist-destinations.html (accessed April 3, 2019).

The Jakarta Post. Indonesia Makes Lombok its Primary Halal Tourist Destination. https://www.thejakartapost.com/travel/2018/12/14/indonesia-makes-lombok-its-primary-halal-tourist-destination.html (accessed April 16, 2019).

The Jakarta Post. Indonesia Aims to be Top 'Halal' Destination. https://www.thejakartapost.com/news/2019/03/14/indonesia-aims-to-be-top-halal-destination.html (accessed April 16, 2019).

The World Bank. Nearly Three Million Indonesians to Benefit from Tourism Project. https://www.worldbank.org/en/news/press-release/2018/05/30/nearly-three-million-indonesians-to-benefit-from-tourism-project (accessed April 3, 2019).

Thomas, J. Malaysia, Indonesia Rivalry Heats Up. *The Asean Post.* https://theaseanpost.com/article/malaysia-indonesia-rivalry-heats (accessed April 20, 2019).

Trew, M. Grab Partners with Indonesia Ministry of Tourism to Boost Inbound Travel to New Destinations. *WiT.* https://www.webintravel.com/grab-partners-with-indonesia-ministry-of-tourism-to-boost-inbound-travel-to-new-destinations/ (accessed April 4, 2019).

World Travel & Tourism Council (WTTC). *Travel & Tourism Benchmarking Reports 2017*; World Travel & Tourism Council: London, 2017.

World Economic Forum (WEF). *The Travel & Tourism Competitiveness Report 2017: Paving the Way for a More Sustainable and Inclusive Future*; World Economic Forum: Geneva, 2017.

CHAPTER 14

HOTEL MANAGERS' VIEWS ABOUT TRADE UNIONS: A TALE OF CONTINUITY AND CHANGE IN NEPAL

SANDEEP BASNYAT

Macao Institute for Tourism Studies (IFTM), Macao, China

E-mail: sandeep@ift.edu.mo

ABSTRACT

In stark contrast to global trends, the Nepalese hospitality industry exhibited a remarkable trend in trade unions' growth, power, influence, and strategic shifts in their strategies related to their interactions with the management since the last two and half decades, particularly between 1990 and 2015. Although from 2015, the power of trade unions seemed to have slowed down faintly, they still exert considerable influence over the hotel management. In the context, where many of those hotel managers are still involved and engaged with trade unions, this chapter investigates their perceptions about trade unions. In doing so, the chapter reviews the roots and nature of employment relationship in the Nepalese hospitality industry, examines political parties' relationships with and influence on trade unions, and recent views of managers on involving trade unions for the effective management of hotel operations.

14.1 INTRODUCTION

In theory, trade unions, as representative organizations, advocate and act on behalf of employees and facilitate the amicable management of labor relations (Blyton et al., 2001; Lucas, 2009). As actors in a dynamic employment relations system (Kaufman, 2008, Wilton, 2010), trade unions operate within broad socioeconomic and political contexts (Benson and Zhu, 2008; Hayward, 1980; Hyman 2001) and strategically employ a variety of methods such as collective bargaining, strike action, and demonstrations to enter into an agreement with employers (Gospel, 2008; Hyman, 1989; Traxler, 1998; Webb and Webb, 1897). Traditionally, the primary aim of trade unions has been to improve wages and the working conditions of employees. In the present day, however, trade unions have aligned themselves with larger social movements, taking on, for example, issues of workplace and gender equality, and family-friendly workplaces for employees (Briskin, 2014; Curtin, 2018; Ravenswood and Markey, 2011) among others.

However, many hospitality managers continue to retain hostile and negative attitudes toward trade unions (LaVan and Katz, 2013; Lucas, 2004; Macaulay and Wood, 1992) and there has been a rise in the number of managers who refuse to recognize trade unions as official representatives of their employees (Baum, 2015; Broughton, 2005; Macfarlane, 1982). As a result, despite the potential benefits of joining trade unions, encouraging employee membership has proven difficult for trade unions in the tourism and hospitality industry (Baum, 2007; Lucas, 2004; Walton, 2012). Consequently, in many countries, trade union densities (in percentages) remain in single figures in the hospitality industry (Baum, 2007; Lucas, 2009). Additionally, the distinctive characteristics of the industry, such as the fragmented nature of the tourism industry, the precarious and menial nature of the work, the irregular working hours and employment of a large number of young people, women and migrant workers is said to prevent workers from fraternizing with one another and hinders the building of solidarity (Brook, 2002; Bryson and Gomez, 2005; Hakim, 1995; Hofmann and Stokburger-Sauer, 2017; Kensbock, Jennings et al., 2013; Lowe and Rastin, 2000; Machin, 2003), resulting in the low level of unionization in the industry. Consequently, barring a few hotels (e.g., Piso, 2003), trade unions have been generally weak in the hospitality industry across the world (Baum, 2007, 2015; Lucas, 2009; Wood, 1997).

In stark contrast to global trends, between 1990 and 2015, the Nepalese hospitality industry saw a remarkable trend in trade unions' growth, power, influence, and strategic shifts in their strategies related to their interactions with the management (Basnyat et al., 2017). With a tremendous increase in the number of trade unions (Thapa, 2004; Upreti et al., 2013), collective bargaining, strike action, participation in and organizing of demonstrations also became prevalent across all sectors of the tourism industry (Adhikari, 2010; Upreti et al., 2013). Although, the activities of trade unions have slowed down since 2015, especially after the two earthquakes that hit the country and aroused the feeling of solidarity among the trade unions in 2015 (see Basnyat et al., 2017 for the softening effect of earthquakes on the Nepalese trade unions), the presence and influence of trade unions in most hotels remain. What do hotel managers perceive about the existence, growth and influence of trade unions in Nepal, a country that exhibits entirely different employment relations in the hospitality industry compared to most other countries? This chapter aims to examine this issue.

This chapter is partially based on the findings of an empirical study on the trade union movement in the Nepalese tourism and hospitality industry. The data for the original empirical study was collected through face-to-face in-depth unstructured interviews with 14 hospitality managers from April 2015 to February 2016 in Kathmandu, the capital city of Nepal. In order to incorporate recent changes and opinions, additional data was collected through Skype interviews with hospitality managers (two previous and three new) between September 2018 and January 2019. Table 14.1 presents the profile of all 17 participants of this study. The results of this study show that while the participating managers have negative perceptions about trade unions for being affiliated with political parties, they unanimously recognized the importance and benefits of trade unions for the hospitality industry. Although the term "employer" is commonly used for a business owner of a hotel, all the participating employers in this study were engaged in managing their employees at the time of data collection of this study or in the past. Therefore, for simplicity, the term "manager" in the current study also refers to the participating business owners of hotels. After a brief background of the roots and nature of the employment relations system in the Nepalese hospitality industry, this chapter presents the findings of the interview data which will be followed by a short concluding remark.

14.2 ROOTS AND NATURE OF EMPLOYMENT RELATIONS SYSTEM IN THE NEPALESE HOSPITALITY INDUSTRY

Nepal is a Himalayan, landlocked, sovereign state located in South Asia. With a population of approximately 27 million (CBS, 2011), it is bordered to the north by China, and in the south, east, and west by India. Since the early 1990s, Nepal has experienced a period of political and natural instability, the effects of which are still evident in the Nepalese society, including in employment relations in the tourism and hospitality industry. The first political movement for the establishment of democracy began in 1989 in Nepal and evolved into

TABLE 14 1 Profile of the Study Participants.

Participants' Nicknames	Position	Age range (years)	Gender	Nature of experience with trade unions	Years of experience with unions
Ganesh	HR and finance manager	45–50	Male	Collective bargaining, dispute settlements, negotiations	23
Balram	HR director	45–50	Male	Collective bargaining, dispute settlements, negotiations	25
Mahendra	Sales and marketing manager	30–35	Male	Collective bargaining, negotiations	12
Rabindra	Hotel employer	45–50	Male	Collective bargaining, dispute settlements, negotiations	14
Anand	Past hotel employer	60–65	Male	Collective bargaining, dispute settlements, negotiations	1975–2007
Bachhan	Past hotel employer	50–55	Male	Collective bargaining, dispute settlements, negotiations	1975–2004
Prahlad	General manager	50–55	Male	Collective bargaining, dispute settlements, negotiations	25
Pratikshya	HR manager	25–30	Female	Collective bargaining, negotiations	4
Surya	HR director	55–60	Male	Collective bargaining, dispute settlements, negotiations	25
Khageshwor	HR manager	35–40	Male	Collective bargaining, dispute settlements, negotiations	15
Bidur	Training manager	30–35	Male	Collective bargaining, negotiations	9
Prerana	Hotel employer	45–50	Female	Collective bargaining, dispute settlements, negotiations	20
Rakshya	Sales and marketing manager	25–30	Female	Collective bargaining, negotiations	4
Sandarbha	Finance manager	40–45	Male	Collective bargaining, negotiations	8
Arjun	Hotel employer	45–50	Male	Collective bargaining, dispute settlements, negotiations	9
Abhinav	General manager	55–60	Male	Collective bargaining, dispute settlements, negotiations	22
Anup	HR manager	30–35	Male	Collective bargaining, negotiations	10

a fully-fledged revolution in 1990 (Basnett, 2009). On April 25, 1990, then King Birendra dismantled the party-less "Panchayat system" and established a multiparty democracy (Thapa, 2004). In 1996, the Communist Party of Nepal (Maoist), one of the political parties in the parliament, moved underground and instigated a *people's war* against the Nepali Congress, the Communist Party of Nepal—Unified Marxist–Leninist (CPN-UML) and other political parties for their inability to address and resolve pressing economic and social issues of the nation's poor and marginalized communities (Basnett, 2009). In the resultant civil war, more than 15,000 people were killed and 200,000 were internally displaced (Upadhayaya et al., 2011). In 2006, the Maoist rebels signed a peace agreement with the political parties and the government and joined the mainstream politics (ILO, 2013). Between 2008 and 2013, mass protests, political disputes, and nationwide strikes were rampant in Nepal (BBC, 2015). Shrestha and Chaudhary (2014) reported that 1621 (36.4%) out of 4451 strikes were organized by political parties between 2008 and 2013. As a result of the political instability, over the period from 2008 to 2015, seven Prime Ministers served the country (Basnyat et al., 2017).

Trade unions played key roles in organizing strikes and protests for the political parties during the time of instability between 2008 and 2013. It is important to note that it was only after the reinstatement of democracy in 1990, and after the promulgation of the Labour Act and the Trade Union Act in 1992, the trade unions formally started to organize in various employment sectors in Nepal including hotels. It is often reported that they gradually became influenced by the political parties which often used trade unions for "doing politics." In fact, all trade unions were affiliated to one or the other political party during the time of data collection for the current study. Table 14.2 presents the list of major trade unions in hotels that were affiliated to various political parties in the year 2016 (as compiled by the author of this current study).

TABLE 14.2 Hotel Trade Unions' Affiliations with Various Political Parties.

Name of trade union	Affiliated political party
All Nepal Hotel, Casino, and Restaurant Workers' Union	United Communist Party of Nepal-Maoist (UCPN (M))
Nepal Tourism, Hotel, Casino and Restaurant Workers' Union	Nepali Congress
Nepal Independent Hotel, Casino, and Restaurant Workers' Union (NIHWU)	Communist Party of Nepal-Unified Marxist-Leninist (CPN-UML)
All Nepal Revolutionary Hotel and Restaurant Workers' Union	Communist Party of Nepal (Maoist)-Biplav

Throughout the period from 2008 to 2013, trade unions not only organized protests and general strikes, but also participated in a number of activities organized by their mother political parties ranging from marching in a protest rally to carrying the political parties' flags and taking part in physical clashes, vandalism and violence against other political parties (Basnyat et al., 2017). Trade unions activities also caused several disputes between trade unions and employers in various tourism establishments including hotels during this period of instability in Nepal (Hoffmann, 2013; Upreti et al., 2013). Table 14.3 lists the most important disputes between trade unions and employers between 2001 and 2015 in the Nepalese tourism and hospitality industry. The list is by no means exhaustive; nonetheless, it does indicate the frequency of the disputes in Nepal. As seen in Table 14.3, the majority of disputes between trade unions and hotel employers occurred between 2006 and 2013. Several employers were physically assaulted, employees were forced to quit their jobs, and a number of hotels either closed down or were forcefully shut down by trade unions during the period. Employers were forced to give an unreasonable amount of money as donations and a significant number of political parties' cadres were employed as hotel employees through trade unions (Basnyat et al., 2017). Analyzing the situation Upreti et al. (2013), once pointed out that a general lack of amicable relationships between management and trade unions in the hotel industry had caused instability in the working environment and posed challenges for the sustainability of the tourism industry. Amidst the ongoing instability, two major earthquakes and subsequent aftershocks hit the nation in 2015 (April 25 and May 12) causing around 9000 deaths, 22,000 injured people and 800,000 damaged houses (fully and partially) (ADB, 2016). The Danish Trade Council for International Development and Cooperation (2019) has estimated that, in 2018, there were about 1.8 million trade union members (approximately 12% of the total labor force and 48% of the wage and salaried workers) associated with nine different trade unions confederations and spread across different employment sectors, including tourism and hospitality, all over Nepal.

Tourism in Nepal, which was largely controlled by the royal family and Kathmandu (capital city) based elites until 1990 (Bhandari, 2010), is now considered an important industry for the generation of employment opportunities. From barely around 9000 in the 1960s when international tourism started in Nepal, the tourist arrivals crossed 250,000 in the 1990s (MOCTCA, 2019). After a series of upheaval between 1996 and 2006 when the country faced the Maoist's civil war, the tourist arrivals gradually improved from 2007 and reached around 800,000 in 2014 until the earthquakes rocked the country in 2015. The earthquakes brought down the tourist arrivals to 500,000 figure (MOCTCA, 2019). However, the country regained the confidence of the tourists fairly quickly. As a result, the tourist arrivals began to improve at a faster pace and more than doubled in less than 3 years after the earthquakes (crossed 1.1 million in 2018) (MOCTCA, 2019). Similar to tourist arrivals, foreign currency earnings (less returned) from international tourists has also experienced fluctuations. The total foreign currency earnings from the tourists in the Fiscal Year 2005/2006 (when the Maoists' civil war was still in place), was roughly around the figure of US Dollar 132 million (MOCTCA, 2019). This figure improved gradually from 2007 and crossed US Dollar 544 million in 2014. After the earthquakes in 2015, the year when the total foreign currency earnings figure came down to US Dollar 392 million, the country has experienced gradual improvements. MOCTCA (2019) reported that the total foreign exchange earnings in 2018 exceeded US Dollar 640 million and is expected to cross US Dollar 750 million marks in 2019. The World Travel & Tourism Council (2018) estimated that the total contribution of travel and tourism to the Nepalese GDP was close to US Dollar 1920 million (7.8%) in 2017 and is forecasted to reach around US Dollar 2950 million (8.2%) in 2028.

TABLE 14.3 Some Important Disputes Between Employers and Trade Unions.

Year	Labor disputes and their impacts on the Nepalese tourism and hospitality industry
2001	Struggle for the introduction of a ten per cent service charge in tourist rated hotels and restaurants in Nepal
2005	Yak & Yeti, a five-star Hotel in Kathmandu shut down its operation
2005	Indian Airlines Office workers' strike
2006	Two five-star hotel managers forced by the trade unions to quit their jobs
2006	Five tourism-related employers' associations closed down their businesses
2006	Trade unions shut down most hotels in two cities Dhulikhel and Nagarkot
2007	Employer of Hotel Woodland was physically assaulted after wage disputes
2008	The employer of Hotel Maharaja was physically assaulted
2008	All hotels in two cities Nagarkot and Daman were closed and tourists' entry was restricted
2009	Demonstration and strike organized at Hotel Beijing China
2010	Long strikes were organized and domestic and international flights were obstructed
2011	Club Himalaya, a three-star resort hotel at Nagarkot, was shut down
2013	Mr. BK Shrestha, first vice president of the Hotel Association Nepal (HAN) and managing director of the Radisson Hotel was physically assaulted
2015	International fast-food chains Kentucky Fried Chicken (KFC) and Pizza Hut shut down
2015	The Everest Hotel, a five-star hotel in Kathmandu, shut down its operation

In 2017, the total contribution (direct and indirect) of the travel and tourism industry to Nepalese employment was 6.6% (102,700 jobs) (World Travel & Tourism Council, 2018), of which hotels portion of contribution was considered significantly larger. MOCTCA (2019) reported that there were altogether 1254 hotels (129 star and 1125 non-star) operating in Nepal in 2018, with a total bed capacity of 40,856 per day. In the last 4 years (Since 2015), the per day bed capacity of hotels has increased altogether by 13%, whereas the total number of hotels increased by 16.5% (MOCTCA, 2018, 2019). In line with the increased per day bed capacity, the hotel occupancy rate has also gone up to above 65% since 2017 (The Kathmandu Post, 2018). The occupancy rate above 60% is usually considered "good" in the hospitality industry, whereas the rate of 70% and above is considered "excellent" (The Kathmandu Post, 2018). The annual growth rate of both hotel and restaurants had dropped significantly in 2015 (−9.68%) due to the earthquakes, albeit, thereafter, it has shown an encouraging sign with above 7% each year and stood at 8.33% in the Fiscal Year 2018/2019 compared to the previous fiscal year (MOCTCA, 2019). Although the comparative official statistics on the hotels' earning is hard to find during these periods, several newspapers have reported that the hotels, in general, have increased their profits by several folds owning to the increased tourist arrivals and occupancy rates (Prasain, 2019; The Kathmandu Post, 2018).

In contrary, the workers' situation in various industries, including tourism and hospitality, is however, challenging as a large number of them, as high as 96% of the total workforce, work in informal sectors and are deprived of social protection benefits such as universal family benefits, social assistance, and social insurance (Danish Trade Council for International Development and Cooperation, 2015; Khanal, 2012). Khanal (2012) has reported that the situation is even worse in the private sector where many employers are not willing to provide any kind of health benefits to workers.

14.3 PERCEPTION OF HOSPITALITY MANAGERS ABOUT TRADE UNIONS

14.3.1 NEGATIVE PERCEPTION ABOUT TRADE UNIONS FOR BEING AFFILIATED WITH POLITICAL PARTIES

As Pratikshya, one of the youngest HR Managers, indicated, most participating managers in the Nepalese hospitality industry believed that trade unions should not have a political party's affiliations. The participating managers argued that trade unions should be organized for the welfare of employees within a particular hotel only, that their loyalty should be dedicated toward the employees in the hotels, and they should not be influenced by the policies of national political parties.

Most managers felt that the way trade union organizations were conceptually developed and formed in early 1990, right after the establishment of the democratic movement, was misleading and provided a biased politicalized foundation for their later developments in hotels. Prahlad explained:

> *The trade union movement in Nepal was initiated as a result of and influenced by political changes in the country. The democratic movement, which was entirely a political movement, also became the foundation for the development of trade unions in the country. Therefore, it is crucial to understand that the concept of the trade union in Nepal did not come because of the concept of trade, but it came as a result of the political change in the country. Political parties formed trade unions for their own benefits in hotels. It was slightly unfortunate because it brought about political nature in trade union organization in Nepal.*

The managers argued that, because trade unions are guided by national political parties' philosophies and ideologies, the whole concept and education about the trade union organization in the country has been misunderstood. Anup commented:

> *The functions of trade unions in our country has been entirely misunderstood and that's because of the influence of political parties. Trade unions are organized here in hotels as a subsidiary or you can say as a wing of the political party they are affiliated with. So, they follow their policies and philosophies. Whatever the political parties ask them to do, they do it here. If the political parties do not like the hotel employer, their affiliated trade unions protest against the management; if the political parties need more donations, they use their affiliated trade unions.*

Consequently, as Anand argued, trade unions have shown less concern about increasing productivity of hotels, and more interests in bargaining with employers and getting as much out of them as possible, resulting in a number of disputes between the trade unions and the management of hotels. Bachhan, one of the senior-most managers, even showed dissatisfaction by commenting that political parties and trade unions "took advantage of their [employers'] open-mindedness as they encouraged the formation of their organization in the early 1990s, and politicized them."

Although the participating managers strongly argued that trade unions should be free of political parties' affiliations, all of them unanimously recognized the importance and benefits of trade unions for the hospitality industry in Nepal. This topic is examined in the next section.

14.3.2 TRADE UNIONS BEING RECOGNIZED AS AN IMPORTANT PART OF THE INDUSTRY

Considering the generally challenging working conditions in most sectors of the hospitality industry, including hotels, the participating hotel managers recognized that trade unions are not only essential for promoting and protecting employees' welfare but also necessary for the effective management of hotels' operations. Additionally, the managers agreed that trade unions are a useful source of information on updated standards of global practices.

Recalling the exploitations that some of the hotel workers had faced before 1990 from their employers, most long-time serving managers believed that many workers who work in the lower ranks in hotels do not have proper education about their organizational, social, and economic rights. Trade unions act as a bridge to educate them. The managers believed that since rights and responsibilities go hand-in-hand, the process of educating rights to employees also help them to be aware of their own responsibilities toward hotels. Some managers also provided examples of how trade unions were protecting and promoting the welfare of employees in their hotel. Ganesh narrated trade unions' efforts this way:

> *A number of improvements such as providing a separate bathroom for staffs, establishing a staff welfare committee and appointing trade union officials as members of the committee, establishing a system for annual salary increments and gratuity, and providing transportation facilities for the staffs have been done in this hotel because of trade unions' suggestions. I think they are good and our management takes it positively.*

The participating managers also believed that trade unions help to maintain a certain level of discipline in hotels, especially on the part of employers. Mahendra noted that managers find it hard to impose their unilateral and discretionary decisions on workers in hotels where there is a presence of an active trade union. Remembering an incident when trade unions aggressively protested against him for mistakenly transferring an employee to a continuous shift, Balram remarked that the presence of trade unions alert managers and does not allow them to avoid or ignore employees' problems.

Most participating managers perceived that trade unions support hotel managers in executing their functions in a number of ways. Balram reported that they invite trade unions during their annual budget meetings where they have been providing useful and constructive suggestions. He further elaborated that trade unions have been often useful in assessing the hotel's competitive reputation in the market as they have links with members in other hotels. Rabindra noted that trade unions' support is essential when the management intends to impose a new policy or take actions against any employee. In such occasions, he arranges a meeting with trade union members and requests them to assist him to expedite the matter. As Rabindra reiterated, trade unions help him to ease his works, particularly when handling employees' related matters.

Abhinav also takes help from trade unions when he wants to impose rules and new policies. However, he uses a different approach. He has established a welfare committee in his hotel wherein trade unions are also the members. He reported that the hotel management easily imposes rules and new policies through this welfare committee because the union members, who are in that committee, convince other employees by themselves. Apart from helping management in imposing new policies and rules, trade unions also assist in organizing social events through welfare committees. Rakshya clarified:

> *We have a 14-member joint welfare committee in our hotel. In this committee, seven members are from the management and the remaining seven are from the trade unions. This committee organizes several social programmes such as annual programs, picnic, and other necessary welfare programs for all the employees. We have seen how helpful the trade unions are to the management in organizing these social programs. They help in promoting programs, finding and evaluating suppliers, and executing the bidding process.*

Furthermore, most managers believed that unlike the 1990s, trade unions in most hotels these days are much more knowledgeable and aware of the existing national and international laws and practices and draw their logics deriving references from a wide array of countries. This has also become possible as a result of the strong networks of union federations which operate globally. As Bidur experienced, for many managers, trade unions are also, therefore, a source of information for the updated international norm, standard and practices on labor relations.

14.3.3 ADOPTION OF REFORMED EMPLOYEE MANAGEMENT PHILOSOPHIES

This study has found that the participating managers in the Nepalese hotels not only recognize the importance and benefits of trade unions, but they have also reformed their employee management philosophies to manage their relationships with trade unions. The managers feel that the management in the Nepalese hotels is not in a position to continue implementing some of the traditional philosophies of dictating, unilaterally controlling, or using divide-and-rule policies among trade unions or other employees. Thanks to the development of Internet and social media, employees including trade unions, customers, and the general public have access to information about activities that happen inside a hotel. Therefore, as the managers reported, it was beneficial for the management to communicate their plans and policies to their employees by themselves and gain their support. Balram explained:

> *There are huge changes in the ways management thinks these days. We think that whatever we think and plan about the employees and the company, the information should reach even to the lower-level employees. These days almost all the employees are well educated, communication facilities are easily available everywhere and the development of the internet has given access to news and information on a national and global level. Therefore, management can't use force against employees. People not only watch and judge and evaluate trade unions but also our activities and performances.*

Some managers felt that sometimes it is even riskier for the management to use traditional divide-and-rule policy for the long-term sustainability of hotels. Abhinav clarified:

> *I understand that some manager may have used divide-and-rule policy in the past among workers to gain control over them. But, in today's world where information is so easily accessible, management should not be using such policies. We are not in a position to take a risk by creating conflicts with trade unions because it might backfire on us. We may not only have to bear the financial costs, but we may also lose our reputation and our potential clients. Cumulatively, this may negatively affect the sustainability of our business in the long run.*

Considering new changes in the employment relationships that are being accepted across the world, the participating managers emphasized that it is important for the management to adopt the concept of employee engagement into their philosophies. The managers believed that along with finance and customer satisfaction, employee satisfaction, and employee engagement have become essential and leading practices for many hotels. Therefore, the hotel management should play creative roles in adopting such practices. Surya reiterated:

> *The management has to play a very creative role while engaging their employees including trade unions because it is a dominating practice nowadays, whether you like it or not. So, instead of saying this is right or this is wrong, the management could guide trade unions properly on how they can approach their demands more effectively. Trade unions and other employees should be satisfied with what we are doing. Probably, it's much easier to resolve their problems if we engage them [trade unions] and also, by doing this, the management would look stronger in the eyes of the trade unions.*

Most managers also perceived that the management needs to take initiative in developing a healthy relationship with trade unions and rest of the employees by letting them know, transparently, what the hotels could or could not offer. In order to achieve this objective, as the managers believed, trade unions should be considered as partners of the management instead of opponents. The managers believed that it is the responsibility of the management to make trade unions feel that they are also a part of the company and their contributions are equally essential for the overall development of hotels. Pratikshya narrated her personal experience of how her beliefs about trade unions shifted:

> *During the early days, I was very negative about trade unions—I used to think that trade unions do not understand the management perspectives. Gradually, I have been realizing that we also think from our, management's, perspective only—whether it's about the profit and loss or managing our staffs. We discuss so many times about staff's unhappiness during our meetings and how unsatisfied we are with whatever we are doing. After having interactions with trade unions, I now realize that it is all related to our own perception. Trade unions periodically remind us to think from the employees' perspective also, not from our own perspective only. When I try to do that, it makes a lot of sense to me.*

Some managers also reported that, instead of cutting down employees' benefits, they were focusing on alternative ways to optimize their costs. For them, spending a portion of their profit for the employees had allowed them to develop cordial relationships with trade unions and it was much easier for them to manage their employees. The managers believed that if the management work cooperatively with employees and trade unions, they would be able to optimize cost, work on leave balances, cut down unnecessarily incurring costs, focus on businesses that have better yields, and focus on quality and consistency of their services.

14.4 CONCLUDING REMARK

This chapter has examined the perceptions of hotel managers about trade unions in Kathmandu, Nepal. Nepal is an appropriate area for analyzing hotel managers' perception about trade unions because of the distinct and contrasting characteristics it has demonstrated for the last two and half decade, particularly between 1990 and 2015. Contrary to the global trend, the strength and influence of trade unions grew rapidly in the hospitality industry within this period in Nepal resulting in an innumerable number of disputes between trade unions and management of hotels. As a result, several hotels were even shut down. Although from 2015, the power of trade unions seemed to have slowed down faintly, they still exert considerable influence over the hotel management. In the context, where many of those hotel managers are still involved and engaged with trade unions, this study attempted to examine their perceptions about trade unions.

In contrast to what most of the earlier studies have found, the results of this study have been remarkable. The findings of this study show that the hotel managers are both positively and negatively disposed about the trade unions. The Nepalese hotel managers perceive that trade unions are an essential part of the employment relations system in the hospitality industry and their presence is important not just for the protection and promotion of employees' welfare but also for the sustainable growth of hotels. They can be a partner for the management and can assist the management to smoothly run their hotels by helping them to manage employees. However, the managers felt negatively for the reason that trade unions are affiliated with political parties and their activities in hotels demonstrate the extent to which political parties control the behaviors of trade unions within hotels. For most managers, the political parties' influence on trade unions' activities was against the fundamental norms for which trade unions are established.

The case of Nepal demonstrates a story of continuity and change in hotel managers' perceptions about trade unions. On the one hand, it represents the continuity of traditional philosophies on the part of management where they are still under the influence of an ideology that hotel employees should exhibit their behavior on the particular way and that external influences are harmful to the overall health of their company. On the other hand, the case of Nepal represents the story of change in the manager's perception owing to their adoption of reformed employee management philosophies. As Hyman (2001) underlines, it is obvious that trade unions inherit their ideology prevalent during their formative years, and they will continue to do so in the future. Therefore, it is advantageous for the hotel management to adapt their perceptions considering this fundamental principle which is embedded in the root of trade unions' formation. At the same time, it is a welcome change that the managerial perceptions incorporate and reflect the new changes that surround the employment relations in the hospitality industry.

KEYWORDS

- **employment relations**
- **hospitality industry**
- **hotel managers' perceptions**
- **Nepal**

REFERENCES

ADB. (2016). Nepal 2015 Earthquake: ADB's Response. *Countries*. http://www.adb.org/countries/nepal/earthquake-adb-response

Adhikari, A. 2010 Turns into 'The Year of Banda' 125 Days of Strikes in 365 Days. *The Kathmandu Post*. http://kathmandupost.ekantipur.com/news/2010-12-31/2010-turns-into-the-year-of-banda-125-days-of-strikes-in-365-days.html (accessed Dec 31, 2010).

Basnett, Y. *From Politicization of Grievances to Political Violence: An Analysis of the Maoist Movement in Nepal*, Vol. 78; London School of Economics and Political Science: UK, 2009.

Basnyat, S.; Lovelock, B.; Carr, N. Political Instability and Trade Union Practices in Nepalese Hotels. *J. Policy Res. Tour. Leis. Events* **2017,** *9*(1), 40–55.

Baum, T. Human Resources in tourism: Still Waiting for Change. *Tour. Manag.* **2007,** *28*(6), 1383–1399.

Baum, T. Human Resources in Tourism: Still Waiting for Change?—A 2015 Reprise. *Tour. Manag.* **2015,** *50*, 204–212.

BBC. Nepal Profile—Timeline. *South Asia,* 2015. http://www.bbc.com/news/world-south-asia-12499391

Benson, J.; Zhu, Y. Trade Unions in Asia: Organization, Strategy and Issues. In *Trade Unions in Asia: An Economic and Sociological Analysis*; Benson, J.; Zhu, Y., Eds.; Routledge: London and New York, 2008; pp 1–10.

Bhandari, K. Tourism in Nepal: Post-Monarchy Challenges. *J. Tour. Cult. Change* **2010,** *8*(1–2), 69–83.

Blyton, P.; Lucio, M. M.; McGurk, J.; Turnbull, P. Globalization and Trade Union Strategy: Industrial Restructuring and Human Resource Management in the International Civil Aviation Industry. *Int. J. Human Res. Manag.* **2001,** *12*(3), 445–463.

Briskin, L. Strategies to Support Equality Bargaining Inside Unions: Representational Democracy and Representational Justice. *J. Ind. Relat.* **2014,** *56*(2), 208–227.

Brook, K. Trade Union Membership: An Analysis of Data from the Autumn 2001 LFS. *Labour Mark. Trends* **2002,** *110*(7), 343–354.

Broughton, A. Industrial Relations in the Airline Sector, 2005. https://www.eurofound.europa.eu/publications/report/2005/industrial-relations-in-the-airline-sector

Bryson, A.; Gomez, R. Why Have Workers Stopped Joining Unions? The Rise in Never-Membership in Britain. *Br. J. Ind. Relat.* **2005,** *43*(1), 67–92.

CBS. *Nepal Living Standards Survey 2010/11: Statistical Report, Vol. 2*; Central Bureau of Statistics: Kathmandu, 2011.

Curtin, J. *Women and Trade Unions: A Comparative Perspective*; Routledge: London and New York, 2018.

Danish Trade Council for International Development and Cooperation. *Labour Market Profile 2015: Nepal*; LO/FTF Council's Analytical Unit: Copenhagen, Denmark, 2015.

Danish Trade Council for International Development and Cooperation. *Labour Market Profile 2019: Nepal*; Danish Trade Union Development Agency, Analytical Unit: Copenhagen, Denmark, 2019.

Gospel, H. Trade Unions in Theory and Practice: Perspectives from Advanced Industrial Countries. In *Trade Unions in Asia: An Economic and Sociological Analysis;* Benson, J.; Zhu, Y., Eds.; Routledge: London and New York, 2008; pp 11–23.

Hakim, C. Five Feminist Myths about Women's Employment. *Br. J. Sociol.* **1995,** *46*(3), 429–455.

Hayward, J. Trade Union Movements and their Politico-Economic Environments: A Preliminary Framework. *West Eur. Politics* **1980,** *3*(1), 1–9.

Hoffmann, M. P. Labour Unionism in a Post-Conflict Context: Maoists, Unionists, and Formerly Bonded Labourers in an Urban Municipality in the Western Tarai *Max Planck Institute for Social Anthropology, Working Paper No. 145*; 2013. https://www.eth.mpg.de/pubs/wps/pdf/mpi-eth-working-paper-0145

Hofmann, V.; Stokburger-Sauer, N. E. The Impact of Emotional Labor on Employees' Work-Life Balance Perception and Commitment: A Study in the Hospitality Industry. *Int.J. Hosp. Manag.* **2017,** *65*, 47–58.

Hyman, R. *Strikes*; Macmillan: Basingstoke and London, 1989.

Hyman, R. *Understanding European Trade Unionism: Between Market, Class and Society*; Sage: London, Thousand Oaks, New Delhi, 2001.

ILO. *Decent Work Country Programme 2013-2017 Nepal*; International Labour Organization: Nepal, 2013.

Kaufman, B. E. *Managing the Human Factor: The Early Years of Human Resource Management in American Industry*; Cornell University Press: Ithaca and London, 2008.

Kensbock, S.; Jennings, G.; Bailey, J.; Patiar, A. The Lowest Rung': Women Room Attendants' Perceptions of Five Star Hotels' Operational Hierarchies. *Int. J. Hosp. Manag.* **2013,** *35*, 360–368.

Khanal, D. R. *Social Security/Social Protection in Nepal: Situation Analysis*; International Labour Organization: ILO Country Office for Nepal, 2012.

LaVan, H.; Katz, M. Current State of Management/Union Relations in Hospitality Sector. *Manag. Hosp. Rev.* **2013,** *30*(2), 73–89.

Lowe, G.; Rastin, S. Organizing the Next Generation: Influences on Young Workers' Willingness to Join Unions in Canada. *Br. J. Ind. Relat.* **2000,** *38*(2), 203–222.

Lucas, R. *Employment Relations in the Hospitality and Tourism Industries*; Routledge: London and New York, 2004.

Lucas, R. Is Low Unionisation in the British Hospitality Industry Due to Industry Characteristics? *Int. J. Hosp. Manag.* **2009,** *28*(1), 42–52.

Macaulay, I. R.; Wood, R. C. Hotel and Catering Industry Employees' Attitudes Towards Trade Unions. *Empl. Relat.* **1992,** *14*(3), 20–28.

Macfarlane, A. Trade Union Growth, the Employer and the Hotel and Restaurant Industry: A Case Study. *Ind. Relat. J.* **1982,** *13*(4), 29–43.

Machin, S. Trade Union Decline, New Workplaces and New Workers. In *Representing Workers: Trade Union Recognition and Membership in Britain*; Gospel, H.; Wood, S., Eds.; Routledge: London and New York, 2003, pp 15–28.

MOCTCA. *Nepal Tourism Statistics 2017*; Government of Nepal, Ministry of Culture, Tourism & Civil Aviation: Kathmandu, 2018.

MOCTCA. *Nepal Tourism Statistics 2018*; Government of Nepal, Ministry of Culture, Tourism & Civil Aviation: Kathmandu, 2019.

Piso, A. Unionisation in the Dublin Hotel Industry. *Int. J. Tour. Res.* **2003,** *5*(3), 211–224.

Prasain, S. Luxury Hotels in Kathmandu Post Record Profits on Higher Arrivals. *The Kathmandu Post,* 2019. https://kathmandupost.com/money/2019/06/04/luxury-hotels-in-kathmandu-post-record-profits-on-higher-arrivals

Ravenswood, K.; Markey, R. The Role of Unions in Achieving a Family-Friendly Workplace. *J. Ind. Relat.* **2011,** *53*(4), 486–503.

Shrestha, M. B.; Chaudhary, S. K. The Economic Cost of General Strikes in Nepal. *NRB Econ. Rev.* **2014,** *26*(1), 1–23.

Thapa, B. Tourism in Nepal: Shangri-La's Troubled Times. *J. Travel Tour. Mark.* **2004,** *15*(2–3), 117–138.

The Kathmandu Post. (2018). Valley Hotels Net Record Profits in 2017. *The Kathmandu Post.* https://kathmandupost.com/money/2018/02/15/valley-hotels-net-record-profits-in-2017.

Traxler, F. Austria: Still the Country of Corporatism. In *Changing Industrial Relations in Europe*, 2nd ed.; Ferner, A.; Hyman, R., Eds.; Wiley-Blackwell: Oxford, 1998; pp 239–261.

Upadhayaya, P. K.; Müller-Böker, U.; Sharma, S. R. Tourism Amidst Armed Conflict: Consequences, Copings, and Creativity for Peace-Building Through Tourism in Nepal. *J. Tour. Peace Res.* **2011,** *1*(2), 22–40.

Upreti, B. R.; Sharma, S. R.; Upadhaya, P. K.; Ghimire, S.; Iff, A. *Making Business Count for Peace: Reflections from Tourism Sector in Nepal*; Kathmandu: South Asia Regional Coordination Office of the Swiss National Centre, 2013.

Walton, J. K. The Tourism Labour Conundrum' extended: Historical Perspectives on Hospitality Workers. *Hosp. Soc.* **2012,** *2*(1), 49–75.

Webb, S.; Webb, B. *Industrial Democracy*; Longmans, Green, and Company: London, New York and Bombay, 1897.
Wilton, N. *An Introduction to Human Resource Management*; Sage Publications: Los Angeles, London, New Delhi, Singapore, Washington DC, 2010.
Wood, R. *Working in Hotels and Catering*; International Thomson Business Press: London, 1997.
World Travel & Tourism Council. *Travel & Tourism: Economic Impact 2018 Nepal*; World Travel & Tourism Council: London, 2018.

INDEX

T

U

V